AUTOMOTIVE ELECTRICITY AND ELECTRONICS

CONCEPTS AND APPLICATIONS

BOYCE H. DWIGGINS and **EDWARD F. MAHONEY**

Prentice Hall
Englewood Cliffs, New Jersey   Columbus, Ohio

Library of Congress Cataloging-in-Publication Data
Dwiggins, Boyce H.
   Automotive electricity and electronics : concepts and applications / B.H. Dwiggins and E.F. Mahoney.
      p.   cm.
   ISBN 0-13-359233-2 (pbk. : alk. paper)
   1. Automobiles—Electric equipment.   2. Automobiles—Motors—Computer control systems.   I. Mahoney, Edward F.   II. Title.
TL272.M24 1995
629.25′4—dc20   95-15895
                                                                    CIP

Cover art: Kevin Hulsey Illustration, Inc.
Editor: Ed Francis
Production Editor: Alexandrina Benedicto Wolf
Cover Designer: Proof Positive/Farrowlyne Assoc. Inc.
Production Manager: Laura Messerly
Marketing Manager: Debbie Yarnell

This book was set in Century Book and Kabel by The Clarinda Company and was printed and bound by Quebecor Printing/Semline. The cover was printed by Phoenix Color Corp.

© 1996 by Prentice-Hall, Inc.
A Simon & Schuster Company
Englewood Cliffs, New Jersey 07632

All rights reserved. No part of this book may be reproduced, in any form or by any means, without permission in writing from the publisher.

Printed in the United States of America

10 9 8 7 6 5 4 3 2 1

ISBN: 0-13-359233-2

Prentice-Hall International (UK) Limited, *London*
Prentice-Hall of Australia Pty. Limited, *Sydney*
Prentice-Hall of Canada, Inc., *Toronto*
Prentice-Hall Hispanoamericana, S. A., *Mexico*
Prentice-Hall of India Private Limited, *New Delhi*
Prentice-Hall of Japan, Inc., *Tokyo*
Simon & Schuster Asia Pte. Ltd., *Singapore*
Editora Prentice-Hall do Brasil, Ltda., *Rio de Janeiro*

This book is dedicated to the automotive technicians of tomorrow, who will need ever-increasing knowledge of electricity and electronics.

PREFACE

The largest single investment one usually makes is for the purchase of a home. The next largest investment, for most, is an automobile. Both of these investments have much in common: they require insurance, periodic maintenance, and needed repairs.

Homeowners require plumbers, electricians, carpenters, painters, and other specialized craftsmen for necessary repairs and maintenance. The same is true for car owners—they require automotive technicians who specialize in tune-ups, transmissions, brakes, air conditioning and cooling systems, and electrical systems. When a vehicle is taken to a specialty shop for air conditioning service, those working on the car have considerable experience in air conditioning diagnosis and repair. When a car is taken to a dealer or general repair facility, it will be worked on by Tom, who is the in-house air conditioning technician. If you were to visit a large garage or service facility, you would find that bays (work areas) are assigned to technicians skilled in a particular service—Curt for brakes, Tim for transmissions, and so on. But skilled technicians are not always males; many are female. Today you may find Lara in the tune-up department and Karen in the electrical service department.

The purpose of this book, then, is to help the serious student, male or female, prepare to enter the automotive profession as an automotive electrician or an automotive electrical technician. We emphasize *serious*, because this text covers automotive electrical theory in depth. It is intended for those interested in becoming a part of the large team of automotive technicians dedicated to keeping today's modern vehicles operationally safe and sound.

B. H. Dwiggins
E. F. Mahoney

BRIEF CONTENTS

UNIT 1
Introduction to Automotive Electricity and Electronics   1

UNIT 2
Electron Theory   11

UNIT 3
Ohm's Law   21

UNIT 4
Electrical Circuits   35

UNIT 5
Magnetism   49

UNIT 6
Electrical Testing   57

UNIT 7
Batteries   73

UNIT 8
Capacitance and Inductance in dc   87

UNIT 9
Relays, Solenoids, and Motors   97

UNIT 10
Electrical Power and Energy   111

UNIT 11
Wiring Circuits   117

UNIT 12
Semiconductor Devices   127

UNIT 13
Semiconductor Integrated Circuits   139

UNIT 14
Alternating Current and the Alternator   147

UNIT 15
Lighting   159

UNIT 16
Safety Systems   177

UNIT 17
Panel Instruments and Warning Devices   189

UNIT 18
Optional and Convenience Systems   205

UNIT 19
Comfort Systems   229

UNIT 20
Sensing and Conversion Devices   241

UNIT 21
Ignition Systems   253

UNIT 22
Computer Controls   263

APPENDIX A
Powers of Ten   273

APPENDIX B
Periodic Table   277

INDEX   279

CONTENTS

UNIT 1
Introduction to Automotive Electricity and Electronics   1
Objectives 1

The Electronic Car 1

A Brief History 2

The Car of Today 4

The Car of Tomorrow 5

The Future of Automotive Electronic Systems 5
  On the Drawing Board, 6   Automatic Highway Control, 6   Automatic Steering Control, 7   Computerized Energy Distribution and Automated Control, 7   Laser Optical System, 7   Radar Braking System, 7   Route Guiding System, 8   Vehicle Proximity Detection System, 8

Tomorrow 8

Summary 8

Review 8

UNIT 2
Electron Theory   11
The Structure of Matter 11

Bohr's Law 12

Insulators 14

Semiconductors 14

The Behavior of Electrons 14
  Static Electricity, 14   Dynamic Electricity, 14

Voltage and Current 15

Analogy of Electricity 15

Open and Closed Circuits 18

Practical Exercise 19

Summary 19

Review 20

UNIT 3
Ohm's Law   21
Objectives 21

Current and Resistance 21

Resistance Values 22

Resistor Color Code 23

Variable Resistors 24

Ohm's Law 24

Using Ohm's Law in Automotive Circuit Current 26
Automotive Circuit Voltage, 27   Automotive Circuit Resistance, 27

Resistance of Wires and Cables 27
Resistance: Mil/Foot, 28   Length of Wire, 29   Stranded Wire, 30   Insulation, 30

Wires and Cable: Summary 31

Practical Exercise 32

Summary 33

Review 33

UNIT 4
Electrical Circuits 35

Objectives 35

Complete Electric Circuits 35

Wire Resistors 35

Switches 39

Resistive Parallel Components 40

Parallel Circuits 41
Product Over Sum Method, 43   Reciprocal Method, 44   Scientific-Calculator Solutions, 44

Series-Parallel Circuits 45

Practical Exercise 46

Summary 47

Review 47

UNIT 5
Magnetism 49

Objectives 49

Magnets and Magnetism 49
Natural Magnets, 49   Artificial Magnets, 49

Magnetic Polarity 50

Magnetic Fields 50

Atomic Arrangement 51

Effects of Magnetic Fields 51
Electromagnets, 52   Coil Strength, 54

Practical Exercise 55

Summary 56

Review 56

UNIT 6
Electrical Testing 57

Objectives 57

Electrical Test Equipment 57
Electric Meters, 57   Analog Meters, 57   Ammeters, 58   Voltmeters, 59   Ohmmeters, 59   Digital Meters, 61

Electrical Testing 62
Voltmeter Use, 62   Ammeter Use, 63   Ohmmeter Use, 64   Other Test Instruments, 65

Practical Exercise 69

Summary 70

Review 71

UNIT 7
Batteries 73

Objectives 73

Internal Resistance 74

Batteries in Series 75

Batteries/Cells in Parallel 75

Maximum Power Transfer 76

The Automotive Battery 76

Battery Safety 76
Neutralizing Sulfuric Acid, 77

Battery Components 77

Battery Chemical Action 78

The Electrolyte 79

Battery Power Ratings 80

Battery Care 80

Battery Testing 80
Specific Gravity Test, 81   Drain Load Test, 81

Loads 81
Parasitic Loads, 81   Sneak or Phantom Loads, 82   Self-Discharge, 82

Capacity Test   82

Testing Maintenance-Free Batteries   82

Charging Rate   83

Jump Starting   83

Conclusion   85

Summary   85

Practical Exercise   85

Review   86

UNIT 8
Capacitance and Inductance in dc   87

Objectives   87

Capacitance   87

Current Flow in Capacitive Circuits   88
    Current, 88   Voltage, 88

RC Time Constant   89

Capacitor Construction   91

Capacitor Ratings   91

Inductance   92

Inductor Ratings   92

Current and Voltage Relationship   92

LR Time Constant   92

Summary   94

Practical Exercise   94

Review   95

UNIT 9
Relays, Solenoids, and Motors   97

Objectives   97

Relays and Solenoids   97
    Relays, 97   bemf Polarity, 98   Solenoids, 99

Motor Principles   100

Torque   101

Starter Motor   102
    Starter Motor Circuits, 103   Starter Control Circuits, 105   Starter Drives, 105   Starter Motor Update, 105

Motor Control   105
    Series Resistive, 106   Pulse Width Modulation, 106

Summary   108

Practical Exercise   108

Review   109

UNIT 10
Electrical Power and Energy   111

Objectives   111

Horsepower   113

Electric Energy   113

Measuring Electric Power   114

Summary   115

Practical Exercise   115

Review   116

UNIT 11
Wiring Circuits   117

Objectives   117

Schematics   117
    Schematic Symbols, 117   Color Coding, 121
    Schematic Interpretation, 121   Schematic Sequence and Arrangement, 123

Wiring Harnesses   124

Summary   126

Practical Exercise   126

Review   126

UNIT 12
Semiconductor Devices   127

Objectives   127

Semiconductors   127
    Silicon Diodes, 127   Testing a Diode, 128
    Diode Uses, 129   Zener Diodes, 129

Transistors   130
    Bipolar Transistors   130   The Transistor as a Switch, 131   Transistor Amplifier, 132   Current Gain, 132   Application of a Signal to the Emitter-

Base Circuit, 133   Maximum Ratings and Electrical Characteristics, 133   Light-Emitting Diodes, 133   Photoconductive Cells, 133   Photovoltaic Cells, 134   Phototransistors, 134   Field Effect Transistors, 135   MOSFETs, 135

Summary   136

Practical Exercise   136

Review   137

UNIT 13
Semiconductor Integrated Circuits 139

Objectives   139

Semiconductor Devices   139

Gate Circuits   140

One-Shot Multivibrator   142

Flip-Flop Circuit (+5 V System)   143

Digital Counting   143

Large-Scale Integration (LSI)   144

Summary   144

Practical Exercise   145

Review   146

UNIT 14
Alternating Current and the Alternator 147

Objectives   147

Generating ac Voltage   147

Generated Electromotive Force (Voltage)   147

Rotating Coils   148

Sine Wave   149

Three-Phase Power   149

Resistive ac Circuits   149

Capacitance in ac Circuits   150

Inductance in ac Circuits   150

Tank Circuits   150

Alternators   151

Induced Polarity   152

The Rotor   152

The Stator   153

Alternator Output   153

Diodes   153

Voltage Regulators   155

Summary   156

Practical Exercise   156

Exercise Summary   156

Review   157

UNIT 15
Lighting 159

Objectives   159

Headlamps   159
Headlamp Circuit Protection, 160   Headlamp Aiming, 160   Headlamp Dimmer Switches, 162   Automatic Headlamp Control, 162   Twilight Sensing, 162   Headlamps Delay Off, 162   Headlamp Switches, 162

Parking Lamps   165

Tail Lamps   165

License Plate Lamp   165

Side Marker Lamp   165

Cornering Lamps   166

Clearance Lamps   166

Backup Lamps   166
Lamps (Bulbs), 166

Turn Signals   167
OFF Position, 167   LEFT Position, 167   RIGHT Position, 167   Front Side Marker Lamps, 168   Turn Signal Flashers, 170

Hazard Lamps   171
Hazard Flashers, 171

Brake Lights   172

Courtesy and Convenience Lamps   173
Dome Lamps, 173   Map Lights, 174   Door Lights, 174   Dash Lights, 174   Rheostats, 174

Fiber Optics   174

Summary   175

Practical Exercise 175

Review 176

UNIT 16
Safety Systems 177

Objectives 177

Windshield Wipers 177
 Nondepressed Park Systems, 177 Depressed Park Systems, 180

Windshield Washers 183
 Electrically Operated Washer Systems, 183 Low Fluid-Level Warning Systems, 183

Rear-Window Wipers 185

Rear-Window Defoggers 185

Rear-Window Deicers 185

Seat Belt Warning Systems 186

Summary 186

Practical Exercise 187

Notes 187

Review 187

UNIT 17
Panel Instruments and Warning Devices 189

Objectives 189

Dash and Display Panel Instruments 189

Voltage Limiters 189

Thermoelectric Gauges 190

Sending Units 191

Fuel-Level System 191

Coolant Temperature System 192
 Gauges, 194
 Lamps, 194

Oil Pressure System 195
 Gauges, 195 Gauges and Lamps, 196

Charging System 196
 Gauges, 196 Lamps, 197 Gauges and Lamps, 197

Horns 197
 Horn Circuit Without Relay, 197 Horn Circuit With Relay, 198 Horn Buttons, 198 Horn Relays, 199

Antitheft Alarm Systems 199

Speedminders 199

Speedometer Calibration Check 199

Key-in Warning Systems 201
 Quick Check, 201

Headlamp-on Warning Systems 202
 Quick Check, 202

Reverse Warning Systems 202

Door-Ajar Warning Systems 202

Hood and Trunk Lid-Open Warning Systems 203

Summary 203

Practical Exercise 203

Review 203

UNIT 18
Optional and Convenience Systems 205

Objectives 205

Front Seats 205
 Power Front Seats, 205 Seat Back Locks, 207 Power Reclining Seat Backs, 208

Power Windows 208
 Safety Precautions, 208 Power Tailgate Windows, 210 Power Window Adjustment, 210 Power Window Motor Testing, 211 Power Window Electrical Testing, 212 Power Window Relays and Interlocks, 212 Sunroofs, 212

Electric Door Locks 213
 Solenoid-Operated Door Locks, 214 Motor-Operated Door Locks, 214

Automatic Door Locks 214
 Keyless Entry Systems, 214 Tailgate Locks, 214

Trunk Lock Release Systems 214

Trunk Lid Closing Systems 215

Power-Operated Rear-View Mirrors 217

xiv CONTENTS

Electronic Fuel-Injection Systems 217
   Electronic Control Unit, 218   Fuel Delivery System, 218   Air Induction System, 219   Sensor System, 219

Fuel Pumps 219

Electric Overdrives 219

Speed Controls 219
   Electropneumatic Speed Controls, 220
   Electrical Speed Controls, 221

Electric Clocks 221

Radios and Sound Systems 221
   AM Radio, 222   FM Radio, 222   FM Stereo Radio, 222   Cassette Player, 222   Compact Disc, 222   CB Radio, 223   Cellular Telephones, 223

Antenna Systems 224
   Antennas, 224   AM/FM Antennas, 224   CB Antennas, 224   AM/FM/CB Antennas, 225   Cellular Phone Antennas, 225

Repairs 225

Summary 225

Practical Exercise 226

Review 226

UNIT 19
Comfort Systems   229

Objectives 229

Heaters 229

Air Conditioning Systems 231
   Mechanical Aspects, 231   Air Conditioning Physics, 231   Electrical Aspects, 232   Protective Switches, 232   Air Conditioner System Schematics, 233

Automatic Control Air Conditioning Systems 234
   Auto Temp II, 237

Summary 238

Practical Exercise 239

Review 239

UNIT 20
Sensing and Conversion Devices 241

Objectives 241

Position Sensing Switches 242
   Potentiometer, 242   Light Sensing, 242

Temperature Sensors 242

Pressure Sensors 243
   Piezoelectric Sensors, 243   Capacitance Sensors, 243

Light Sensors 244

Speed Sensors 244
   Magnetic Pickup Sensors, 244   Hall Effect Sensors, 244

Oxygen Sensors 245

Air-Flow Sensors (Hot Wire) 245

Crash Sensor 246

Digital Conversion Circuits 246
   Analog-to-Digital Converters, 247   dc Voltage to Digital, 247   Frequency-to-Digital Conversions, 248   Digital-to-Analog Converters, 248

Summary 250

Practical Exercise 250

Notes 250

Review 250

UNIT 21
Ignition Systems   253

Objectives 253

Motor Vehicle Ignition 253
   Conventional Ignition, 254   Ignition Waveform, 254

Distributors 255
   Conventional Distributor, 255

Transistorized Ignition System 255
   Advantages, 256   Breakerless Ignition, 256   Reluctor Pickup, 256   Hall Effect Ignition, 256   Distributorless Ignition System (DIS), 258   Single-Sensor Electronic Ignition, 259

Summary 260

Practical Exercise 260

Review 260

UNIT 22
Computer Controls 263

Objectives 263

Multiplexing 263

A Simple Multiplexing System 264
    Operation, 264   Brake and Turn Signals, 265

Multiplexing of Digital Engine Data 266
    Memory Cell, 267   Operation, 267

Parallel-to-Series and Series-to-Parallel Converters 269
    Parallel-to-Series, 269   Series-to-Parallel, 269

Summary 270

Practice Exercise 270

Review 270

APPENDIX A
Powers of Ten 273

APPENDIX B
Periodic Table 277

INDEX 279

◆ UNIT 1 ◆

INTRODUCTION TO AUTOMOTIVE ELECTRICITY AND ELECTRONICS

OBJECTIVES

On completion of this unit you will be able to:

- Understand the early development of the car.
- Discuss the technology found in today's car.
- Anticipate some of the plans for the car of the future.

The increased use of electricity and electronics for control of systems and subsystems is evident in almost every industry. Production equipment is rapidly becoming automated through the use of digital computers tied through analog systems to handle repetitive processes.

The American automobile is becoming increasingly controlled by the use of electronic sensors and computers. Electronics is used in the blending of engine and transmission controls, advanced power train controls, spark timing, manifold boost, and fuel mixture. In today's world if it is mechanical and requires positioning control, it probably includes computer electronics.

The purpose of this text is to help the serious student prepare to enter the automotive profession as an automotive electrician or automotive electrical technician. We emphasize "serious," because this text covers automotive electrical theory in depth. It is intended for those interested in becoming part of the elite team of automotive technicians dedicated to keeping today's car operationally safe and sound.

THE ELECTRONIC CAR

Electronic, by narrow definition, implies systems in which components are "purely electronic," such as integrated circuits (ICs) and various other solid-state devices. A broader definition, when applied to the automotive field, must include all types of hybrid electrically operated components, including those with pneumatic, hydraulic, or mechanical subsystem functions. A good example of this combination of components is the antilock brake system. When the brake pedal is applied (mechanical) by power assist (pneumatic), the on-board computer (solid-state electronic) "reads" the braking effort at each wheel and proportions individual wheel-braking effort (hydraulic) accordingly. Consequently, the term *automotive electronics* applies to any automotive electrical system or subsystem with pneumatic, mechanical, or hydraulic application (Figure 1–1).

When a key is turned in the ignition switch of a vehicle, the engine is always expected to start. Few realize, however, that starting a car's engine requires more

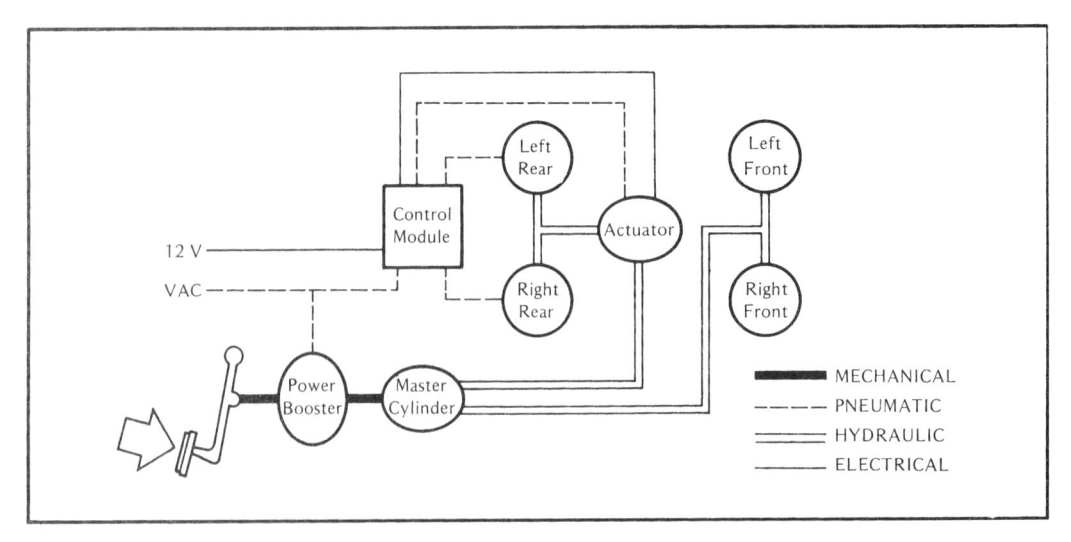

FIGURE 1–1 Antiskid braking system: when the brake pedal is applied by power assist, the on-board computer (control module) "reads" the braking effort and proportions individual wheel braking action

electrical power than starting many home air conditioners—2500 to 3000 watts. The engine starter requires this great amount of power for only a short period of time, however. Once started, the maximum electrical power that is required, even when driving at night with lights and air conditioner, is about 500 watts or a little more. This power, taken for granted by most motorists, is supplied by the alternator charging system and battery—only two components of the total automotive electrical system. The automotive electrical system, because it is taken for granted, is one of the most neglected systems of the car.

A BRIEF HISTORY

The "grandfather" of automotive electricity, Count Alessandro Volta (1745–1827), an Italian physicist-chemist, lived before the invention of the automobile and certainly did not know what his discovery would lead to when, in 1780, at the age of 55, he developed the first source of "constant current electricity" (later to be known as direct current), the voltaic pile. He, and other scientists, used the voltaic pile to decompose water by electrolysis, electroplate precious metals, and form electromagnets. For his achievement, in France, Volta was made a count and a senator of the kingdom of Lombardy, and he was awarded the Cross of the Legion of Honor. The greatest tribute for his discovery came more than 50 years after his death: the unit of electromotive force, the *volt*, was named in his honor in 1881.

Other physicists and scientists share credit for today's modern electronics and electricity. Charles Augustine de Coulomb (1736–1806), a French scientist and inventor, discovered the principles for measuring the force of magnetic and electrical attraction. In 1884 the International Congress of Electricians named the electrical unit "coulomb" in his honor. A *coulomb* identifies the quantity of electrons (6.25×10^{18}) produced by a current of one ampere for one second.

Andre Marie Ampere (1775–1836), a French physicist, noted that two parallel conductors attract each other when current is passed through them in one direction and oppose each other when current is passed through them in the opposite direction. He discovered the unit of electrical strength, the *ampere* (often called *amp*), among other discoveries.

James Watt (1736–1819), a Scottish engineer and inventor, originated a method of determining steam engine power output in 1769. He named it "horsepower" because in Watt's day power was determined by the work ability of the average draft horse. Today horsepower relates to the amount of energy required to raise 550 pounds (249.5 kg) one foot (30.5 cm) in one second. The electrical unit, watt, is named in his honor. Incidentally, one horsepower is equal to 745.7 (746) *watts*.

Georg Simon Ohm (1787–1854), a German physicist, discovered a principle known as Ohm's law. Ohm's law establishes the relationship between the current flow (ampere), the potential difference (voltage), and the resistance of individual components in a circuit (ohm). The original ohm, discovered in 1827, and the international ohm, established in 1893 by the International Electrical Congress, are almost identical. The absolute ohm is equal to 0.999505 international ohm. The unit of resistance is the *ohm*.

These five men—Volta, de Coulomb, Ampere, Watt, and Ohm—are largely responsible for modern automotive electronics and the methods by which we can understand, design, troubleshoot, and repair automotive electrical systems. Together their lives spanned less than 120 years (1736–1854); the last of them died just six years before a Belgian, Etienne Lenoir, patented the first gasoline-powered internal-combustion engine, in France in 1860.

Perhaps Charles Franklin Kettering (1876–1958) should be credited with the beginning of the electronic car. In 1908 he patented the first battery ignition system, the forerunner of the conventional ignition system. Kettering's system consisted of a battery, a coil, and a set of points (Figure 1–2). Or perhaps the credit should be shared by Henry Martyn Leland (1843–1932). With Kettering's help, Leland developed the electric starter available for the first time on 1912 Cadillacs. Regardless of who gets the credit, by 1912 the modern car was fast carving a notch in history, only 16 years after a Duryea "motor wagon" was featured in a Barnum and Bailey Circus parade as a novelty (April 2, 1896).

The car of the early 1900s was a product of the development begun by a German engineer, Karl Benz (1844–1929). In 1885 he built a three-wheeled vehicle using an engine designed by Gottlieb Daimler (Germany) and a carburetor designed by Fernand Forest (France). Benz added a magneto electrical system of his own design and produced the first known motor vehicle with an electrical system. Although several motor vehicles were built before Benz's effort, he sparked the continuing development of the motor vehicle.

After reading about Benz's achievements in *Scientific American*, two bicycle mechanics, brothers Charles E. and J. Frank Duryea, developed a two-cylinder car. The Chicago *Times-Herald* sponsored a 55-mile (88.5-km) race in 1895; of the Duryea car, a Benz car, and four other cars entered, only the Duryea and Benz cars finished. Duryea was the victor with an average speed of 5 miles per hour (8 km/hr) for the course. Average driving speed was actually 7 miles per hour (11.3 km/hr) if the time required for repairs is deducted. In the 11-hour race, more than three hours was required for repairs—considered average for those days.

The first electric battery manufacturing concern, known as the Electric Storage Battery Company, was formed by an American lawyer, Isaac L. Rice (1850–1915). Rice soon merged with the motor carriage division of the Pope Manufacturing Company to form the Electric Vehicle Company. Pope Manufacturing, the first electric motorcar factory, was founded in 1897 in Hartford, Connecticut. Its founder, Colonel Albert A. Pope, theorized that combustion-powered vehicles would not sell well because the gasoline engine was located under the driver's seat. As he rationalized, "You can't get people to sit on top of an explosion." He also built gasoline-powered motor cars. Pope's theory held for the first two years of operation—he sold almost twelve times as many electric cars as gasoline-powered cars during that period. Incidentally, although Pope manufactured both electric- and gasoline-powered vehicles, he prudently continued to manufacture bicycles under the trade name Columbia. The Columbia bicycle, produced into the modern era, outlasted both his electric- and gasoline-powered machines.

Important events in the development of the car are shown in Table 1–1.

The war effort in the early 1940s greatly accelerated technological advancement and production methods and primed the automotive industry for today's level.

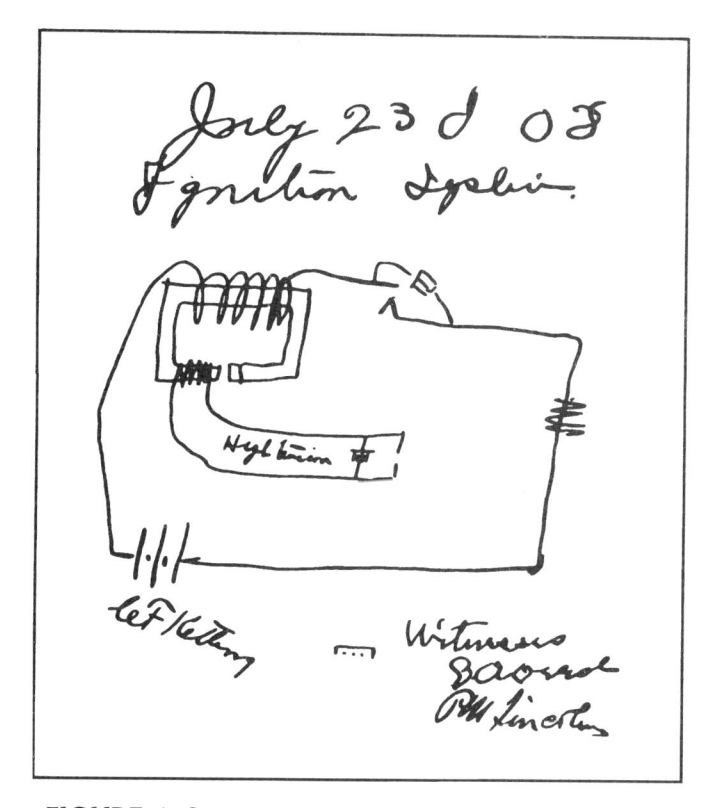

FIGURE 1–2 A copy of the drawing submitted by C. F. Kettering to obtain the first U.S. patent for an ignition system

TABLE 1–1 Major events in the development of the car

1897	First commercial production of an automobile
1902	American Automobile Association (AAA) founded
1903	Henry Ford (1863–1947) founds the first of the "big three," the Ford Motor Company
1905	The Society of Automobile Engineers (SAE) founded (The name was later changed to Society of Automotive Engineers.)
1908	The Ford Motor Company introduces the Model T, known as the "tin Lizzie." Charles F. Kettering (1876–1958) patents the first battery ignition system. William C. Durant (1860–1947) founds General Motors Corporation.
1911	Henry M. Leland (1843–1932), with the help of Kettering, develops the electric starter first used on 1912 Cadillacs
1914	The first production line is started at Ford Motor Company. Henry Ford raises employees' pay to $5 per day, about twice the average pay of other manufacturers at the time.
1916	The Federal Road Act is passed
1921	The Federal Highway Act is passed
1923	Ethyl gasoline is developed
1925	Walter P. Chrysler (1875–1940) founds the Chrysler Corporation
1928	The average hourly wage for automotive workers is raised to $0.75 per hour, about 27 percent higher than the average wage of other manufacturers
1935	The United Auto Workers (UAW) union is chartered
1940	The first turnpike is opened in Pennsylvania
1942	Civilian vehicle production is halted. All manufacturers produce military vehicles for use in World War II.
1943	The fiftieth anniversary of the Duryea car, now in the possession of the Smithsonian Institution, passes with little notice because of the war
1945	Production of civilian automobiles is resumed by all automotive manufacturers
1947	The fiftieth anniversary of commercial automobile manufacturing is celebrated in Hartford, Connecticut

THE CAR OF TODAY

Improved highways and automotive mechanical engineering technologies paved the way for faster speeds that, in turn, require precise and accurate timing of automotive functions. This, with the greater demand for comfort, requires much more sophisticated electronics than the car of yesterday.

Today's average car has over 1600 feet (488 meters) of wire made up into about 50 individual wiring harnesses. There are more than 30 light bulbs in the modern car, depending on the make and model. The air conditioning system has sufficient capacity to easily cool a small two-bedroom home.

The electrical system of the car consists of many subsystems. The power system provides primary power with a battery and secondary power with an alternator that keeps the battery charged. The starting system provides a means for starting the engine, and the ignition system provides low- and high-voltage power circuits necessary to keep the engine running. The lighting system includes several subsystems such as head/tail lamps, parking lamps, turn signal/brake lamps and courtesy/convenience lamps. The power system includes window, sun roof, and seat circuits. The safety system includes windshield wipers/washers, hazard lighting, seat restraint systems, and equipment failure warning lamps. Other electrical systems include heating/air conditioning, radio/sound equipment, and other accessory equipment (see Figure 1–3).

Electronics warn of low brake fluid level, worn brake linings, and other braking defects. Unsafe conditions of the wheels, the steering system, and the front-end parts are audibly and visually signaled to the driver. Other fail-safe devices may warn of low engine coolant and oil level. Safety problems are fed into an on-board

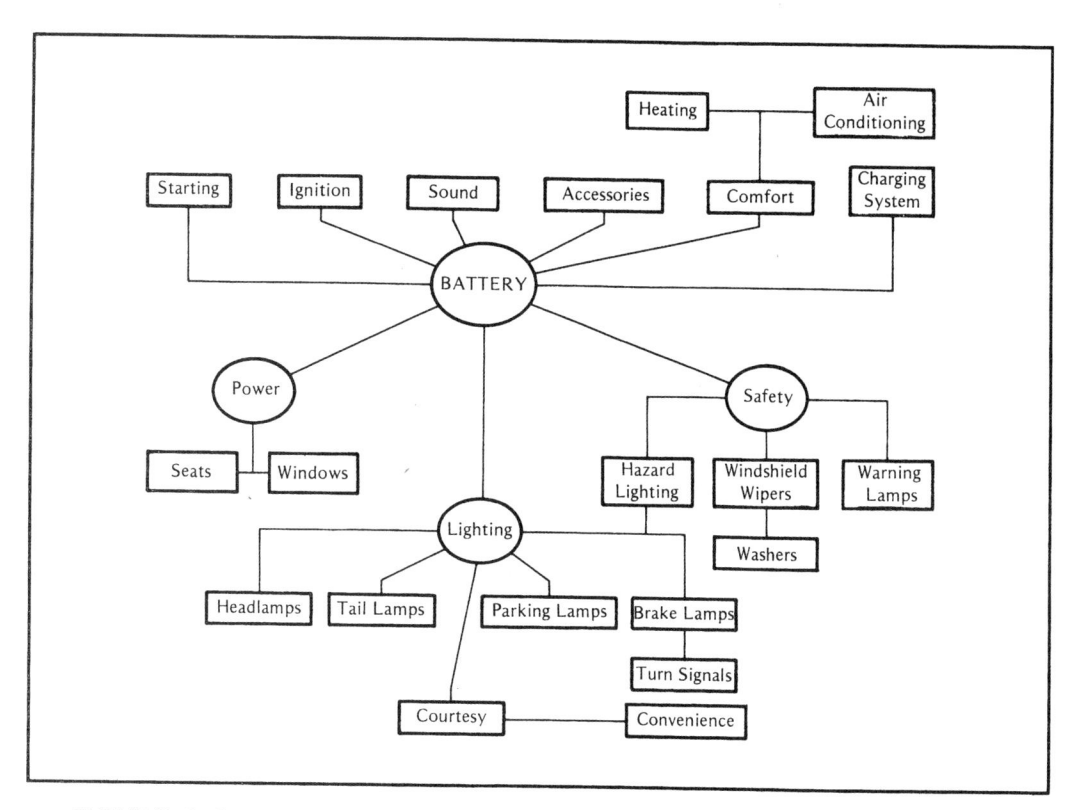

FIGURE 1-3 Many subsystems make up the total automotive electrical system

logic circuit that allows ample warning time to get off the road safely before stopping the engine. Today's on-board computer monitoring systems "scan" individual components as often as 10 times each second and make complex adjustments automatically to ensure optimum performance and safety under a wide variety of conditions.

THE CAR OF TOMORROW

The car of tomorrow (Figure 1–4) may make present-day cars look as old fashioned as antique cars now seem. Sophisticated on-board electronics will make tomorrow's car safer by taking much of the human error involved in response time and the neglect factor away from the driver.

More complex electronics will include all-wheel steering and anticollision devices that will limit the driver's ability to approach and pass another car. This system will prevent passing unless there is ample car-to-car clearance based on variables such as road conditions, speed, and traffic. The car of tomorrow may require only an annual fuel cell for power. It may be programmed over certain routes to ensure swift and safe travel with a minimum of driver intervention. Some of these devices are now in experimental cars; others are on the drawing board.

THE FUTURE OF AUTOMOTIVE ELECTRONIC SYSTEMS

Most automotive electronic control systems are factory adjusted and preset. Many are not field adjustable, essentially because the average automotive technician does not yet have the technical knowledge to adjust them. The new breed of automotive electronic technician will have the knowledge required to test, adjust, and repair complex automotive electronic systems and subsystems. Meanwhile, the trend is for component replacement at the manufacturer's or customer's expense, depending on warranty status. Present Detroit thinking is, "No can tinker, no can foul up." Also, electronic system performance analyzers (Figure 1–5) are a major investment that will separate the amateurs from the true professionals. Many analyzers cost as much as or more than the average luxury car. The analyzer, however, is an essential aid to determine which (if any) subsystem is defective.

FIGURE 1-4 The car of tomorrow

At present totally electronic automotive systems are not being installed in all production cars. Most production cars, for the next few years, will have partially electronic systems, as in earlier years. These systems are not, however, suitable to meet the emission and economy standards proposed for the close of the twentieth century. Of development since 1978, one California test engineer notes, "What we are after is better fuel economy, lower emissions, improved performance, and consistency. That is what electronics can give us."

On the Drawing Board

Many systems for the future automobile are in development by automotive engineers. These systems, as complex as their names imply, include automatic highway control (AHC), automatic steering control (ASC), computerized energy distribution and automated steering control (CEDAC), laser optical system (LOS), radar braking system (RBS), route guidance system (RGS), and vehicle proximity detection system (VPDS). These systems and others, such as the automatic battery condition checker, oil level checker, low fluid-level (cooling system, brake system, and windshield washer) warning system, and the lighting system monitor, will be covered briefly.

Automatic Highway Control

The automatic highway control (AHC) automatically maintains a safe distance between vehicles, depending on speed and road conditions. This system is ideal under heavy traffic conditions; it eliminates the physically tiring control by the driver of brake pedal and accelerator pedal movement. Near-ideal spacing of all

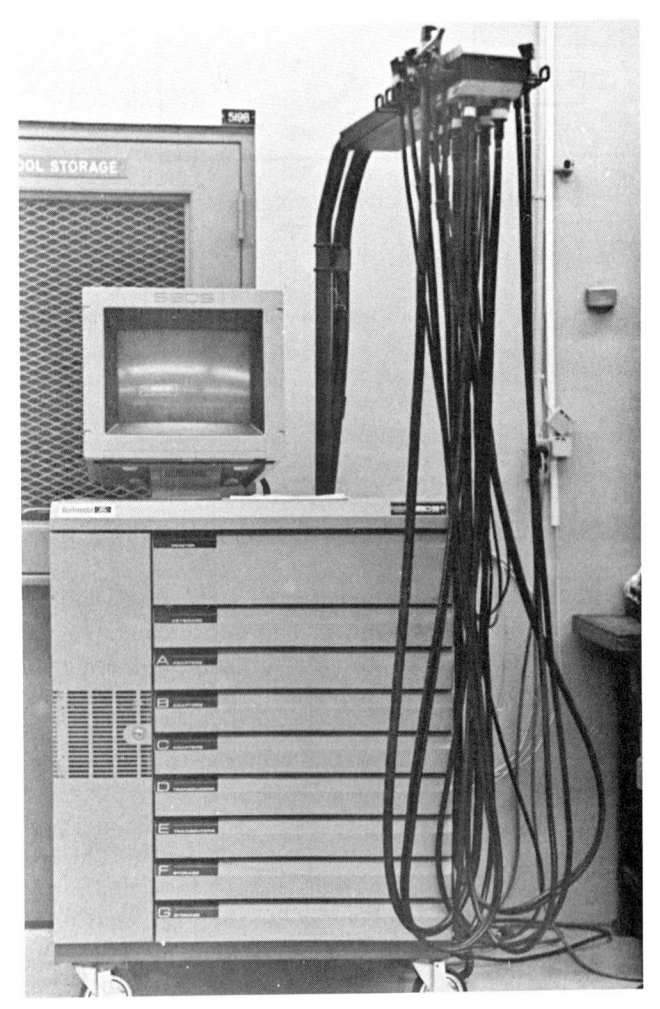

FIGURE 1-5 Diagnostic computer used to check systems and subsystems (*Source:* BET, Inc.)

vehicles also aids in traffic control. If all vehicles move at a constant and safe speed (for a given road condition), more vehicles will pass through any particular area in a given time period. This system will virtually eliminate rear-end collisions because it eliminates poor driver judgment, which is often caused by fatigue.

Automatic Steering Control

There are a number of automatic steering controls (ASC) being developed; the most promising is the wire-guidance type. Regardless of the type, however, all serve the same general function. The ASC is an automated one-way communication computer system that maintains continuous tracking contact with automated vehicles. It provides continuous steering guidance, sign information, and voice and digital-logic communications. Among other functions, speed, braking, steering, and headway controls may also be provided. Also, these controls may be provided by operator (voice) or automatic programmer (digital).

Computerized Energy Distribution and Automated Control

The computerized energy distribution and automated control (CEDAC) ties together all on-board electronic subsystems to form a totally automatic operation (Figure 1–6). CEDAC is an electrofluidics control system with a central computer with a single energy distribution and control harness. Perhaps by scan, the control computer of CEDAC may be used to monitor and control every function of a vehicle from antilock braking to lamp failure monitoring—even climate control (air conditioning) temperature and humidity conditions. Some of the subsystems that the on-board logic computer will remember and control under all conditions include the following:

1. Lighting: on–off, bright–dim, brake, turn signals
2. Climate control: temperature and humidity
3. Fuel: air and fuel ratio mixtures
4. Ignition: timing, low- and high-voltage control
5. Transmission: shifting or variable torque ranges
6. Braking: antilock and anticollision

Because CEDAC utilizes logic memory, there will be a reduction in both number and complexity of switches and levers on the dashboard. Many controls that are now driver operated will operate automatically by preprogrammed memory computer logic in the car of tomorrow.

Laser Optical System

The laser optical system (LOS) is designed to see the "blind spot" behind, under, or near the rear bumper of a car. The LOS provides an audible and/or visual warning of any object within 10 feet (3 meters) of the rear bumper. The driver must then determine whether it is safe to back up before putting the vehicle in reverse gear. The main purpose of the LOS is to make the driver aware that an obstacle is present. This is particularly important for backing out of driveways in residential areas where small children play.

Radar Braking System

The radar braking system (RBS) provides automatic braking when closing at an excessive speed on a slower-moving vehicle. In addition, it will completely and automatically stop a vehicle 8 to 10 feet (2.4 to 3.0 meters) from any obstacle. Braking effort is semiautomatic, depending on road and traffic conditions. The RBS has two modes: city and highway. The city mode cuts in the brakes at 30–40 feet (9.1–12.2 meters), whereas the highway mode cuts in the brakes at about 150 feet (45.7 meters).

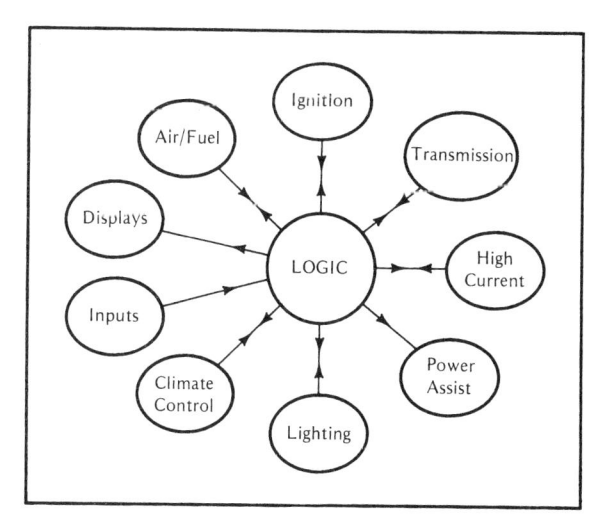

FIGURE 1–6 Computerized energy distribution and automatic control system

Route Guidance System

The route guidance system (RGS) is similar in many respects to the automatic steering control (ASC). It consists of an in-car (mobile) and an on-highway (fixed) receiver-transmitter that establishes a two-way communication link. Before starting a trip, the driver enters a code into the mobile unit for the destination address. Appropriate route and maneuvering instructions are then returned to the mobile unit. This system differs from the ASC in that the driver is in complete control of the vehicle at all times. The RGS only suggests the route to take.

Vehicle Proximity Detection System

The vehicle proximity detection system (VPDS) detects the presence of a vehicle approaching, tailing, or overtaking the VPDS-equipped vehicle. It effectively warns of another vehicle in the right- or left-rear blind spots. Information on the speed, distance, and location of the approaching vehicle is displayed on the dash. The system responds as long as the approaching vehicle is traveling at least 35 mph (56.3 km/h) and is within 25 feet (7.6 meters). This system is particularly helpful for driving in rain and fog.

Other Systems

Other systems, already available on certain models, display a visual or audible dashboard warning of problems such as the following:

1. Battery: under-/overcharging and/or low fluid level
2. Engine oil: fluid level and pressure
3. Windshield washer: low fluid level
4. Engine coolant: fluid level and temperature
5. Brake system: fluid level, temperature, and wear
6. Lighting system: on–off, or burned-out bulbs
7. Road icing conditions

TOMORROW

The shape, style, and electronic equipment of the car of tomorrow is anybody's dream. However, it is certain that future cars will have more complex on-board electronics. Although this development may require less skill and attention on the part of the driver, it will demand more skill and training on the part of automotive electronics technicians.

SUMMARY

- Tomorrow's technician will have to possess the skills and knowledge necessary to work on increasingly sophisticated automobiles.
- The automotive technician will probably need to be more of a diagnostician and less of a mechanic in the future.
- Assemblies and subassemblies will be replaced as a component, rather than be repaired.
- Electrical and electronic diagnostics will be a prerequisite to all automotive repairs.

REVIEW

1. SAE is an acronym for:
   a. Society of Automotive Electricians
   b. Safety of Automotive Engineering
   c. Society of Automotive Engineers
   d. None of the above is correct.

2. The electric starter was first used by:
   a. Cadillac   c. Ford
   b. Chrysler   d. Leland

3. The grandfather of automotive electricity is:
   a. Volta   c. de Coulomb
   b. Ampere   d. Kettering

4. The first auto race, won by Duryea, averaged _____ mph.
   a. 5   c. 11
   b. 7   d. 55

5. Today's average car has _____ feet of wire.
   a. 160
   b. 1600
   c. 16,000
   d. 50

6. Technician A says that the electronic system of a car may be checked with a standard multimeter if the specifications are known. Technician B says that the electronic system of a car must be checked with sophisticated test equipment.
   a. Technician A is correct.
   b. Technician B is correct.
   c. Either technician A or B may be correct.
   d. Neither technician A nor B is correct.

7. Of particular help to "see" the blind spot behind the vehicle, the _____ is being developed.
   a. route guidance system (RGS)
   b. laser optical system (LOS)
   c. vehicle proximity detection system (VPDS)
   d. look back detection system (LBDS)

8. Electronic control includes _____ subsystems.
   a. hydraulic
   b. mechanical
   c. pneumatic
   d. all the above

9. The Columbia bicycle was manufactured by:
   a. Henry Ford
   b. Isaac Rice
   c. Albert Pope
   d. Gottlieb Daimler

10. The American Automobile Association (AAA) was founded in:
    a. 1900
    b. 1902
    c. 1904
    d. 1906

UNIT 2

ELECTRON THEORY

OBJECTIVES

On completion of this unit you will be able to:

- Relate electricity to things that are common in your environment.
- Picture, in your mind, electrons and their movement.
- Define the terms *voltage* and *current*.
- Recognize the difference between an open and a closed circuit.

The information presented in this unit is theoretical. *Theory* is a line of reasoning that is assumed to be correct. Theory may change from time to time as new information comes to light and suggests new lines of reasoning.

Electrical theory is highly developed at the present time and requires a knowledge of higher mathematics for complete understanding. However, mathematical explanations of electricity are not the concern of this text; rather, we present an elementary approach to those electrical concepts, theories, and formulas that are essential to the technician.

We suggest that this unit be read as a story. The material provides good background information that will assist the learner in understanding the nature of electricity. A technician with a good grasp of the nature of electricity will find troubleshooting a much easier task.

THE STRUCTURE OF MATTER

Everything in the known universe is made from 96 stable elements, along with a few unstable short-term elements (see Appendix B), in their pure forms or in combination with other elements. A combination of elements is called a *compound* or *alloy*. In their pure forms, the elements are made up of electrons, protons, and neutrons in a special arrangement for each element. The negatively charged, tiny object called an electron must be considered in order to understand the behavior of electricity. This requires an investigation into the structure of matter: of what are things really made?

Imagine looking at a piece of copper (Cu) under a microscope as you increase the magnification to an extremely high magnitude. You will see the structure of copper as it really is. With the microscope at a magnification of 10 (10 ×), the Cu looks like Figure 2–1. As the magnification increases to 100 ×, the rough crystalline structure is seen (Figure 2–2).

It takes a large jump in magnification to 10,000,000 × before the beginning of atomic structure becomes evident, as seen in the bumpy surface in Figure 2–3. Finally, at a magnification of 100,000,000 ×, individual atoms can be seen (Figure 2–4).

What is there to see? The center of a Cu atom, the nucleus, consists of positively charged and neutral particles—protons and neutrons. Negatively charged particles—electrons—whirl in four orbits around the center. It is on these negatively charged particles that electron theory is based. Figure 2–5 shows an accepted representation of a Cu atom.

In the study of electricity, the electrons whirling about the nucleus are the center of interest. The ability

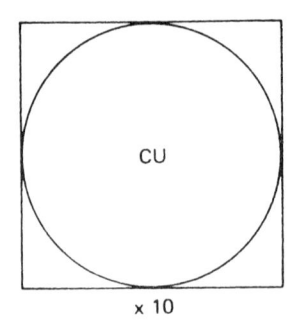

FIGURE 2–1 Plain copper (Cu)

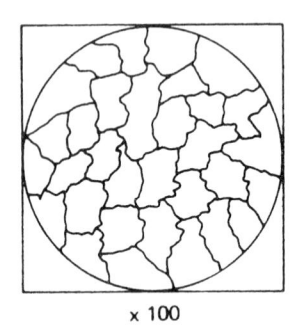

FIGURE 2–2 Copper (Cu) crystal

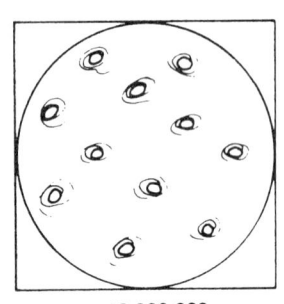

FIGURE 2–3 Atomic structure

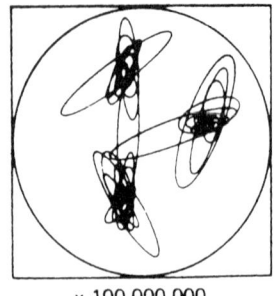

FIGURE 2–4 Copper (Cu) atoms

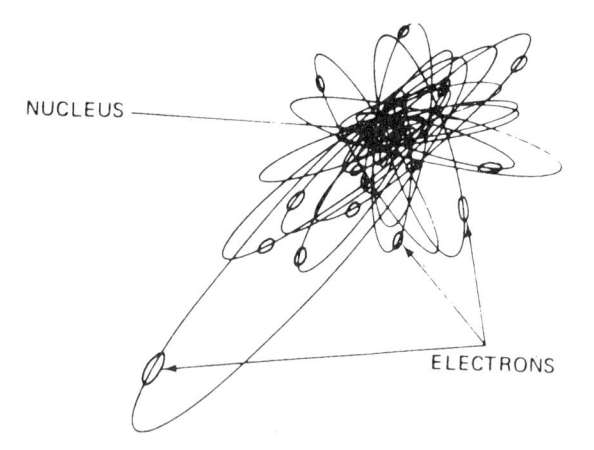

FIGURE 2–5 Copper (Cu) atom

to move these electrons on command produces electricity that can be used to do work.

BOHR'S LAW

As scientists experimented with things of an electrical nature, they learned about the characteristics of electricity and electrical phenomena. Physicist Niels H. D. Bohr (1885–1962) developed a model of electron, proton, and neutron arrangements to represent everything in the universe; this became known as Bohr's quantum theory of atomic structure. According to Bohr, an atom of hydrogen (H), the lightest known element, consists of one proton in the nucleus and one electron rotating about it (Figure 2–6). Bohr's theory of the hydrogen atom is regarded as the basis of modern atomic theory.

Heavier materials move up the atomic scale, having a greater number of particles in the nucleus and a greater number of electrons whirling about that center.

As the theory developed and is now understood, the arrangement of electrons in rings (orbits) around the nucleus is fixed. The following rules have been found to prevail:

1. In the first orbit around the nucleus, there can be no more than 2 electrons.
2. In the second orbit, the maximum is 8 electrons.

FIGURE 2–6 Hydrogen (H) atom

3. The third orbit has a maximum of 18 electrons.
4. The fourth orbit has a maximum of 32 electrons.

As this list indicates, the number of electrons in each orbit around the nucleus of an atom is fixed. The number of electrons in orbit, as well as the number of protons at the center, determines the type of material.

Copper, for example, has a nucleus of 29 protons and 34 neutrons with 29 electrons rotating around it. According to the fixed arrangement of electrons, there are 2 in the first orbit, 8 in the second orbit, and 18 in the third orbit. This accounts for only 28 electrons. The remaining electron orbits all by itself in the fourth ring (Figure 2–7). Since this electron is all by itself in the fourth ring, it is not held as strongly to the nucleus as are the other electrons. The electron is said to be "free"—free to move whenever an outside force acts on it. Copper is a good conductor of electricity because each of the billions of tiny atoms in a copper conductor has one free electron ready to move when a force is applied to it. A reasonably small copper wire will pass a relatively large amount of electricity.

Another example of a good conductor is aluminum (Al). The Al atom has a total of 13 electrons in three orbits. Two are in the first orbit, eight in the second orbit, and three in the third orbit (Figure 2–8). There is a

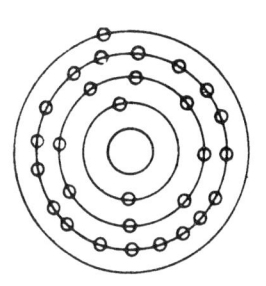

FIGURE 2–7 Copper (Cu) atom

capacity for 18 electrons in the third orbit. Thus the three electrons in the third orbit are not tightly bound to the nucleus, and the third orbit can momentarily accommodate extra electrons. Aluminum, therefore, is a good conductor of electricity, but not as good as copper. If the same size of copper and aluminum wire were used, copper would conduct better. If weight and cost are factors to be considered, aluminum, a lightweight metal, is an excellent choice. With wire of equal length and weight (not size), aluminum will conduct electricity twice as well as copper. Aluminum wire of equal weight, however, will be much larger.

Although aluminum offers a weight advantage and overall vehicle weight is an important consideration, aluminum wire is not used for automotive electrical service. The use of multiplex wiring systems, covered in Unit 22, negates the advantages of aluminum wiring.

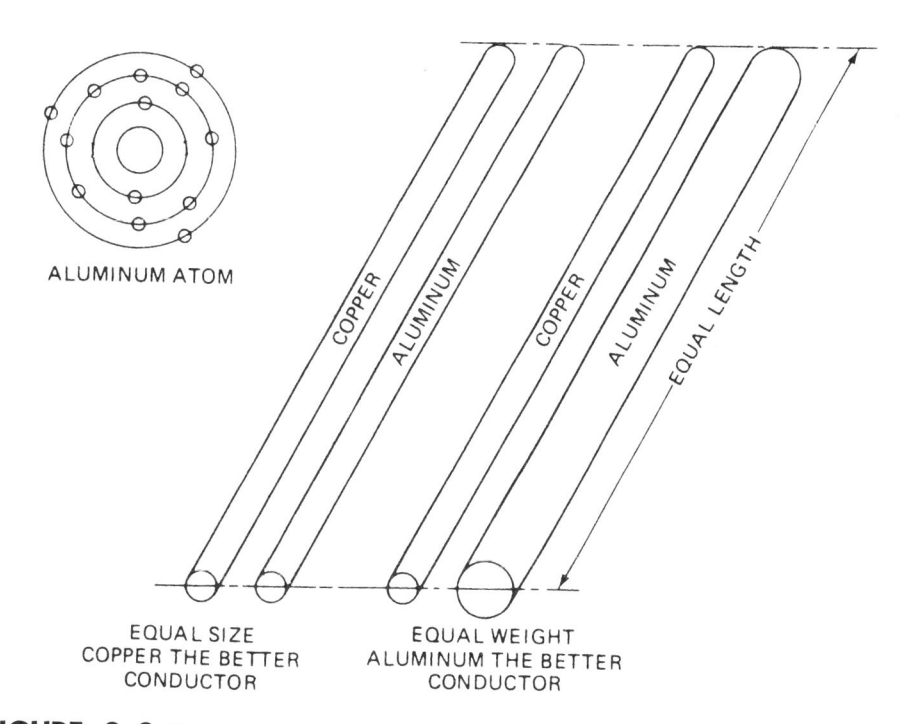

FIGURE 2–8 Equal size and equal weight copper (Cu) and aluminum (Al) conductors

INSULATORS

Insulators are made of materials that have few, if any, free electrons. Materials such as glass, rubber, and most plastics are good insulators. Actually, according to Ohm's law (covered in Unit 3), everything in nature conducts electricity. It is reasonable to note, however, that the conductivity rate for insulators is so minute it is not worth consideration.

SEMICONDUCTORS

Semiconductors are materials that fall midway between good conductors and poor conductors. Silicone (Si) and germanium (Ge) are good examples of elements that are semiconductor material. Semiconductors are covered in Unit 12.

THE BEHAVIOR OF ELECTRONS

After walking across a wool rug or sliding out of a carseat, you sometimes have the strange ability to cause an electric spark when you touch some other object. When this happens you are "electrically charged"—you carry an actual electric charge. You can observe this same phenomenon by running a plastic comb through dry hair. The comb will then attract lightweight objects, such as small scraps of paper.

Static Electricity

The electricity that is obtained by rubbing some objects together is called *static* electricity. It is called static because electrons are picked up by an item, and the electrical charge remains on the item until it touches some other object that has an unlike charge.

As they performed experiments on static electricity, scientists found that two types of charges could be obtained, depending on the nature of the materials rubbed together. If two glass rods are rubbed with a silk cloth, they are charged alike and repel each other; if two hard rubber rods are rubbed with wool, they are also charged alike and also repel each other.

Like charges repel each other

If a charged glass rod is approached by a charged rubber rod, the glass rod and rubber rod will be attracted to each other. The charges on the rods are unlike each other, so they attract.

Unlike charges attract each other

Benjamin N. Franklin (1706–1790), an American scientist, suggested the terms *positive* and *negative* to classify these charges. In our example the glass rods have a positive charge and the rubber rods have a negative charge.

Static electricity is electricity at rest

Special care must be exercised when handling metal oxide semiconductor (MOS) transistors and chips. They are susceptible to damage from static electricity discharges. Many automotive electronic devices contain MOS transistors. When handling MOS devices, make sure that you are always grounded (Figure 2–9).

Dynamic Electricity

Static electricity is not a form of electricity that can be used to produce power. A practical form of electricity, the form that is used to provide power and energy all over the world, is dynamic electricity.

Dynamic electricity is electricity in motion

In an automobile, the battery provides dynamic electricity, as does the alternator (generator). For many power companies, water (H_2O) at high pressure behind dams moves large alternators that provide dynamic electricity. In areas where there are no rivers or dams, large steam plants are often used to turn alternators that generate dynamic electricity.

A newer kind of power plant, which uses atomic energy to provide the heat needed for steam turbine operation, has been constructed in many areas of the world. One such atomic plant is located at Turkey Point, Florida (Figure 2–10). The Turkey Point plant, located near Miami, produces much of the electricity used in the South Florida area.

Another energy source that may be of greater importance in the future, geothermal, is presently being investigated. In order to tap this energy source, a deep hole must be drilled into the earth to an area of intense heat. Water introduced into the hole is converted to superheated steam. The steam is returned to the earth's surface to operate turbines that rotate the electric alternators, providing electricity.

Regardless of the energy source, dynamic electricity is produced by generators or alternators that convert mechanical energy into electrical energy.

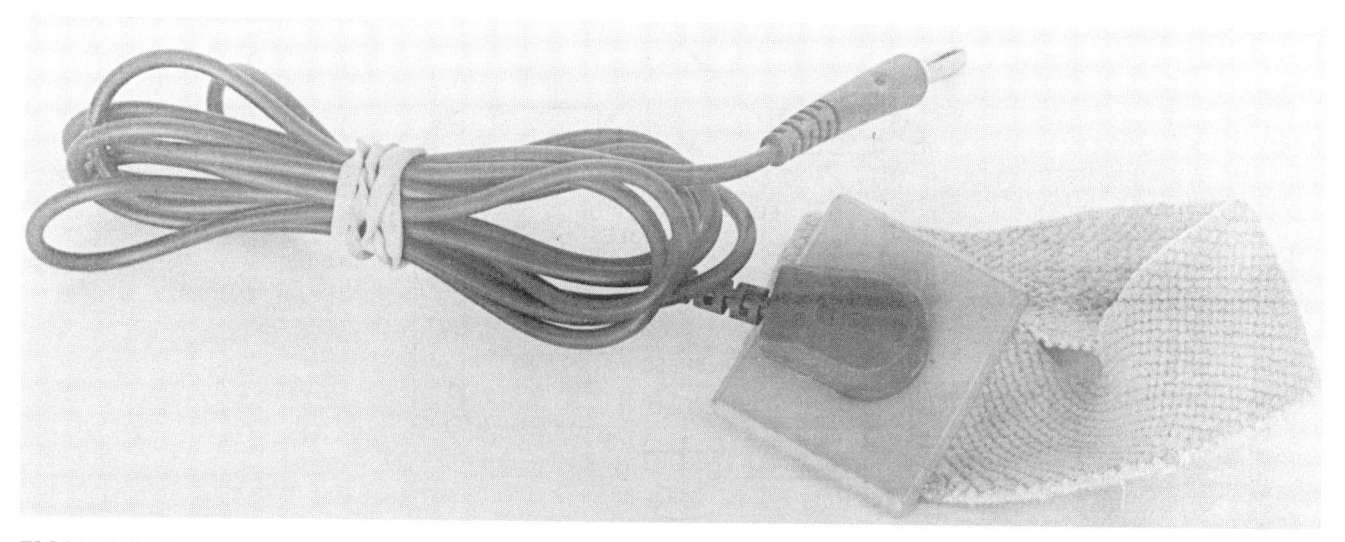

FIGURE 2–9 Be sure you are grounded when handling delicate solid-state devices (Courtesy of BET, Inc.)

VOLTAGE AND CURRENT

A well-developed understanding of the basic terms relating to dynamic electricity and their relationship with each other is necessary in order to troubleshoot electrical systems effectively.

Voltage and *current* are terms associated with dynamic electricity. We cover voltage and current together because it is impossible to obtain the one without some of the other. *Voltage* is the pressure that causes electrons to move. *Current* is the movement of electrons.

ANALOGY OF ELECTRICITY

As is often the case, it is easier to understand a new subject when you can relate it to some past experience. Let us consider the electrical phenomenon in relation to air—more specifically, to an air-moving system.

In the automobile a fan is provided in order to move air through the radiator and air conditioning condenser. The rotating fan blade pulls air molecules through the radiator and condenser.

This can be compared to an electrical system. The fan or air mover must be placed in a closed system (Fig-

FIGURE 2–10 Turkey Point nuclear power plant (*Source:* Florida Power and Light Company)

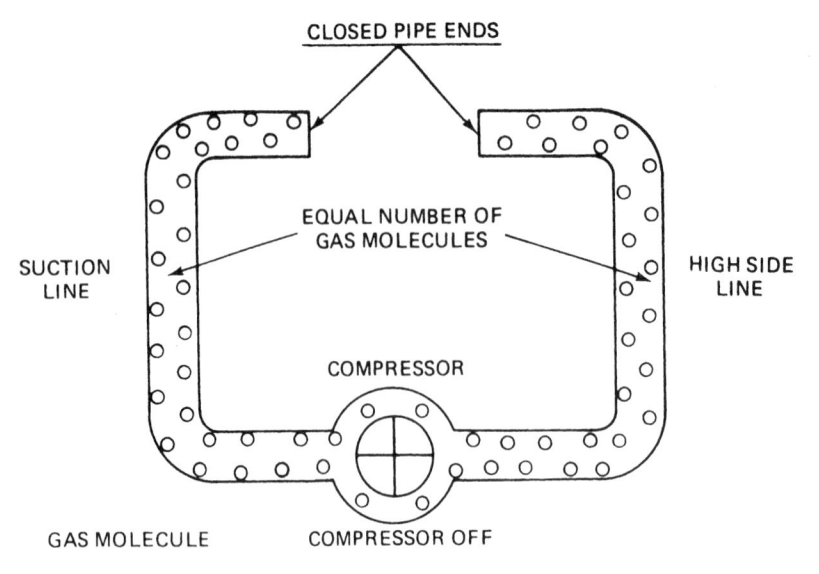

FIGURE 2-11 Closed air system in balance (fan off)

ure 2-11) in order to relate to electricity, since electricity is mainly confined to flow in wires.

The fan in Figure 2-11 is enclosed and has pipes connected at its input and output sides. The pipes are closed at their ends. When the fan is not running there will be an equal number of air or gas molecules on either side of the system. When this occurs the system is said to be *in balance*. This is a natural state in nature that all systems try to reach.

Figure 2-12 shows the fan in operation. Some of the gas molecules are moved by the fan from the suction line (on the left) to the discharge side (on the right). There is a relative excess of gas molecules in the discharge line (compression), causing a pressure difference between the two lines.

If a small tube, as shown in Figure 2-13, were connected from the discharge line to the suction line, some of the gas molecules would move through the tube. There would be a continuous movement of gas molecules through the system as long as the fan was operating. The suction side of the system would continue to be under a vacuum, and the discharge side would continue at high pressure as long as the small tubing offered some restriction (resistance) to the movement (flow) of gas molecules through the system. In an electrical system the action is very similar to that of the fan-air system.

In Figure 2-14 an electric generator has been substituted for the compressor, and the air lines have been replaced with copper wire. The circuit is open, since

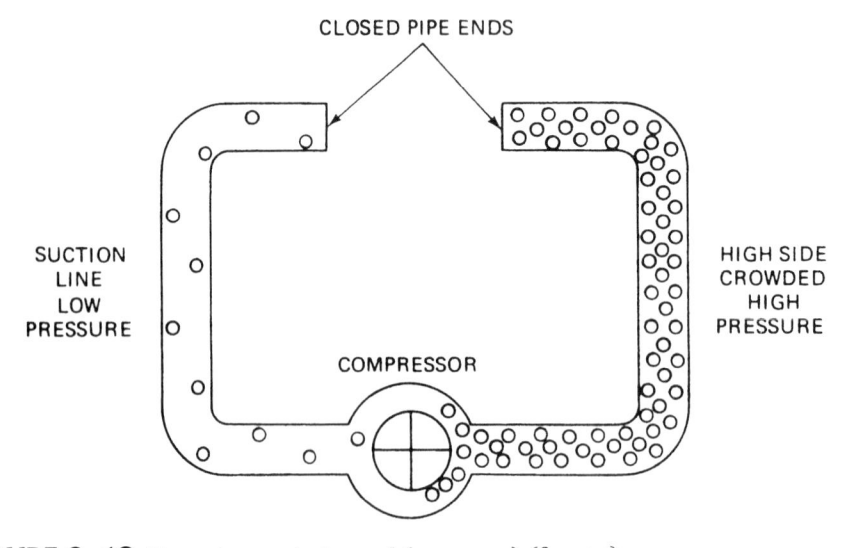

FIGURE 2-12 Air system unbalanced (pressure) (fan on)

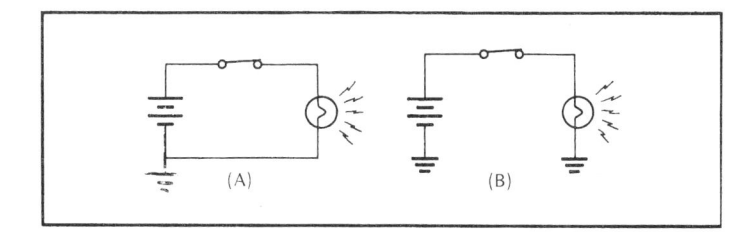

FIGURE 2–18 Circuit with ground wire (a), and circuit with common ground (b)

conductors. Although the steel frame and body parts do not have the same conductivity characteristics as copper, the mass of the steel used in the automobile as a ground conductor provides a low-resistance path for current.

PRACTICAL EXERCISE

The first exercise requires a plastic or rubber pocket comb and paper. The second exercise requires a motor vehicle.

EXERCISE 1

1. Tear a small piece of paper into smaller pieces—say, a half-inch (12–13 mm) square.
2. Pass the comb through your hair several times.
3. Place the comb near the pieces of paper.

QUIZ

1. Did the comb attract the pieces of paper? _____
   a. If no, determine why not before proceeding.
   b. If yes, proceed to question 2.
2. What do you think caused this attraction? _____
3. Is this exercise an example of static or dynamic electricity? _____ Explain. _____

EXERCISE 2

1. Open the hood of the motor vehicle. DO NOT start the engine.
2. Locate the battery.
3. Locate the cable connecting the battery negative (−) terminal to the engine ground.
4. Locate the cable at the battery positive (+) terminal.

QUIZ

1. Where does the other end of the positive (+) cable terminate? _____
2. Battery cables are the largest wires that are found in the vehicle. Why do you think this is necessary? _____
3. Some vehicles may have smaller wires included with the cables. Why do you think smaller wires are included with the ground (−) cable? _____
   Why do you think smaller wires are included with the positive (+) cable? _____
4. Does the battery supply power in the form of static or dynamic electricity? _____ Explain your answer. _____

SUMMARY

- Whenever there is an imbalance in an electrical system a pressure will exist trying to balance the system.
- In an electrical system the imbalance, or pressure between two points in the system, is the electromotive force (emf), most often called voltage. Remember that voltage is measured between two points.

- If a circuit is provided between the unbalanced points, current (a movement of electrons) will flow through the circuit.
- If a circuit is not connected between the unbalanced points (open circuit), no current will flow.

REVIEW

Answer the following questions True *(T)* or False *(F)*.

(T) (F) 1. Most cars have a negative (−) ground.

(T) (F) 2. In a closed circuit, direct current (dc) electron flow is from negative (−) potential to positive (+) potential.

(T) (F) 3. The simplest atom is the hydrogen (H) atom.

(T) (F) 4. Electrons in the innermost layer of an atom are referred to as loosely bound.

(T) (F) 5. Insulators have poor conductivity for the movement of free electrons.

(T) (F) 6. Dynamic electricity is electricity in motion.

(T) (F) 7. Like charges repel each other.

(T) (F) 8. Voltage is a pressure that causes electrons to move.

(T) (F) 9. Current is the movement of electrons.

(T) (F) 10. The natural state is for everything to be in a balanced condition.

(T) (F) 11. Current is measured in amperes.

(T) (F) 12. Electromotive force, emf, is measured in volts.

(T) (F) 13. Given equal diameter and length of conductors, copper (Cu) is a better conductor than aluminum (Al).

(T) (F) 14. There are 2 electrons in orbit about the nucleus of many atoms.

(T) (F) 15. There are 4 electrons in the second orbit of a hydrogen (H) atom.

(T) (F) 16. Heavier materials have a greater number of electrons in each atom than lighter materials.

(T) (F) 17. The heavier the material, the better the conductor.

(T) (F) 18. The number of electrons in the third and higher orbits is a random number.

(T) (F) 19. One electron per second equals one ampere.

(T) (F) 20. Once an object is charged with static electricity, it cannot be discharged.

◆ UNIT 3 ◆

OHM'S LAW

OBJECTIVES

On completion of this unit you will be able to:

- Recognize resistors as components in electrical and electronic circuits.
- Determine the resistance value of a color-coded resistor.
- Recognize the schematic symbol for fixed resistors.
- Recognize the schematic symbol for variable resistors.
- Use Ohm's law to determine the third of three values (voltage, current, or resistance) when two of the values are known.

CURRENT AND RESISTANCE

In this unit we introduce the mathematics that relates to automotive electricity and electronics. Be assured that the mathematics required for automotive electricity-electronics is minimal. To solve problems in this area the student should be able to perform basic addition, subtraction, multiplication, and division. See Appendix A, Electrical/Electronic Math, which provides some basics in this area.

In Unit 2 we covered voltage and current. There is another factor that is of equal importance in the study of electricity and electrical circuits: *resistance*. In the discussion of voltage and current, we observed that when a generator is operating, electrons are pulled from the wire on one side and forced out on the wire on the other side; this creates an imbalance called *voltage*. This action is shown in Figure 3–1. As long as the generator operates, it will keep the imbalance, and a voltage will exist between the two wires.

Note that in the circuit of Figure 3–1 there is no wire connected between the output wires of the generator. The resistance of air is extremely high and may be considered infinite for low voltages. With such a high resistance, no current will flow.

If a small wire were connected between the ends of the original wires, a current would flow in the circuit, as shown in Figure 3–2. The amount of current that would flow in the circuit would depend on the type of material of which the wire is made, the size (diameter or circular mil) of the wire, and its length. In the circuit in Figure 3–2, the small wire is said to offer resistance to the current flow. Although resistance is offered, current will continue to flow.

The connecting wire in the circuit in Figure 3–3 is the same length and of the same material. However, the cross-sectional area is twice the size of the wire used in Figure 3–2. More current flows in the circuit in Figure 3–3 than in the circuit in Figure 3–2 because resistance is reduced. Actually, twice as much current flows when a wire twice the size is used.

The resistance of a wire is dependent on four things:

1. The material used to make the wire (available free electrons).
2. Cross-sectional area of the wire (size).
3. Length of the wire.
4. Temperature (normally, as the temperature increases, resistance increases).

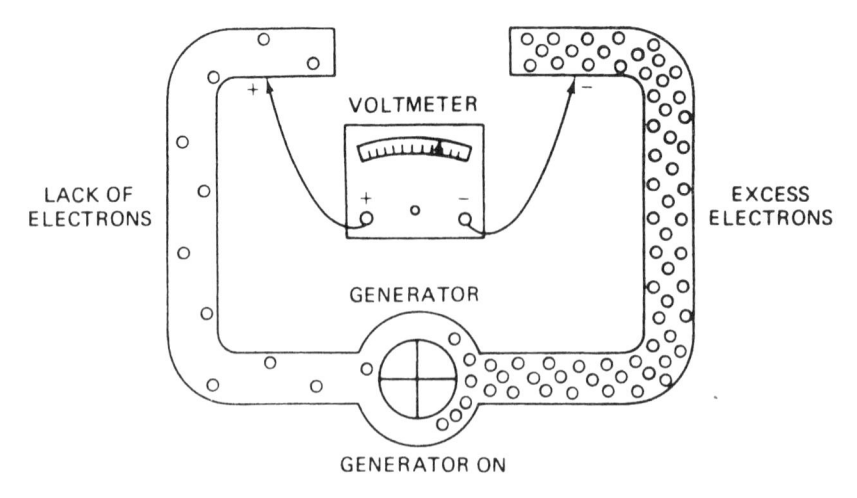

FIGURE 3–1 Electron distribution—power on

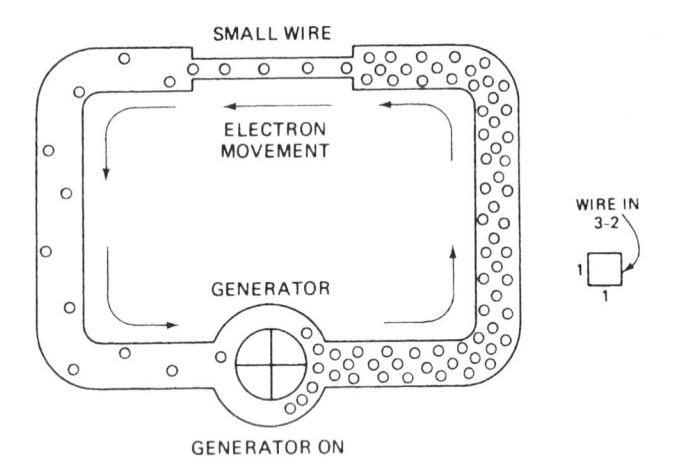

FIGURE 3–2 Current flow—high resistance

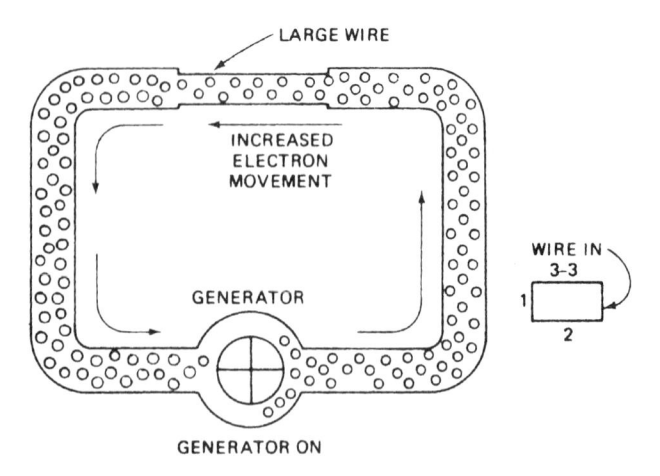

FIGURE 3–3 Current flow—low resistance

Resistors come in many shapes and sizes and configurations. Six different types of resistor elements are shown in Figure 3–4.

Resistance is measured in ohms. The Greek letter *omega* (Ω) is used as the symbol for ohm. One ohm of resistance is the value that will allow one ampere to flow when one volt is impressed on it.

RESISTANCE VALUES

Resistors used in electrical and electronic circuits vary in value from less than one ohm (Ω) to values in million ohms (Meg Ω). Some resistors have their value printed on their face. For example, a 200-Ω, 25-watt (W) resistor would probably have "200Ω 25W" printed right on it. Other resistors of smaller wattage may have a coded indication of the resistor value. The first resistor in Figure 3–4 is a standard, color-coded 1/4-watt carbon deposit resistor. Resistor two is a color-coded 1/2-watt carbon composition resistor, and resistor three is a 25-watt wire-wound power resistor.

The fourth resistor is a multielement resistor. These are special resistive circuits in which resistors are connected in combination depending on the user's need. You would need to refer to the schematic diagram of the element in order to determine what the component consists of. This schematic is sometimes printed on the component.

The fifth resistor is a leadless, surface-mount resistor found on printed circuit boards. The sixth is a molded power resistor, also found in printed circuit

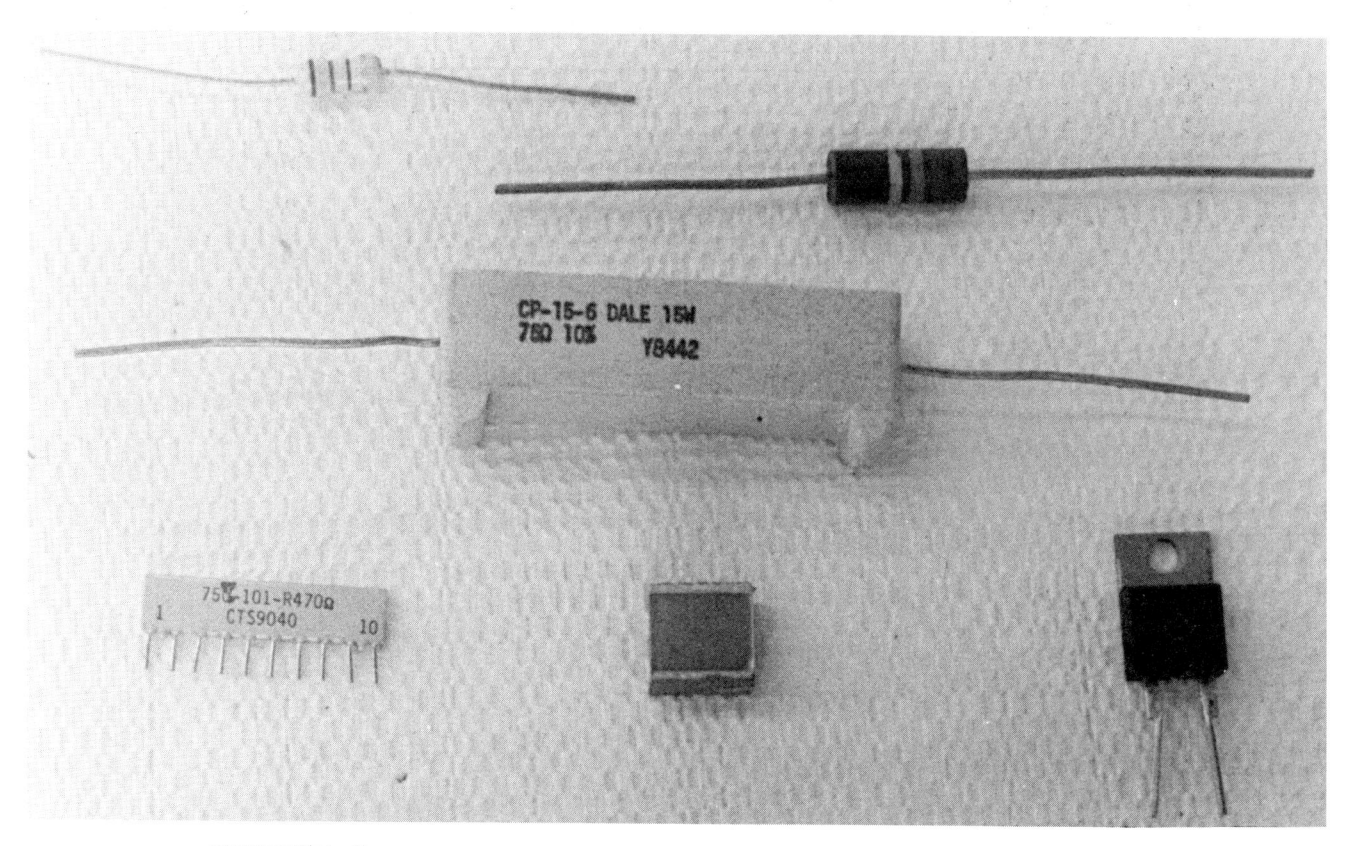

FIGURE 3-4 Resistors

board applications. One metal tap of this resistor is sometimes connected to a heat sink in order to assist in radiating heat generated in the resistor. A *heat sink* is a piece of metal, usually aluminum, that conducts heat readily.

RESISTOR COLOR CODE

For many years the value of resistors has been identified by the use of color coding. The color-code system uses bands of color on the resistor to indicate the resistor's value and tolerance. The color-code numbering system is shown in Table 3–1.

Standard resistor color coding calls for a minimum of three and a maximum of six color bands. The fifth and sixth bands are of engineering and statistical significance and so not of concern to the automotive technician. They provide information as to temperature sensitivity and failure rate. The first four color bands are, however, of particular significance to automotive technicians. The coded bands are called out in Figure 3–5.

EXAMPLE 1

What is the resistance value of a resistor coded yellow, violet, orange, gold?

Solution

1. The first figure is 4.
2. The second figure is 7.
3. The number of zeros is 3.
4. The tolerance is 5 percent.

The resistance value is 47,000 Ω plus or minus ($\pm$) 2350 Ω. Another way of stating the resistance value is to say that the manufacturer indicates the resistor is between 44,650 Ω and 49,350 Ω.

EXAMPLE 2

What is the value of a resistor coded brown, black, blue?

TABLE 3–1 Resistor color-code system

Color	1st Band	2nd Band	3rd Band (Times or Add Zeros)		4th Band
Black	0	0	10^0	1	None	1%
Brown	1	1	10^1	10	1	2%
Red	2	2	10^2	100	2	
Orange	3	3	10^3	1,000	3	
Yellow	4	4	10^4	10,000	4	
Green	5	5	10^5	.	5	
Blue	6	6	10^6	.	6	
Violet	7	7	10^7	.	7	
Gray	8	8	10^8	.	8	
White	9	9	10^9	.	9	
Gold			10^{-1}	0.1		5%
Silver			10^{-2}	0.01		10%
No color						20%

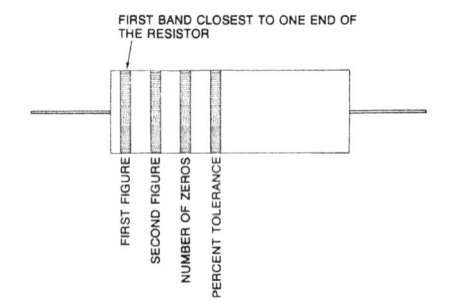

FIGURE 3–5 Color-coded resistors

Solution

1. The first figure is 1.
2. The second figure is 0.
3. The number of zeros is 6.
4. The tolerance is 20 percent.

The resistor is 10,000,000 Ω (10 Meg Ω). The tolerance is ± 20 percent, or 2 Meg Ω. Actual value of the resistor is 8 to 12 Meg Ω.

Standard resistors are color coded in wattage sizes from 1/8-watt to 2-watt values.

VARIABLE RESISTORS

Variable resistors are used in circuits whenever there is a requirement for changing the magnitude of voltage or current fed to other components in the circuit. If the variable resistor is connected to provide a variable voltage, it is called a *potentiometer* (Figure 3–6), generally referred to as a "pot." If the variable resistor is connected in the circuit to provide a variable current, it is called a *rheostat*. Electronic technicians tend to use the term *pot* in both situations.

Variable resistors typical of those used in the trade today are shown in Figure 3–7. The shape and size of variable resistors is controlled only by need and manufacturing capability.

OHM'S LAW

The relationship of voltage, current, and resistance may be determined by a relatively simple formula known as *Ohm's law*. It was devised by a German physicist, George Simon Ohm, in 1848. He discovered that the current in any circuit is directly proportional to the voltage applied and is inversely proportional to the resistance.

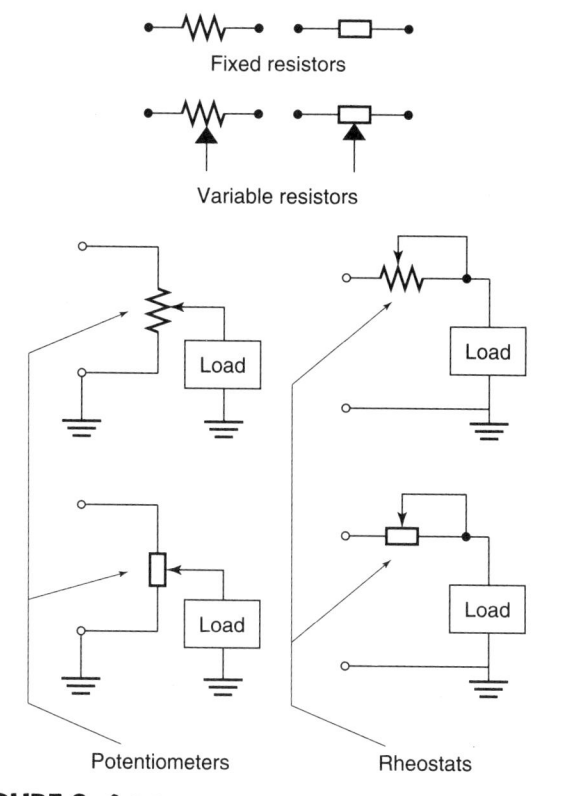

FIGURE 3–6 Schematic symbols for resistors

$$I = E/R$$
or
$$R = E/I$$
or
$$E = I \times R$$

From Ohm's law, the following can be derived:

1. If the resistance is kept the same and the voltage is increased, the current will increase.
2. If the voltage is kept the same and the resistance is increased, the current will decrease.

In honor of Ohm's discovery of this useful law, the unit of resistance measurement is the ohm, indicated, as you will recall, by the Greek letter *omega* (Ω). The symbol for resistance is the letter R. The symbol for voltage is the letter E (from electromotive force, emf) or V for voltage. The modern tendency is to use V. In this text we use both E and V, reflecting what is found in automotive reference materials. The symbol for current is the letter I (from current intensity). By using one of the three forms of Ohm's law, you can always find the third factor in a circuit when two factors are known.

FIGURE 3–7 Variable resistors

EXAMPLE 1 (*I is the unknown*)

A 12-Ω resistor has 24 volts across it (Figure 3–8). How much current is flowing through it?

Solution

$$I = E/R$$
$$I = 12 \text{ V}/6 \text{ }\Omega$$
$$I = 2 \text{ amperes}$$

Mathematically, the answer is 2 amperes. If an ammeter were connected in series with the 6-Ω resistor, as in Figure 3–9, the meter would indicate 2 amperes.

EXAMPLE 2 (*E is the unknown*)

A resistor of 6 Ω is connected across an electrical power source. An ammeter connected in series with

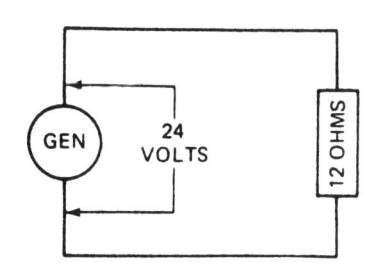

FIGURE 3–8 Using Ohm's law to determine current

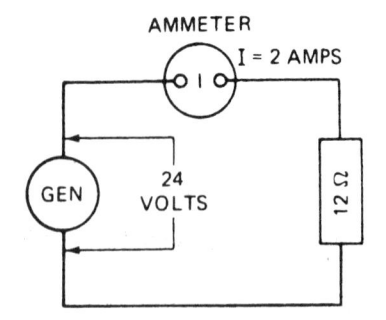

FIGURE 3-9 Connecting ammeter to indicate current

the resistor indicates 2 amperes (Figure 3-10). What is the voltage of the power source?

Solution

$$E = I \times R$$
$$E = 2 \times 6$$
$$E = 12 \text{ volts}$$

Mathematically, the answer is 12 volts. If a voltmeter were connected across the power source, it would indicate 12 volts (Figure 3-11).

EXAMPLE 3 *(R is the unknown)*

A resistor is connected across a 12-volt power source. An ammeter connected in series with the resistor indicates 10 amperes (Figure 3-12). What is the value of the resistor?

Solution

$$R = I/E$$
$$R = 12/10$$
$$R = 1.2 \text{ }\Omega$$

Mathematically, the answer is 1.2 Ω. If a sensitive ohmmeter (Figure 3-13) were connected to the resistive device, it would indicate 1.2 Ω.

In normal automotive service, the technician would not be required to solve Ohm's law problems. An understanding of Ohm's law does, however, provide the means to a better understanding of electricity. A good understanding of electricity is important to the technician when troubleshooting equipment.

USING OHM'S LAW IN AUTOMOTIVE CIRCUIT CURRENT

To find amperes, or current *(I)*, when the voltage *(E)* and resistance *(R)* are known, divide the voltage by the resistance. For example, determine the amperage required of the following devices in a 12-volt electrical system. The formula:

$$I = E/R$$

1. Blower motor (1.5 Ω)

$$12 \text{ V}/1.5 \text{ }\Omega = 8 \text{ A}$$

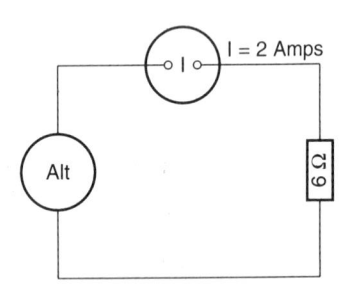

FIGURE 3-10 Using Ohm's law to determine voltage

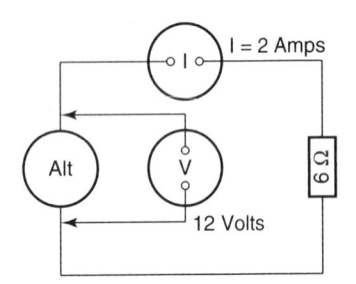

FIGURE 3-11 Measuring voltage with a voltmeter

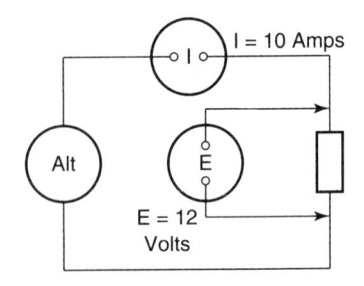

FIGURE 3-12 Using Ohm's law to determine resistance

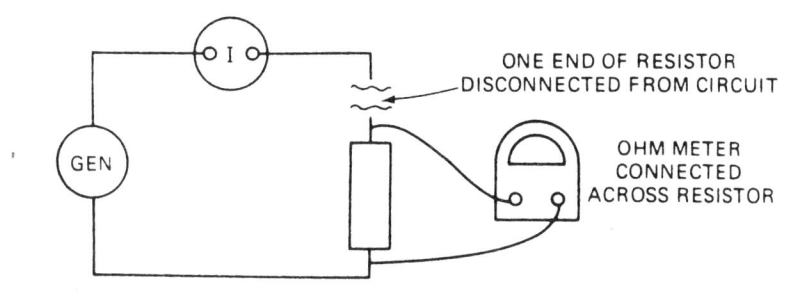

FIGURE 3-13 Using an ohmmeter to measure resistance

2. Compressor clutch (2 Ω)

$$12 \text{ V}/2 \text{ Ω} = 6 \text{ A}$$

3. Accessory load (3 Ω)

$$12 \text{ V}/3 \text{ Ω} = 4 \text{ A}$$

Automotive Circuit Voltage

To find the voltage E when the resistance R and amperage I are known, multiply the current I by the resistance R. By rearranging the previous example, we can determine the voltage required to pass the desired amount of current through a known resistance as follows. The formula:

$$E = I \times R$$

1. Blower motor (1.5 Ω at 8 A)

$$1.5 \times 8 = 12 \text{ V}$$

2. Compressor clutch (2 Ω at 6 A)

$$2 \times 6 = 12 \text{ V}$$

3. Accessory load (3 Ω at 4 A)

$$3 \times 4 = 12 \text{ V}$$

Actually, in automotive service, this part of the Ohm's law formula is of little use. Automotive electrical systems, for the most part, are 12-volt systems.

Automotive Circuit Resistance

To find the resistance R of a circuit when the voltage E and current I are known, divide the voltage by the current. Again, using the previous examples, the circuit resistance can be determined as follows. The formula is:

$$R = E$$

1. Blower motor (8 A at 12 V)

$$12 \text{ V}/8 \text{ A} = 1.5 \text{ Ω}$$

2. Compressor clutch (6 A at 12 V)

$$12 \text{ V}/6 \text{ A} = 2 \text{ Ω}$$

3. Accessory load (4 A at 12 V)

$$12 \text{ V}/4 \text{ A} = 3 \text{ Ω}$$

RESISTANCE OF WIRES AND CABLES

Automotive wires and cables are sized by the manufacturer to carry a given load. Because automotive manufacturers are cost conscious and in a highly competitive market, a wire or cable is usually sized no larger than necessary—the larger the wire, the greater the cost. Wire sizes are stated in gauge numbers. The higher the gauge number, the smaller the wire size. Although battery cables may be as large as 00 or 0 gauge, the most common gauges for automotive use are 8 through 18 (see Figures 3–14 and 3–15).

The current-carrying characteristic (resistance) of a wire is determined by three factors: its composition,

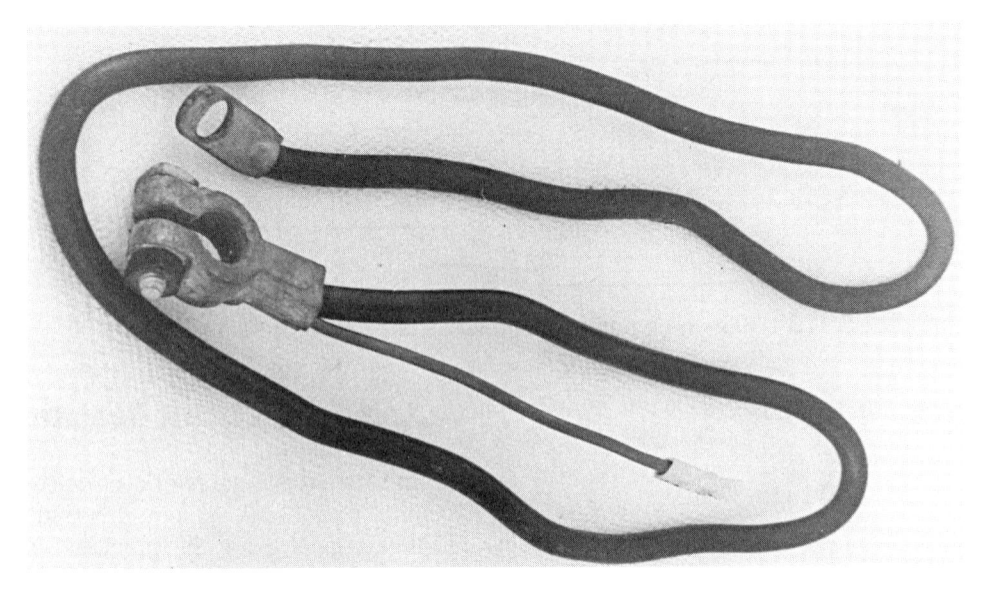

FIGURE 3–14 Battery cable

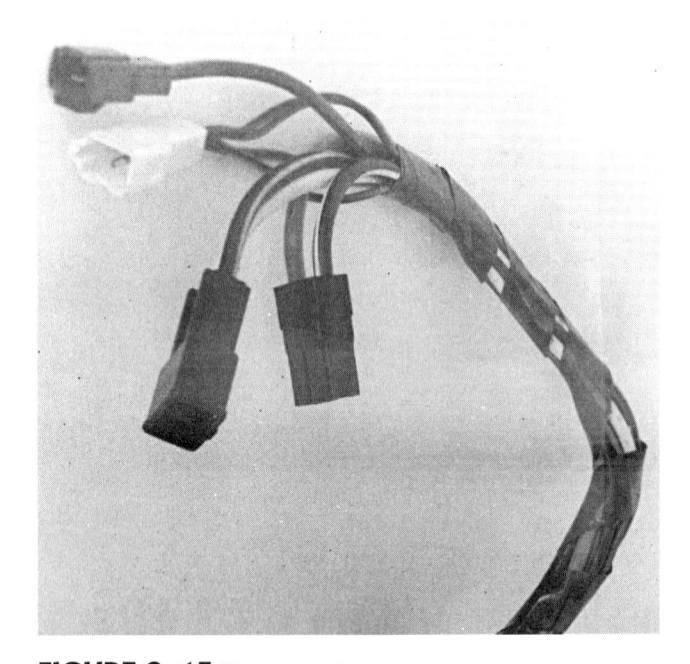

FIGURE 3–15 Harness wiring

length, and diameter. Often, when repairing a damaged wiring harness or adding wiring for an accessory, it is necessary to calculate the approximate resistance of a wire for proper sizing. If a wire is too small, too much electrical energy will be lost in the form of heat, and an electrical fire can develop. When in doubt, always use a larger wire. The slight additional cost can be good insurance against damage caused by an electrical fire.

To find the total resistance of a wire or cable, the following formula is used:

$$R_T = K \times L/cm$$

In this formula, RT is the total resistance; K is the ohm (Ω) resistance per mil/foot; L is the length of the wire in feet; and cm is the circular mil area of the wire. To apply this formula, examine each of the factors. Although a service technician does not normally have to solve the formula, knowledge of it reinforces understanding of automotive electricity.

Resistance: Mil/Foot

In the preceding formula, the letter K represented a standard of one mil/foot. A wire one mil in diameter and one foot long is called a mil/foot. The resistance of one mil/foot of wire depends on its composition. Table 3–2 shows the resistance for some popular materials used in automotive wiring. One mil/foot of aluminum (Al) wire, for example, has 17 Ω of resistance; ten mil/feet has 170 Ω; 100 mil/feet has 1700 Ω; and so on. Nichrome (Ni/Cr) wire, often used in wire-wound resistors, has the greatest resistance per mil/foot. The copper (Cu) wire commonly used in automotive service has the least resistance per mil/foot.

The cross-sectional area of a wire is given as a gauge number that is based on the diameter of the wire (Figure 3–16). The diameter is measured in

RESISTANCE OF WIRES AND CABLES

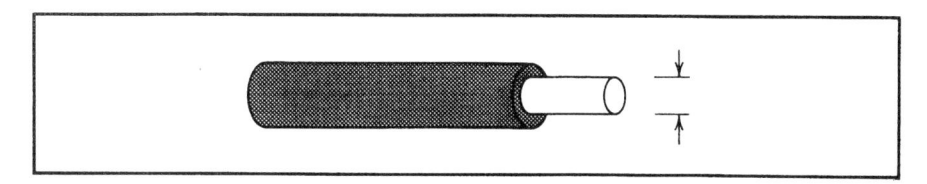

FIGURE 3–16 The diameter of a wire, minus its insulation, is measured in thousandths of an inch, or mils

TABLE 3–2 Ohm Resistance per mil/foot of some popular automotive wiring materials

Material	Ohm Resistance per Mil/Foot
Aluminum	17
Copper	10.4
Iron	60
Nichrome	600
Steel	75
Tungsten	33

TABLE 3–3 Diameter and circular mil area for the most popular sizes of automotive wiring

Gauge	Diameter Mils	Circular Mils
8	128	16384
10	102	10404
12	81	6561
14	64	4096
16	51	2601
18	40.3	1642

thousandths of an inch, or mils. One thousandth of an inch (1/1000 or 0.001) is one mil. The English mil is not to be confused with the metric mil. The English mil (0.001 in.) converts to the metric measure of 0.0254 millimeters. To avoid confusion, the discussion of wire sizes in this section will be in English measure only.

The cross-sectional area of a wire is measured in circular mils. A circular mil is the square of the diameter mil. The square of a number is obtained by multiplying a number by itself. For example, the square of 2 is 4 (2 × 2 = 4); the square of 8 is 64 (8 × 8 = 64); the square of 10 is 100 (10 × 10 = 100); and so on. The diameter of a 14-gauge wire (see Table 3–3) is 64 mils or 4096 circular mils (64 × 64 = 4096). The diameter and circular mil area for the most popular sizes of automotive wiring is given in Table 3–3.

Length of Wire

The length of a wire, in feet, is a factor in computing total resistance. If, for example, a 16-gauge copper wire had a resistance of 0.004 Ω per foot, 50 feet would have a total resistance of 0.2 Ω (0.004 × 50 = 0.2).

For example, determine the total resistance of 50 feet of 16-gauge copper wire, using the following formula:

$$R_T = K \times L/\text{cm}$$

First, find K from Table 3–2. We find that copper has a K factor of 10.4 Ω per mil/foot. The formula now becomes:

$$R_T = 10.4 \times 50/\text{cm}$$

Now, find cm. Table 3–3 indicates that 16-gauge wire has a circular mil area of 2601. (Do not confuse circular mils with diameter mils at this point.) Insert the number 2601 into the formula. The formula now becomes:

$$R_T = 10.4 \times 50$$

Now, solve the formula:

$$10.4 \times 50/2601 = 520/2601 = 0.1999 \ \Omega \ (0.2 \ \Omega)$$

Stranded Wire

Stranded wire has about the same current-carrying characteristics as solid wire. But stranded wire is more flexible than solid wire and is therefore more desirable for automotive use. The size and number of strands that make up a wire or cable determine its flexibility and current-carrying characteristics. For example, a 00-gauge wire with a circular mil of 133,225 can be made up of seven strands of 7-gauge wire, 20 strands of 12-gauge wire, or 37 strands of 15-gauge wire. The more strands, the more flexible the wire. But regardless of the strand gauge and number of strands, the circular mil area should be 00 gauge.

1 strand, 00 gauge = 133,225 cm

7 strands, 7 gauge (7 × 19000) = 133,000 cm

20 strands, 12 gauge (20 × 6561) = 131,220 cm

37 strands, 15 gauge (37 × 3600) = 133,200 cm

Stranded wire may be either bunch or concentric stranded. *Bunch stranding* (Figure 3–17) consists of a number of wires twisted together in no particular order. *Concentric* or *conical stranded* wire (Figure 3–18) consists of a single strand (core) surrounded by one or more layers of wire. Each layer consists of six more wires than the previous layer. Also, each layer is twisted in the opposite direction from the previous layer.

Insulation

Wires must be insulated in some manner to prevent them from shorting to each other or to ground. The most common insulation material is polyvinylchloride (PVC) plastic. PVC is among the least expensive of insulation materials and offers good insulation characteristics.

The insulation of wires is color coded as an aid in identifying circuits. However, the color of the wire is not an indication of its size. Because there are only so many colors, striping of the wire is a common practice. This gives two colors, expanding the color-coding capabilities. For example, consider the nine colors: yellow,

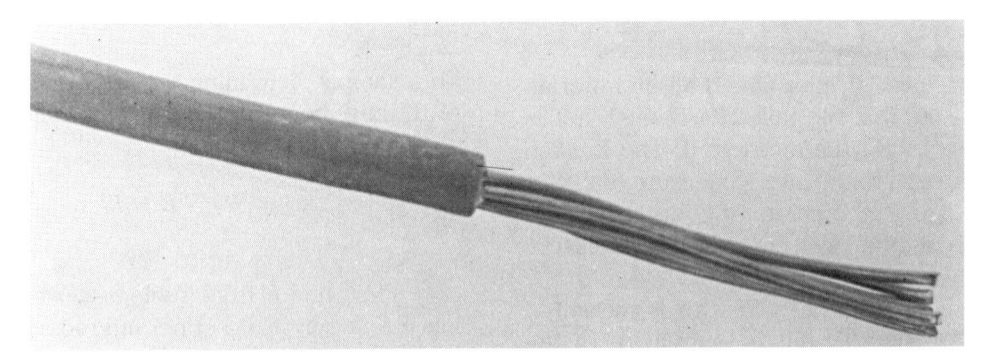

FIGURE 3–17 Bunch stranding of wire

FIGURE 3–18 Conical or concentric stranding of wire

blue, green, red, black, white, orange, brown, and tan. By adding one stripe, the total possible combination of these colors is expanded to 81. Additional expansion is made possible by using a combination of light and dark colors. Color coding and identification of wires and cables is also discussed in Unit 11.

WIRES AND CABLES: SUMMARY

Most of the more than 1600 feet (488 meters) of wire in the average car is made of copper (Cu) and is stranded. In some applications, aluminum (Al) or an exotic alloy of nichrome (Ni/Cr) or chromel (Cr/Al) may occasionally be found. The larger the wire size (diameter), the smaller the gauge number. For 12-volt applications, gauge numbers of 4, 10, 12, 16, and 18 are commonly used. Battery cables are usually 4 gauge, although some discount automotive supply houses market 6-gauge cables as replacements. Power distribution cables for battery-to-ignition switches, fuse panels, headlamp switches, seat/window circuits, air conditioning systems, and other heavy current-consuming devices are usually 10 or 12 gauge. Exterior and interior lighting circuits are usually 16 or 18 gauge. Automotive wires are usually covered with plastic for insulation; the color of the plastic cover, either solid or striped, aids in circuit identification. Since striping may be concentric, longitudinal, solid, or dashed, several hundred wires may be identified with the use of only a few colors (Figure 3–19).

Wires are sized according to gauge number for the maximum amount of current they are to carry. If a wire is undersized (Figure 3–20), electrical resistance will cause a drop in voltage, resulting in poor performance of the device for which the current is being supplied.

If, on the other hand, a wire is oversized, no harm is done; there will be less voltage drop, and the device will function at its full performance rating. Whenever in doubt, use the next larger wire gauge size. For example, a 10-percent reduction in battery voltage at the headlamps will result in a 30-percent reduction of lamp brilliance. Reduced voltage at motors not only reduces their efficiency, but also contributes to their early failure.

CONNECTORS AND CONNECTIONS

There are about 500 wiring connections in the average car, and they must also be considered when servicing an automobile electrical system. The most frequent problem with these connector devices is corrosion and oxidation, which prevents good electrical contact. The most common and well-known automotive connector with this problem is the one that connects the electrical system to the battery. Each year, according to the Battery Manufacturers Association (BMA), thousands of batteries are needlessly replaced simply because of dirty or corroded battery terminal connections. Consequently, always clean and recheck terminal connectors before replacement. Often you may find that the terminal is causing the problem and the connector is sound. Wires and wiring connectors are also discussed in other units of this text, especially Unit 11.

FIGURE 3–19 Some of the methods used to color code wires and cables

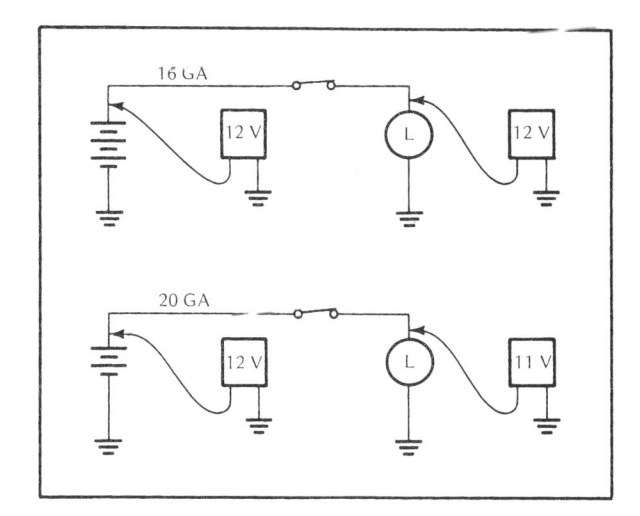

FIGURE 3–20 Using a smaller size (gauge) wire will result in reduced voltage at the load (L)

PRACTICAL EXERCISE

This exercise requires 10 assorted color-coded resistors of various types and styles, and a late-model vehicle.

EXERCISE 1

1. Observe some standard color-coded resistors. Determine their value by their color codes. Refer to the color-code chart in Table 3–1 and Figure 3–5.
2. Enter your values in the work sheet in Figure 3–21.
3. Ask your instructor to verify your findings.

EXERCISE 2

1. Open the hood of the vehicle. DO NOT start the engine.
2. Locate exposed electrical wiring at different points of the wiring harness. DO NOT pull wires from harness unless specifically instructed to do so.
3. Identify the colors.
   a. How many different color combinations do you find? _____
   b. How many combinations and variations of color identification can you find (solid, striped, etc.)? _____

First Band Color	Second Band Color	Third Band Color	Fourth Band (If There)	Resistor Value

FIGURE 3–21 Color code chart

4. On an electrical schematic, what colors are depicted by the following abbreviations?

   a. WHT _____
   b. BLK _____
   c. GRN _____
   d. YEL _____
   e. ORG _____
   f. BLU _____
   g. TAN _____
   h. PUR _____
   i. BRN _____

SUMMARY

- The mathematics required to solve electrical/electronic problems is simple addition, subtraction, multiplication, and division.
- In every electrical circuit there is a mathematical relationship between current, voltage, and resistance. This relationship is called Ohm's law.
- The three forms of Ohm's law are:
  - $I = E/R$
  - $R = E/I$
  - $E = I \times R$

REVIEW

(T) (F) 1. Most automotive wiring is made of aluminum (Al).

(T) (F) 2. Exterior lighting circuits are usually 16 or 18 gauge.

(T) (F) 3. Undersized wiring may result in early component failure.

(T) (F) 4. Most battery cables are 6 gauge.

(T) (F) 5. A 30-percent reduction in battery voltage results in a 10-percent reduction in headlamp brilliance.

6. The resistance of a piece of wire is dependent on four factors. What are they? _____

7. In an electric circuit, if the resistance is kept the same and the voltage is increased, the current will (increase) (decrease).

8. In an electric circuit, if the voltage is kept the same and the resistance is increased, the current will (increase) (decrease).

9. If a 6-Ω resistor has 20 amperes flowing through it, the voltage across the resistor will be _____ volts.

10. If 24 volts is measured across an 8-Ω resistor, how much current must be flowing through it?

11. An ammeter connected in series with a resistor indicates 12 amperes. A voltmeter connected across the resistor indicates 12 volts. What is the value of the resistor?

12. What voltage is needed to cause 4 amperes to flow through a 110-ohm resistor?

13. If the voltage across a 15-ohm resistor is measured at 90 volts, how much current is flowing through the resistor?

14. If the voltage applied to a resistor is doubled, the current will (double) (decrease to half) the original value.

15. What size resistor will cause a current of 3 amperes to flow when 12 volts are applied to it?

16. What is the biggest problem that may occur when a technician uses a wire size smaller than that required for the current it is to carry?

17. Using Tables 3–2 and 3–3, find the total resistance of 75 feet of 18-gauge copper (Cu) wire.

18. What is the circular mil of a wire with a diameter mil of 50?

19. Briefly describe concentric stranding of wires.

20. What is the total resistance of 6 inches (6″) of 18-gauge Nichrome (Ni/Cr) wire?

◆ UNIT 4 ◆

ELECTRICAL CIRCUITS

OBJECTIVES

On completion of this unit you will be able to:

- Recognize a complete circuit.
- Recognize a series circuit by component connections.
- Solve series-circuit problems.
- Recognize a parallel circuit by its component connections.
- Solve parallel circuit problems.
- Recognize a series-parallel circuit.
- Solve series-parallel circuit problems.

In order for electricity to accomplish things of value for us (that is, to do work) the electricity must be controlled. We control electricity by providing paths that the electricity must follow. These paths are called *circuits*. Understanding how the different electrical components function and how the systems function when the components are connected together in circuits provides the automotive electrical/electronics technician with the ability to maintain the modern automobile.

In Unit 3 we discussed Ohm's law and its application to simple electric circuits. In this unit we expand the discussion to Ohm's law as it relates to electrical circuits.

COMPLETE ELECTRIC CIRCUITS

In order for current to flow in a circuit, the circuit must be complete. Another way of stating this is that current must be able to flow from the source through the load and back.

In Figure 4–1, an alternator is the energy source, and a lamp is the load. As long as the circuit is complete, current will flow from the alternator (source) through the lamp (load) and back through the alternator. There is a continuous path for current to flow: It is a complete electric circuit.

If a switch is included in the circuit, the same current that flows through the lamp flows through the switch. Two conditions could exist. When the switch is closed, the circuit is complete and current can flow (Figure 4–2). When the switch is in the open position, the circuit is not complete (Figure 4–3). Current cannot flow through the open switch; therefore, current cannot reach the lamp.

In Figure 4–4, two circuits are shown. Note the position of the switches. In both circuits, the switches and the lamp are connected in series. In the circuit in Figure 4–4a, if either switch A-1 or A-2 is open, the circuit is not complete: The lamp will not light. Similarly, in the circuit in Figure 4–4b, if either switch B-1 or B-2 is open, the circuit is not complete and the lamp will not light. Electrically, both circuits are the same; they are both series circuits. The same current must flow through each component of the circuit.

WIRE RESISTORS

A common material used in the construction of wire-wound resistors is nichrome, an alloy of nickel (Ni) and chromium (Cr). Nichrome wire has a relatively high resistance for short lengths. For example, 100 feet of

36   Unit 4 ELECTRICAL CIRCUITS

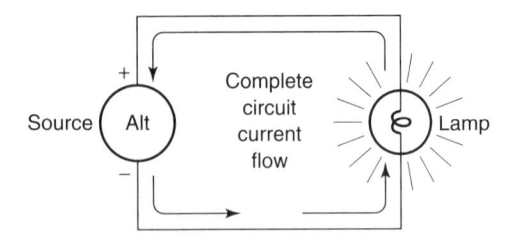

FIGURE 4–1 Complete series circuit

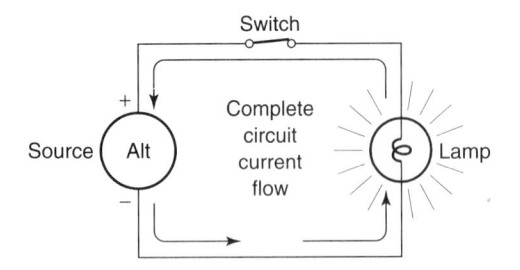

FIGURE 4–2 Series circuit with switch closed

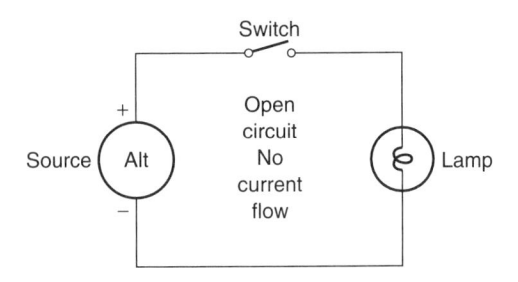

FIGURE 4–3 Series circuit with switch open

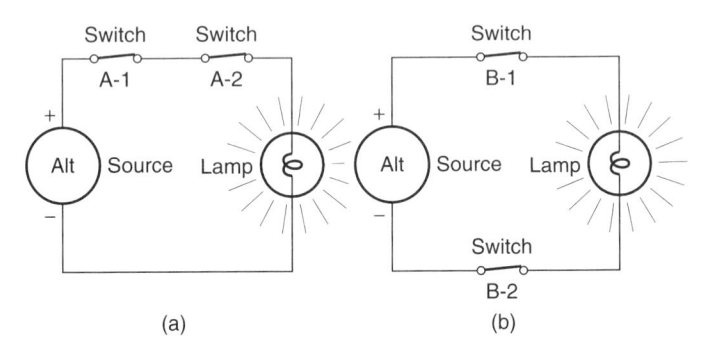

FIGURE 4–4 Series circuits with two switches

nichrome wire could have a resistance of about 100 Ω. A practical, usable resistor can be created when nichrome wire is wound on a core of ceramic with terminals at either end (Figure 4–5).

The circuit in Figure 4–6 consists of a source (an alternator), connecting wires, and a load (a 100-Ω wirewound resistor). According to Ohm's law, if the alternator were producing 100 volts, one ampere of current would flow in the resistor. A voltmeter connected across the resistor would read 100 volts.

$$I = E/R$$
$$I = 100/100$$
$$I = 1 \text{ ampere}$$

If the resistor were broken at the exact center, we would have two resistors of 50 Ω each (two 50-foot lengths of the nichrome wire) (Figure 4–7). The two broken ends of wire could be clamped together to remake the 100-Ω resistor (Figure 4–8). In other words, the two 50-Ω resistors, tied in series, make the 100-Ω resistor: 50 + 50 = 100.

The two reconnected 50-Ω resistor sections may be connected into the original circuit (Figure 4–9). This circuit is a series circuit. The same current, 1 ampere, flows through each of the 50-Ω resistor sections. Ohm's law may be used to calculate the voltage across R_1, the first resistor section.

$$E = I \times R$$
$$E = I \times 50$$
$$E = 50 \text{ volts}$$

Ohm's law is also used to calculate the voltage across R_2, the second resistor section.

$$E = I \times R$$
$$E = 1 \times 50$$
$$E = 50 \text{ volts}$$

Note that the sum of the voltages across R_1 and R_2 equals the supplied voltage, 50 V + 50 V = 100 V.

$$E_T = E_{R_1} + E_{R_2}$$

Also note that the total resistance, 100 Ω, is the sum of the individual resistors:

$$50 \text{ Ω} + 50 \text{ Ω} = 100 \text{ Ω}.$$
$$R_T = R_1 + R_2 + \ldots$$

We can now state the laws of series circuits.

1. The same current flows through each component of a series circuit.
2. The total voltage in a series circuit is equal to the sum of the voltages across the individual components.
3. The total resistance of a series circuit is equal to the sum of the resistances of the individual components.

EXAMPLE 1

A circuit consists of a 25-Ω resistor and a 50-Ω resistor connected in series across a 150-volt alternator (Figure 4–10). Determine the current flow in the circuit and the voltage across each resistor.

Solution

The total resistance is equal to the sum of the individual resistances.

$$R_T = R_1 + R_2$$
$$R_T = 25\ \Omega + 50\ \Omega$$
$$R_T = 75\ \Omega$$

Use Ohm's law on the whole circuit:

$$I = E/R$$
$$I = 150\ V/75\ \Omega$$
$$I = 2\ \text{amperes}$$

Because the same current flows through each component, the voltage across each component may be determined by using Ohm's law:

$$E_{R_1} = I \times R \qquad E_{R_2} = I \times R$$
$$E_{R_1} = 2 \times 25 \qquad E_{R_2} = 2 \times 50$$
$$E_{R_1} = 50\ \text{volts} \qquad E_{R_2} = 100\ \text{volts}$$

The supply voltage is equal to the sum of the voltages across the individual components.

$$E_T = E_{R_1} + E_{R_2}$$
$$E_T = 50 + 100$$
$$E_T = 150$$

EXAMPLE 2

Determine the value of each resistor, the voltage across each resistor, and the total resistance of the circuit shown in Figure 4–11. The alternator supplies 12 volts; the ammeter indicates 4 amperes.

Solution

Solve for R_T, using Ohm's law for the whole circuit:

$$R_T = E_T/I$$
$$R_T = 12/4$$
$$R_T = 3\ \Omega$$

Solve for R_2:

$$R_2 = E_{R_2}/I$$
$$R_2 = 4/4$$
$$R_2 = 1\ \Omega$$

Solve for E_{R3}:

$$E_{R_3} = I \times R_3$$
$$E_{R_3} = 4 \times 0.5$$
$$E_{R_3} = 2\ \text{volts}$$

Solve for R_4:

$$R_4 = E_{R_4}/I$$
$$R_4 = 2/4$$
$$R_4 = 0.5\ \Omega$$

Solve for R_1, using the total resistance law:

$$R_T = R_1 + R_2 + R_3 + R_4$$
$$3 = R_1 + 1 + 0.5 + 0.5$$
$$R_1 = 3 - 2$$
$$R_1 = 1\ \Omega$$

Check the solution, using total voltage:

$$R_1 = 1\ \Omega$$
$$E_{R_1} = I \times R$$

$E_{R_1} = 4 \times 1$

$E_{R_1} = 4$ volts

$E_T = E_{R_1} + E_{R_2} + E_{R_3} + E_{R_4}$

$12 = 4 + 4 + 2 + 2$

$12 = 12$

The solution checks.

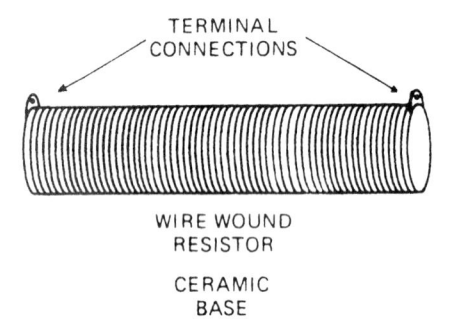

FIGURE 4–5 Wire-wound resistor

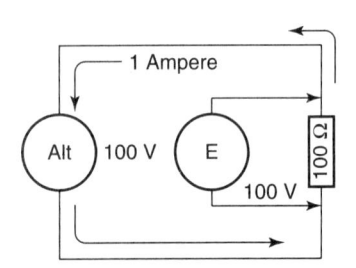

FIGURE 4–6 Single-resistor circuit

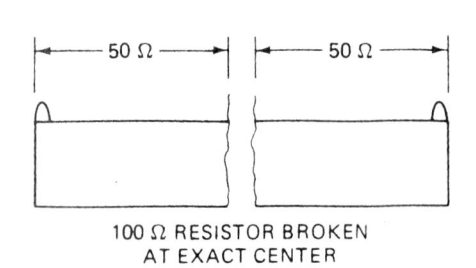

FIGURE 4–7 A 100-Ω resistor broken at exact center

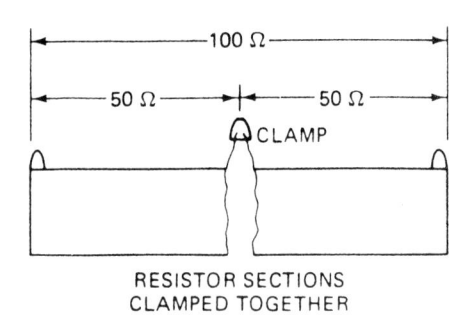

FIGURE 4–8 Two 50-Ω resistors joined together

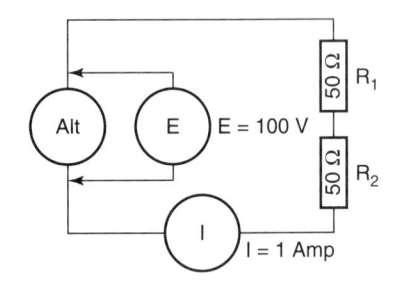

FIGURE 4–9 Two-resistor series circuit

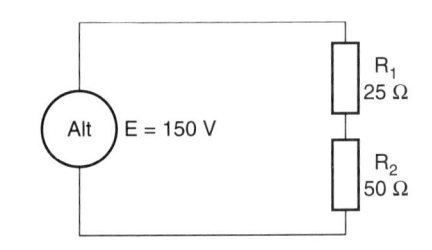

FIGURE 4–10 Circuit: 25 Ω and 50 Ω in series

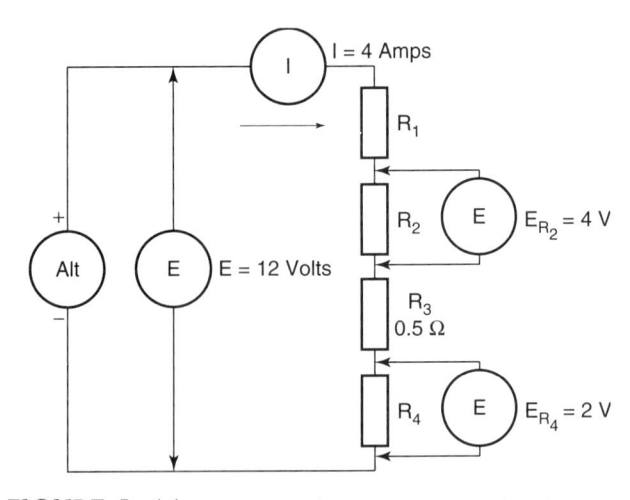

FIGURE 4–11 Using Ohm's law in series circuits

SWITCHES

In automotive electrical systems, switches are used to control the operation of devices such as fan motors, heaters, relays, and indicator lights. These switches are connected in series, parallel, or series parallel (as required) in order to accomplish the desired circuit action. Some examples of switch circuit combinations are shown in Figure 4–12.

The lamps shown in these circuit diagrams could easily have been a compressor clutch coil or fan motor. The switches could have been circuit breakers, temperature controls, or pressure controls. The important point to remember is how switches control various electric circuits.

Whenever switches are connected in series parallel, the possibility of sneak circuits exists. A *sneak circuit* is one that permits the operation of a component at a time when operation is not wanted. Sneak circuits are not always obvious at the time a circuit is designed. Some sneak circuits cause operation only under odd conditions and are therefore difficult to discover.

Consider the circuit in Figure 4–13. The path through the switch is normally considered to start from the left top and proceed through the switches to the lamp. Different paths and switches might be considered, but most people would follow a path from left to right. The circuit designer no doubt drew the circuit while considering paths from left to right.

A sneak circuit might exist in a path through S2 from right to left. If switches S4, S2, S5, and S6 are closed, there is a complete path for current flow, as shown in Figure 4–14. The circuit path may be unwanted, causing lamp L1 to be on when it should not be. When a sneak circuit is discovered, redesign is required.

Most sneak circuits would not be as obvious as the one shown in this example. In some cases, a considerable amount of circuit investigation is necessary before the problem can be discovered. Keep in mind that, when a sneak circuit is causing a problem, the

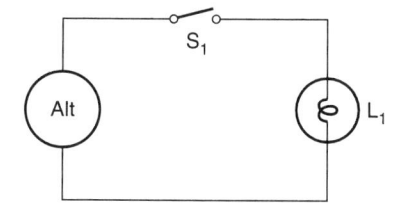

(a) Switch S_1 must be closed if lamp (L_1) is to light.

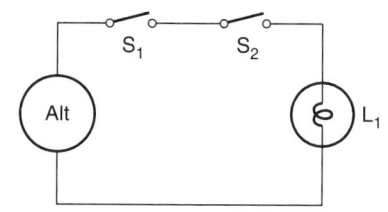

(b) Both switches, (S_1) and (S_2), must be closed if lamp (L_1) is to light.

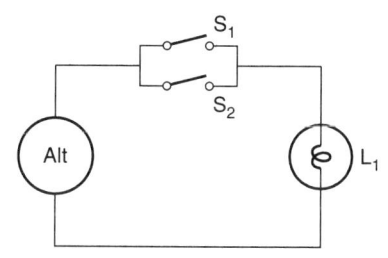

(c) Either S_1 or S_2 must be closed if lamp (L_1) is to light.

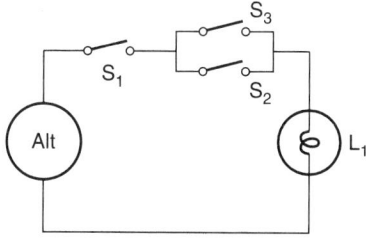

(d) Switch (S_1) and switch (S_3) or (S_4) must be closed if lamp (L_1) is to light.

FIGURE 4–12 Simple switch circuits

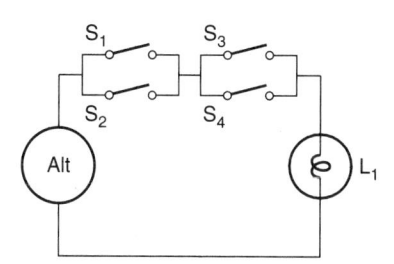

(a) Switches (S_1) or (S_2) and switches (S_3) or (S_4) must be closed if lamp (L_1) is to light.

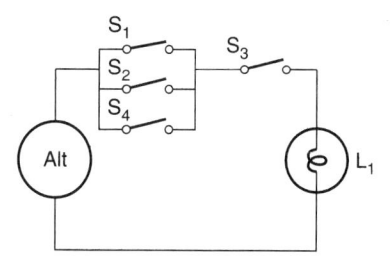

(b) Switch (S_4) or switches (S_1) and (S_3) or switches (S_2) and (S_3) must be closed if lamp (L_1) is to light.

FIGURE 4–13 Multiswitch circuit

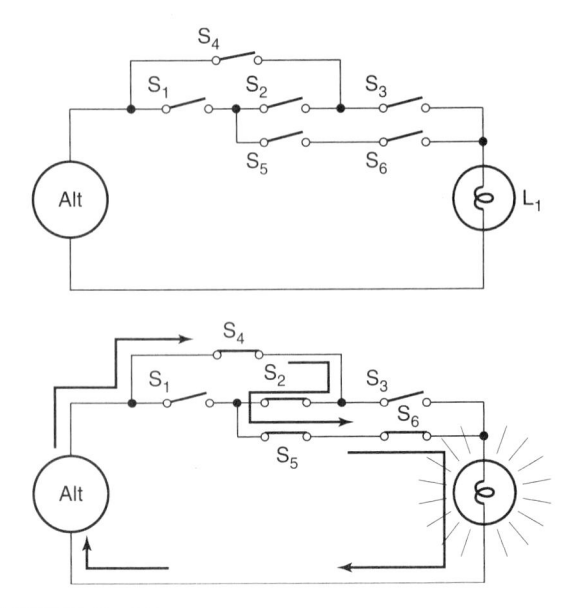

FIGURE 4–14 Sneak circuit

technician investigating the problem normally assumes that something is malfunctioning, when, in fact, every component is operating correctly.

Sneak circuits are normally in-plant manufacturing problems, not field problems. However, they do sometimes occur in the field and technicians must be aware of them.

RESISTIVE PARALLEL COMPONENTS

Consider a paper two inches (50.8 mm) wide and four inches (101.6 mm) long. A very thin layer of carbon material is deposited on this paper, forming a resistor, as

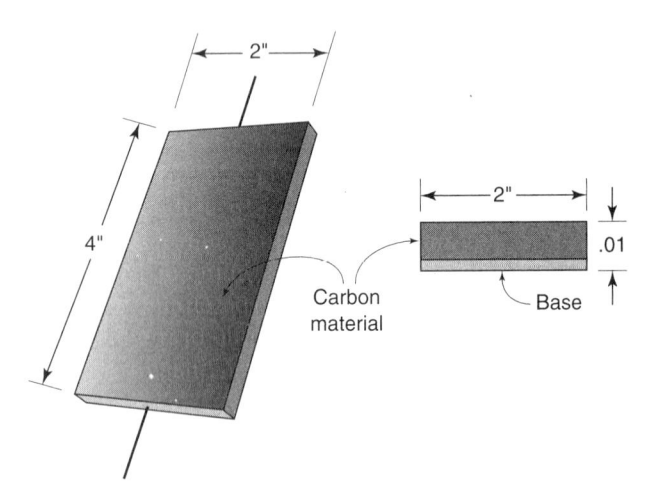

FIGURE 4–15 Homemade resistor

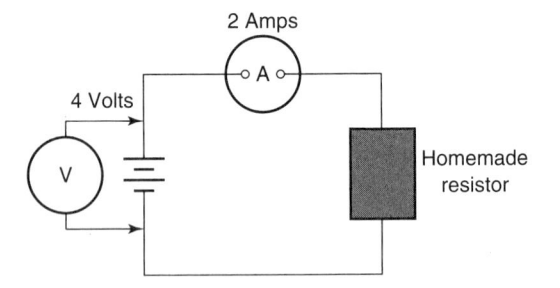

FIGURE 4–16 Using a homemade resistor in a circuit

shown in Figure 4–15. The path for current, then, is two inches (50.8 mm) wide by 0.01 inch (0.254 mm) high.

A voltage of four volts is impressed across the resistor and a current of two amperes is seen to be flowing through it, as shown in Figure 4–16.

By applying Ohm's law the resistance of the homemade resistor may be calculated:

$$R = V/I$$
$$R = 4/2$$
$$R = 2\,\Omega$$

Next, a very sharp instrument is used to cut our homemade resistor in half lengthwise without removing any of the carbon deposit (Figure 4–17). Since none of the carbon deposit was removed the total circuit was not changed electrically. With four volts impressed a current of two amperes flows. The current path is still a total of 2.0 inches (50.8 mm) wide by 0.01 inches (0.254 mm) high.

Since there are now two paths for current to flow, the current will split. Since the two paths are of the same material and of equal dimensions, the current will split equally. One ampere will flow through the resistor on the left and one ampere will flow through the resistor on the right.

Using Ohm's law the value of each resistor may be found. The battery is connected directly across the re-

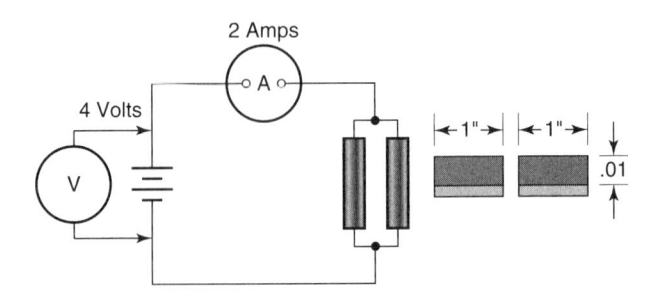

FIGURE 4–17 Homemade resistor divided in half

sistors. Four volts is impressed on each resistor. The current flow through each resistor is one ampere. According to Ohm's law,

$$R = E/I$$
$$R = 4/1$$
$$R = 4 \, \Omega$$

Note that each resistor is 4 Ω, although the total resistance of the circuit is only 2 Ω.

The two resistors form what is called a parallel circuit. A *parallel circuit*, then, is a circuit where there is more than one path for current to flow.

Some other interesting facts may be determined using that same parallel circuit:

1. The same voltage is impressed across each component in a parallel circuit. Many components in circuits may have equal amounts of voltage impressed across them even though they may not be in parallel. If in parallel, it *must* be the same voltage.
2. The total current is the sum of the currents in the individual branches of the parallel circuit.
3. The effective resistance is smaller than that of the smallest resistor in a parallel circuit.

PARALLEL CIRCUITS

Parallel circuits are found when the same source voltage is required across two or more components. A typical parallel circuit is shown in Figure 4–18. If the alternator were producing 12 volts at its output terminals, there would be 12 volts across each lamp. Current would flow through each lamp. If each lamp had a resistance of 6 Ω, the current would be 2 amperes in each, according to Ohm's law:

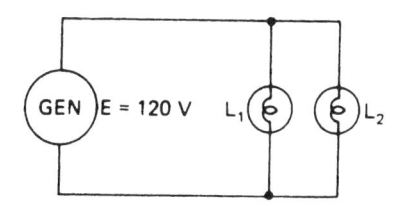

12 volts/6 Ω = 2 amperes

FIGURE 4–18 Parallel lamp circuit

The current for both lamps would come through the alternator, the power source. There would be 4 amperes flowing through the alternator.

$$(2 + 2 = 4)$$

Current flow in a parallel circuit is similar to water flow in a pipe system. In the water-pipe junction shown in Figure 4–19, 2 gallons (7.57 L) of water (H_2O) per minute are flowing in pipe A, entering from the left. In pipe B, 5 gallons (18.9 L) of water per minute are flowing, also entering from the left. There must be 7 (26.5 L) gallons of water per minute flowing out of pipe C. The water cannot disappear. The sum of the water flows into the junction must equal the water flow out of the junction.

In an electric circuit, the sum of the currents entering a junction must equal the current leaving the junction. In Figure 4–20, the lower junction of Figure 4–18 is shown.

If 2 amperes are flowing to junction A from lamp L1 and 2 amperes are flowing to it from lamp L2, there must be 4 amperes leaving junction A.

Sometimes it is necessary to use this law to find a branch current value. A *branch* is one of the paths for current flow in a parallel circuit. The circuit in Figure 4–21 is a portion of a total circuit. Ammeter 1 indicates 3 amperes, and ammeter 2, in the lamp circuit, indicates 2 amperes. Since there are 3 amperes leaving junction B and only 2 amperes coming to junction B through the lamp, there must be 1 ampere flowing through resistor R_1 to junction B. The total circuit of Figure 4–21 could contain another branch, as in Figure 4–22.

The current entering junction A, coming from junction B, is 3 amperes. The current leaving junction A, going to the generator, is 6 amperes. Where did the other 3 amperes come from? There must be 3 amperes flowing into junction A through R_2. Given that R_2 is an 8-Ω resistor, other unknown values of the circuit may

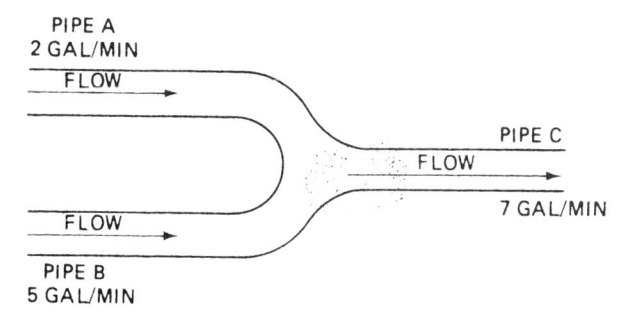

FIGURE 4–19 Parallel water pipe system

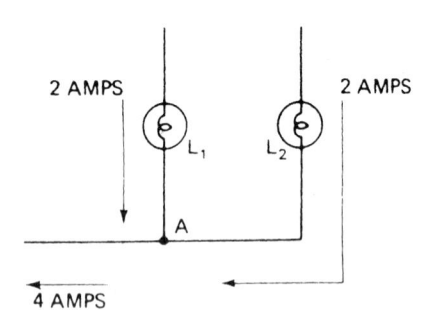

FIGURE 4–20 Current flow in junction

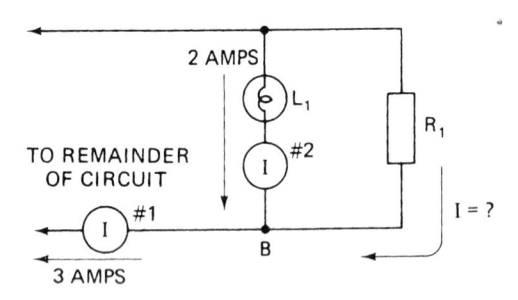

FIGURE 4–21 Branch current determination

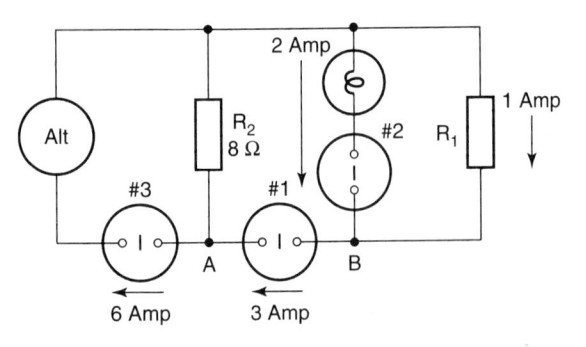

FIGURE 4–22 Using junction current rule

be determined. Determine the voltage across R_2; use Ohm's law for E:

$$E = I \times R$$
$$E = 3\,A \times 8\,\Omega$$
$$E = 24\,V$$

The voltage across each component in parallel is the same.

The voltage across R_2 is 24 volts.
The voltage across L1 is 24 volts.
The voltage across R_1 is 24 volts.

Determine the resistance of lamp L1; use Ohm's law for R:

$$R_{L1} = E/I$$
$$R_{L1} = 24/4$$
$$R_{L1} = 12\,\Omega$$

To determine the resistance of R_1:

$$R_1 = E/I$$
$$R_1 = 24/1$$
$$R_1 = 24\,\Omega$$

To determine the resistance of the circuit, the total current and voltage must be used. The total current as indicated by ammeter 2 (Figure 4–22) is 6 amperes. The total voltage is 24 volts.

$$R_T = E/I$$
$$R_T = 24/6$$
$$R_T = 4\,\Omega$$

The total resistance of the circuit is 4 Ω. This is less than the resistance of any of the three individual components:

$$R_1 = 24\,\Omega$$
$$L1 = 12\,\Omega$$
$$R_2 = 8\,\Omega$$

NOTE: The total resistance of a parallel circuit is less than the smallest branch resistance.

There are other interesting relationships in parallel circuits. The circuit in Figure 4–23 contains an alternator producing 24 volts, wires from the alternator to a plastic box, and an ammeter indicating 6 amperes. There is no indication as to what is in the box. Solve for R_T of the circuit in the box:

$$R_T = E/I$$
$$R_T = 24/6$$
$$R_T = 4\,\Omega$$

PARALLEL CIRCUITS

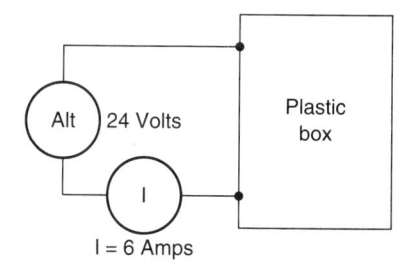

FIGURE 4–23 Black box analogy

When the box is opened, a 12-Ω resistor is found connected in parallel with a 6-Ω resistor (Figure 4–24). The resistance of the parallel combination could also be found using the product over sum method as in the formula:

$$R_T = \frac{R_1 \times R_2}{R_1 + R_2} \quad \text{(The product of } R_1 \text{ and } R_2\text{)}$$
$$\text{(The sum of } R_1 \text{ and } R_2\text{)}$$

For the parallel combination in Figure 4–24:

$$R_T = \frac{12 \times 6}{12 + 6}$$

$$R_T = \frac{72}{18}$$

$$R_T = 4 \, \Omega$$

This is the same total resistance value that was calculated using Ohm's law.

Product Over Sum Method

If more than two resistors are connected in parallel, the product over sum method may be used on two resistors at a time. An example of this procedure is given using Figure 4–25. What is the total resistance of the circuit? It has been shown that the combination of R_1 and R_2 is equal to a 4-Ω resistor. Figure 4–26 shows the equivalent circuit after resistors R_1 and R_2 are combined. The product over sum method may now be used to determine R_T of the total circuit.

$$R_T = \frac{R_x \times R_3}{R_x + R_3}$$

$$R_T = \frac{4 \times 4}{4 + 4}$$

$$R_T = \frac{16}{8}$$

$$R_T = 2 \, \Omega$$

Thus, the R_T of two resistors of equal value connected in parallel is equal to the value of one resistor divided by the number of resistors. If three 9-Ω resistors were connected in parallel, the R_T would be equal to 9 (ohms) divided by 3 (resistors) (Figure 4–27).

Another method that may be used to solve for total resistance of a parallel circuit is to assume a voltage across the circuit. For example, in Figure 4–25, 4-Ω, 12-Ω, and 6-Ω resistors are connected in parallel. If a supply voltage of 12 volts is assumed, the current through each resistor by Ohm's law would be:

$I_{R_1} = 12 \text{ V}/R_1 \quad I_{R_2} = 12 \text{ V}/R_2 \quad I_{R_3} = 12\text{V}/R_3$

$I_{R_1} = 12 \text{ V}/6 \quad I_{R_2} = 12 \text{ V}/12 \quad I_{R_3} = 12 \text{ V}/4$

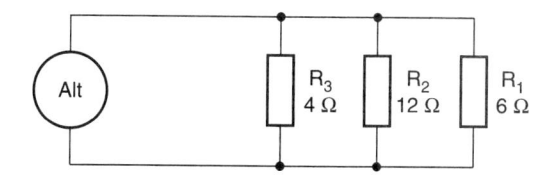

$I_{R_1} = 2 \text{ amperes} \quad I_{R_2} = 1 \text{ ampere} \quad I_{R_3} = 3 \text{ amperes}$

FIGURE 4–25 Product over sum, three-resistor circuit

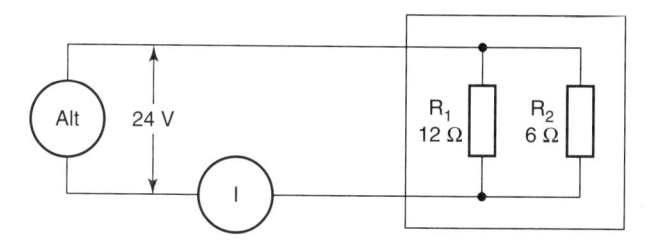

FIGURE 4–24 Parallel circuit (6 Ω and 12 Ω).

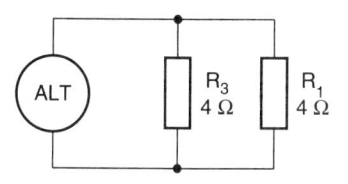

FIGURE 4–26 Equivalent circuit

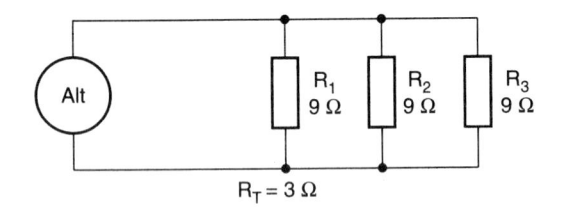

FIGURE 4–27 Three equal-value resistors in parallel

$$I_T = I_1 + I_2 + I_3$$

$$I_T = 2\,A + 1\,A + 3\,A$$

$$I_T = 6\,amps$$

The total current is the sum of the currents in the branches.

To solve for the complete circuit using Ohm's law:

$$R_T = E_T/I_T$$

$$R_T = 12\,V/6\,A$$

$$R_T = 2\,\Omega$$

Because the applied voltage is an assumed value, the currents in the branches are also assumed values. The calculated total resistance is the correct value. Supply voltage does not influence the resistance of resistors or the total circuit resistance.

The use of assumed voltage values leads to another important formula for parallel circuits. The formula is:

$$R_T = \frac{1}{1/R_1 + 1/R_2 + 1/R_3 \ldots 1/R_N}$$

Reciprocal Method

Consider again the circuit of Figure 4–24, in which 4-Ω, 12-Ω, and 6-Ω resistors are connected in parallel. If one volt is assumed across the circuit, the current in the individual resistors becomes:

$$I_{R_1} = 1\,V/R_1 \quad I_{R_2} = 1\,V/R_2 \quad I_{R_3} = 1\,V/R_3$$

The total current is the sum of the currents in the branches.

$$I_T = 1\,V/R_1 + 1\,V/R_2 + 1\,V/R_3$$

The total resistance is the supply voltage (1 V) divided by the total current.

$$R_T = \frac{1\,V}{1\,V/R_1 + 1\,V/R_2 + 1\,V/R_3}$$

Because 1 divided by a number is called the *reciprocal* of the number, the formula for R_T is called the reciprocal of the sum of the reciprocals. This formula provides the most important method for solving parallel circuits, particularly since scientific calculators have come into general use.

Scientific-Calculator Solutions

A scientific-calculator solution of a parallel-circuit problem follows.

PROBLEM

Determine the equivalent resistance of four resistors—8 Ω, 12 Ω, 9 Ω, and 17 Ω—in parallel using the reciprocal of the sum of the reciprocals method.

The resistance of the circuit is 2.6463 Ω.

STEP	ENTER	PRESS	DISPLAY
1	8	1/x	0.125
2		+	0.125
3	12	1/x	0.0833
4		+	0.2083
5	9	1/x	0.1111
6		+	0.3194
7	17	1/x	0.5882
8		=	0.3782
9		1/x	2.6463

Here is a scientific calculator solution of a parallel circuit problem containing resistors in the kilohm (kΩ) range. This requires the use of the exponent (EXP) key for proper entry.

PROBLEM

What is the equivalent resistance of a parallel circuit containing a 2-kΩ, a 6-kΩ, an 11-kΩ, and a 7-kΩ resistor?

$$R_T = \frac{1}{1/R_1 + 1/R_2 + 1/R_3 + 1/R_4}$$

$$R_T = \frac{1}{1/2 \text{ k}\Omega + 1/6 \text{ k}\Omega + 1/11 \text{ k}\Omega + 1/7 \text{ k}\Omega}$$

Step	Enter	Press	Display
1	2	EXP	2.00
2	3	1/x	0.0005
3		+	0.0005
4	6	EXP	6.00
5	3	1/x	0.000166
6		+	0.000166
7	11	EXP	11.00
8	3	1/x	0.0000909
9		+	0.0007575
10	7	EXP	7.00
11	3	1/x	0.0001428
12		=	0.009004
13		1/x	1110.58

The effective resistance of the circuit is 1110.576923 Ω. This problem could have been solved leaving out the exponent, since all of the resistors were given in the kilohm (kΩ) range. If the numbers 2, 6, 11, and 7 were simply entered in the calculator, the final answer would have been 1.11. The k factor could then have been reinserted, providing 1.11 k as the answer. In many problems, the resistors will be mixed—ohms (Ω) with kilohms (kΩ) or kilohms (kΩ) with megohms (MΩ). It is good practice to enter the actual resistor value into the calculator using the EXP key when needed. With practice you will soon develop individual techniques for calculating. At this time, however, you should understand:

1. The same voltage appears across each component of a parallel circuit.
2. The total current in a parallel circuit is equal to the sum of the currents in the branches.
3. The total resistance (R_T) of a parallel circuit is always smaller than the smallest resistor connected in the parallel combination.

SERIES-PARALLEL CIRCUITS

A series-parallel circuit is made up of a combination of components. Some are connected in series, and others are connected in parallel. When working with series-parallel circuits, it is necessary to use the laws of series circuits on the series elements and the laws of parallel circuits on the parallel elements. The circuit in Figure 4–28 is a series-parallel circuit. The 10-Ω resistor, R_3, is connected in series with the parallel combination R_1 and R_2. In order to solve for the current flow in the circuit, the resistors must be combined. Resistors R_1 and R_2 are in parallel.

$$R_T = \frac{R_1 \times R_2}{R_1 + R_2}$$

$$R_T = \frac{15 \times 30}{15 + 30}$$

$$R_T = 450/45$$

$$R_T = 10 \ \Omega$$

The equivalent circuit (Figure 4–29) is a series circuit.

$$R_T = R_1 + R_2$$

$$R_T = 10 + 10$$

$$R_T = 20 \ \Omega$$

The alternator is producing 20 volts. According to Ohm's law, the current through the alternator is:

$$I = E/R$$

$$I = 20/20$$

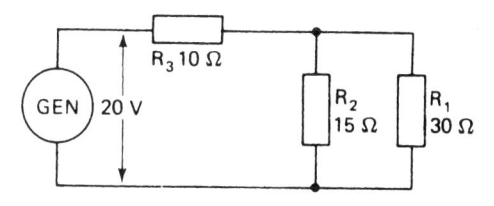

$$I = 1 \text{ ampere}$$

FIGURE 4–28 Series-parallel circuit

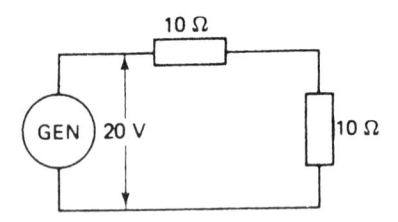

FIGURE 4–29 Equivalent circuit

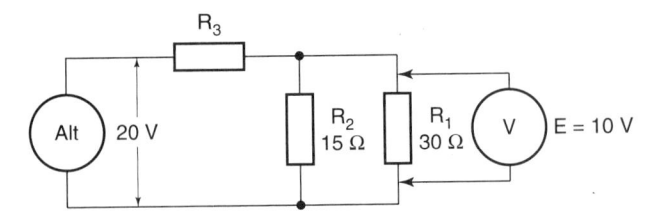

FIGURE 4–30 Series-parallel circuit with unknown resistance value

Going back to the original circuit, Figure 4–28, the current through the series 10-Ω resistor R_3 is 1 ampere, and there are 10 volts across it. With 10 volts across R_3, there are 10 volts left (from the 20 volts supplied) to appear across R_1 and R_2.

Since the same voltage appears across resistors in parallel, the current flow through R_1 is:

$$I = E/R$$
$$I = 10/30$$
$$I = 1/3 \text{ or } 0.333 \text{ ampere}$$

The current through R_2 is

$$I = E/R$$
$$I = 10/15$$
$$I = 2/3 \text{ or } 0.666 \text{ ampere}$$

The total current entering the junction (1 ampere) is equal to the total current in the branches.

$$1/3 \text{ ampere} + 2/3 \text{ ampere} = 1 \text{ ampere}$$

or

$$0.333 \text{ ampere} + 0.666 \text{ ampere} = 0.999 \text{ or } 1 \text{ ampere}$$

There are times when the information known about a circuit is other than the resistance values. For example, the circuit in Figure 4–28 might be encountered, but the known information could be that in Figure 4–30: The alternator is producing 20 volts; the value of resistor R_1 is 30 Ω; the value of resistor R_2 is 15 Ω; and a voltmeter across R_1 indicates 10 volts. The unknowns may be found by using the laws of series circuits and parallel circuits, and Ohm's law.

The current flow through R_1 could be found using Ohm's law. It has already been shown that with 10 volts across 30 Ω, the current is 1/3 (0.333) ampere. According to the laws of parallel circuits, the voltage across R_2 is the same as across R_1, 10 volts. The current through R_2 can be determined: 10 volts across 15 Ω provides 2/3 (0.666) ampere. The current entering junction point A is equal to the sum of the currents leaving it. Therefore, 1 ampere (1/3 A + 2/3 A = 1 A) enters junction point A.

This 1 ampere is the current flowing through the unknown series resistor R_3. Since the supply voltage is 20 volts and there are 10 volts across the parallel resistors R_1 and R_2, there must be 10 volts across R_3. The supply of 20 volts minus the 10 volts across the parallel branch leaves 10 volts (20 − 10 = 10).

According to Ohm's law:

$$R_3 = E/I$$
$$R_3 = 10/1$$
$$R_3 = 10 \text{ ohms}$$

PRACTICAL EXERCISE

This exercise requires a late-model automobile and an appropriate wiring diagram for the park lamp circuits.

1. Turn the park lamps ON.
2. Observe the lamps on both sides of the vehicle.
   a. Are the front and rear lights ON on the left side?
   b. Are the front and rear lamps ON on the right side?

   NOTE: If any lamp is burned out, replace the lamp before proceeding with the exercise.

3. Turn the park lamps OFF.
4. Disconnect the left front lamp.

5. Turn the park lamps ON.
   a. Is the left front lamp ON?
   b. Is the left rear lamp ON?
   c. Are the right front and rear lamps ON?
6. Turn the park lamps OFF.
7. Reconnect the left front lamp.

QUIZ

1. Are the park lamps wired in series or in parallel with each other? _____
2. Is the park lamp switch wired in series or in parallel with the parking lamp circuit? _____

EXERCISE

Complete the sketch of a typical parking lamp circuit in Figure 4–31.

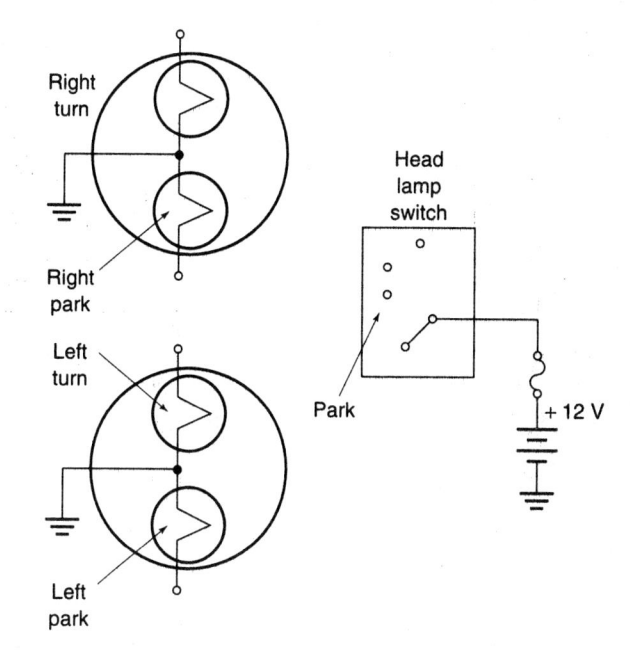

FIGURE 4–31 A typical park lamp circuit (Exercise)

SUMMARY

- A complete circuit is necessary if current is to flow.
- In a series circuit the sum of the voltage across the components is equal to the applied voltage.
- The same current flows through each component in a series circuit.
- The total current in a parallel circuit is equal to the sum of the currents in the branch circuits.
- The same voltage appears across each component of a parallel circuit.
- In a series-parallel circuit the laws of series circuits apply to the components connected in series, and the laws of parallel circuits apply to the components connected in parallel.

REVIEW

Draw a sketch of each of the circuits before attempting the solution.

(T) (F) 1. In parallel-resistive circuits, the total current is the sum of the currents in the branches.

(T) (F) 2. The same voltage appears across each component in parallel.

(T) (F) 3. The total resistance may be found by using the product over sum method.

(T) (F) 4. In series-parallel circuits, the laws of series circuits apply to the portions of the circuit connected in series.

(T) (F) 5. In series-parallel circuits, the laws of parallel circuits apply to the portion of the circuit connected in parallel.

(T) (F) 6. Ohm's law may be applied to the whole circuit or any part of the circuit.

7. A circuit consists of four 20-Ω resistors connected in series. The total resistance of the circuit is _____ ohms.

8. A circuit consists of an 8-Ω resistor and a 4-Ω resistor connected in series. The supply voltage to the circuit is 24 volts.
   a. How much current will flow in the circuit?
   b. How much voltage would be measured across each resistor?

9. A circuit consists of a 10-Ω resistor and a 2-Ω resistor connected to a 24-volt power source. The 10-Ω resistor has 20 volts across it. How much current is flowing in the 2-Ω resistor?

10. A circuit consists of three resistors connected in series to a power source. Each resistor has 6 volts across it. What is the voltage of the power source?

11. A series circuit consists of two resistors connected to a power source of 24 volts. The first resistor has 16 volts across it; the second resistor is a 4-Ω resistor. What is the total resistance of the circuit?

12. The (same) (different) current flows through each resistor in a series circuit.

13. Two 10-Ω resistors are connected in a series circuit. The voltage across the circuit is 40 volts.

    The current through each resistor is _____ amps.

14. A current flow of 3 amperes is flowing in a two-resistor series circuit. The supply voltage is 24 volts. One of the resistors has a value of 6 Ω.

    The other resistor is _____ ohms.

15. An 8-kΩ resistor and a 12-kΩ resistor are connected in parallel with a third resistor. The supply voltage is 24 volts. The line current is 6 mA. What is the value of the third resistor?

16. Twelve milliamperes (MA) are flowing through a circuit made up of an 8-kΩ and a 6-kΩ resistor in series. What is the applied voltage?

17. A three-resistor circuit has 36 volts across it. The current in the circuit is 1/2 mA. One of the resistors is 8 kΩ and another is 3 kΩ in value. What is the value of the third resistor?

18. Four 40-Ω resistors are connected in parallel. What is the total resistance of the combination?

19. A 10-Ω resistance is needed in a control circuit. Three 30-Ω resistors are available. How could these resistors be connected in order to obtain the required resistance?

20. A resistance is needed in a control circuit. The resistance must be more than 2 Ω but less than 4 Ω. Three 2-Ω resistors are available. How would you connect the 2-Ω resistors in order to obtain the required resistance?

◆ UNIT 5 ◆

MAGNETISM

OBJECTIVES

On completion of this unit you will be able to:

- Identify natural magnets and artificial magnets.
- Understand magnetic polarity and magnetic fields.
- Discuss electromagnets and magnetic strength.

Magnets and magnetism are incorporated in many components of automotive electrical circuits and systems. Electric motors, ignition coils, air conditioner clutches, sensing devices, and many instruments operate using magnetic principles. Modern automobile electrical systems could not function properly without magnets and magnetism. A good understanding of the basic principles of magnets and magnetic circuits is essential for all automotive technicians.

MAGNETS AND MAGNETISM

There are three types of magnets: natural magnets, artificial magnets, and electromagnets. All have one common characteristic: the ability to attract or repel metal. The behavior of magnets is most important in the operation of many electrical devices. Therefore, it is important that the automotive technician possess a general understanding of the basic laws of magnetism.

Natural Magnets

We do not know for certain when the first magnet was discovered. However, it is known that magnetic iron ore was the object of superstitious worship by early humans. The Greek philosopher Plato (427–347 B.C.) regarded "magnetic virtue" as divine.

The first known practical use of a natural magnet, called lodestone or loadstone, was late in the twelfth century. Mariners found that a lodestone suspended from a piece of wood and floated in water could be used as a compass, since it always assumed a north-south (N-S) orientation. Lodestone is a magnetic rock with a high iron ore content, scientifically known as magnetite (Fe_3O_4). Magnetite is a black, native oxide of iron that is strongly attracted by a magnet. Early discoveries of large quantities of magnetite were found near Magnesia in Asia Minor; the word *magnet* comes from the Greek word *magnes*, "stone of Magnesia."

Artificial Magnets

Artificial magnets can be made from any material that has a high permeability, such as hard steel, cobalt (Co), nickel (Ni), or a combination of these elements with metals and rare-earth elements. An example of an excellent magnet material is *alnico*, a combination of aluminum (Al) and nickel (Ni). Aluminum is a nonmagnetic material, but in combination with nickel provides a better magnetic material than nickel alone.

An artificial magnet can be created in any one of four ways:

1. By striking the material with a sharp, hard blow while it is aligned parallel to the direction of the Earth's magnetic field.
2. By bringing the material into contact with another magnet, or by stroking the material with the pole of another magnet.
3. By heating the material and allowing it to cool while its principal axis is in the direction of the Earth's magnetic field, or another magnetic field.
4. By placing the material into the field of a powerful electric current.

Because of their importance in automotive electronics, we cover electromagnets in detail later in this unit.

MAGNETIC POLARITY

If a piece of metal such as a steel bar is magnetized, the magnetic effects concentrate at the ends of the bar. These points of magnetic concentration are called *poles* of the magnet. Away from the ends of the bar an invisible force is present, the effect of which can be seen if small bits of iron (Fe) or steel are placed near the magnet. The invisible force around the magnet is called a magnetic *field*.

The planet Earth is a permanent magnet. One end of this giant magnet is near the north pole and the other end is near the south pole (Figure 5–1). It is Earth's invisible magnetic lines of force that cause a compass to function.

MAGNETIC FIELDS

A magnet is a piece of material that has the ability to attract steel, iron, and other ferrous materials (those that contain iron). A magnet will not attract nonferrous materials, such as copper (Cu) and aluminum (Al). Also, magnets will not attract or be attracted to nonmetallic objects such as plastic. Magnets are available in a variety of sizes and shapes. We will use one of the most common magnets, the bar magnet (Figure 5–2), to explain magnetic fields.

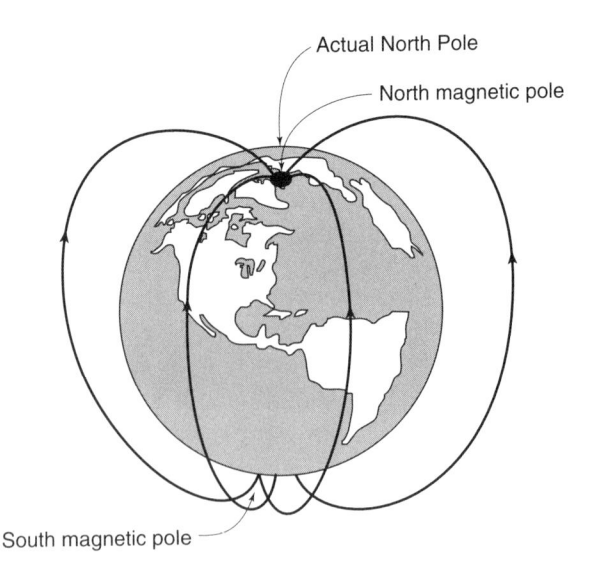

FIGURE 5–1 The Earth as a magnet

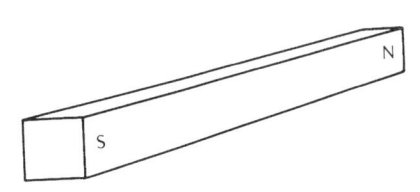

FIGURE 5–2 Square bar permanent magnet with north (N) and south (S) poles

A magnet has two poles: a north (N) pole and a south (S) pole. These poles are located at either end of a bar magnet. If a bar magnet were suspended from a string or provided with another means to pivot at its exact center, one of the poles would turn toward the Earth's magnetic north pole. The bar would then be aligned with the Earth's magnetic field; the end facing north would be the north pole of the magnet. This is the basis of compass construction. A compass needle is merely a lightweight strip of magnetized steel delicately pivoted at its exact center. Its north pole will always seek Earth's magnetic north (Figure 5–3) unless some opposing field prevents it from doing so (Figure 5–4).

It was discovered long ago that the north (N) pole of any magnet will repel the north (N) pole of any other magnet (Figure 5–5). Conversely, the north (N) pole of any magnet will attract the south (S) pole of any other magnet (Figure 5–6). Similarly, south (S) poles repel south (S) poles, but attract north (N) poles. This phenomenon is summarized as the law of magnetic attraction and repulsion:

Like poles repel—Unlike poles attract

EFFECTS OF MAGNETIC FIELDS

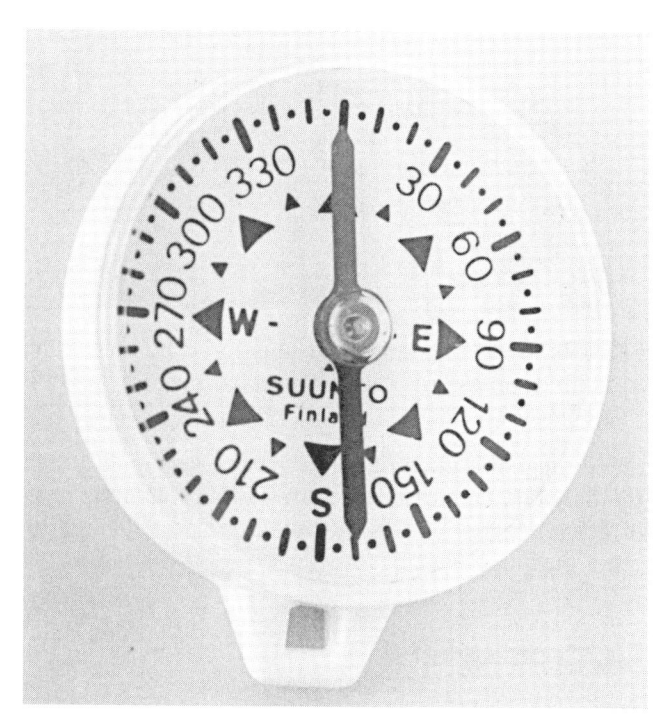

FIGURE 5-3 A compass needle, under normal conditions, will seek Earth's magnetic north

ATOMIC ARRANGEMENT

In an unmagnetized piece of ferrous material the atoms are arranged in haphazard fashion, with atomic N-S poles pointing in any direction, as shown in Figure 5–7(a). The magnetizing process aligns these atomic N-S poles so that they are all facing the same direction (Figure 5–7(b)). The piece of magnetic material then has one north pole and one south pole. As illustrated in Figure 5–8, if a single magnet were broken into two pieces, two separate magnets would result, each with its own north and south pole. If it were broken into four pieces, four separate magnets would result, and so on.

EFFECTS OF MAGNETIC FIELDS

The force (magnetic field) in the space surrounding a magnet can be seen by studying a pattern of iron filings sprinkled on a card placed over a magnet (Figure 5–9). Each small particle of iron becomes a temporary magnet with a north and south pole. Accordingly, the iron filings align with the magnetic field of the magnet. This invisible magnetic field is also called *lines of force*, *lines of flux*, or *magnetic flux*.

Magnetic lines of force are known to penetrate all materials, and there is no insulator that will stop them.

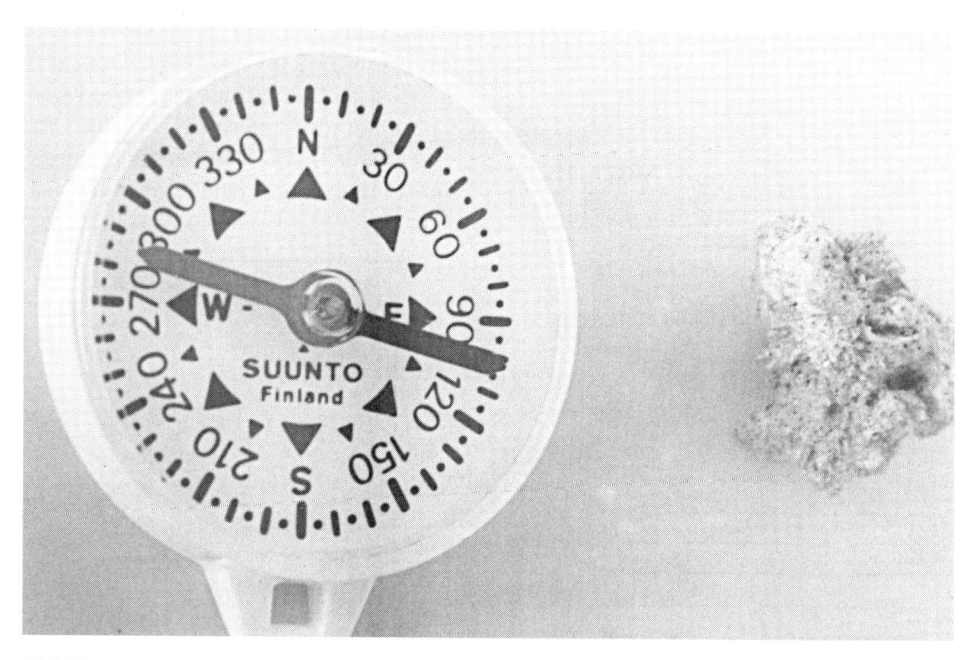

FIGURE 5-4 A compass needle is deflected by an opposing field, here by a lodestone

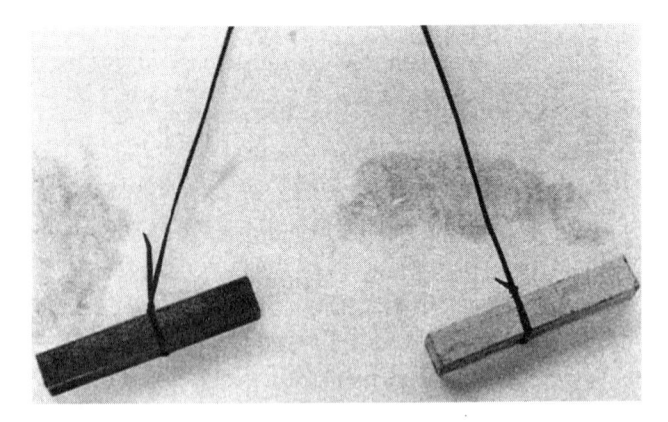

FIGURE 5–5 Like poles repel

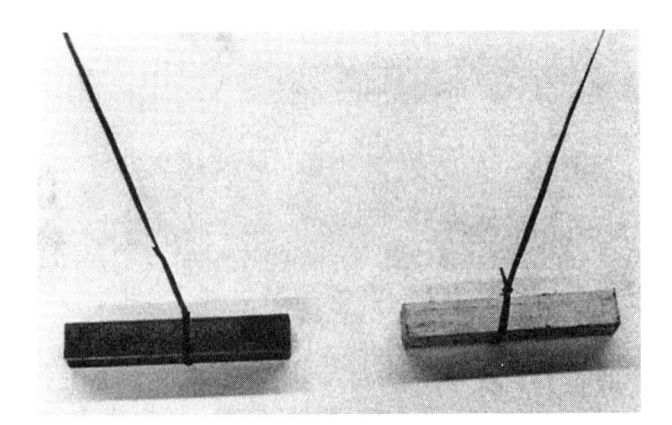

FIGURE 5–6 Unlike poles attract

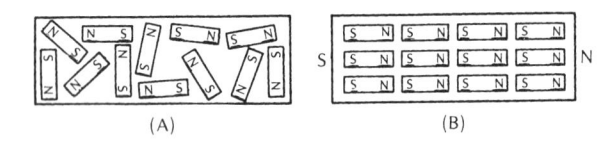

FIGURE 5–7 (a) Unmagnetized and (b) magnetized ferrous metal bars

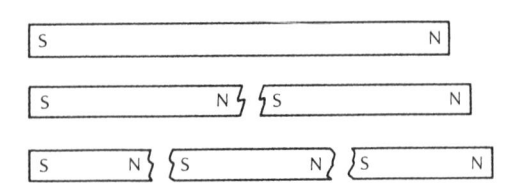

FIGURE 5–8 Regardless of the number of pieces a permanent magnet is broken into, each piece will have a north (N) and south (S) pole

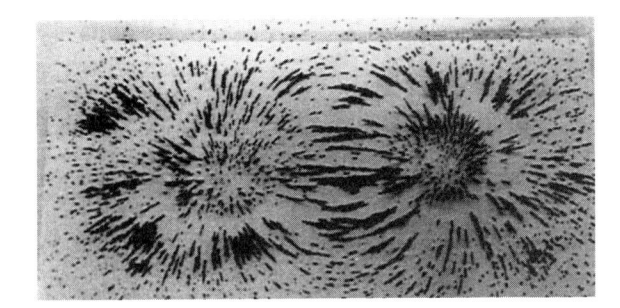

FIGURE 5–9 Magnetic field illustrated by sprinkling iron filings on a card placed over a bar magnet

Since magnetic lines of force never cross each other, the only known barrier to them is other magnetic lines of force. Magnetic lines of force are also deflected by other magnetic lines of force.

Electromagnets

An electromagnet also has a magnetic field. Basically, any wire carrying an electrical current has magnetic lines of force (Figure 5–10). These magnetic lines of force differ from those of permanent magnets, however. Instead of a north and south pole arrangement, these lines of force form concentric circles around the wire. The number of lines of force around a current-carrying wire is in direct proportion to the amount of current in the conductor. The greater the current, the greater the number of lines. The number of lines of force, per unit, is called *flux density*. The flux density is greatest at the surface of a conductor and diminishes with distance from the conductor (Figure 5–11).

When a current-carrying wire is formed into a loop (Figure 5–12), the lines of force must pass through

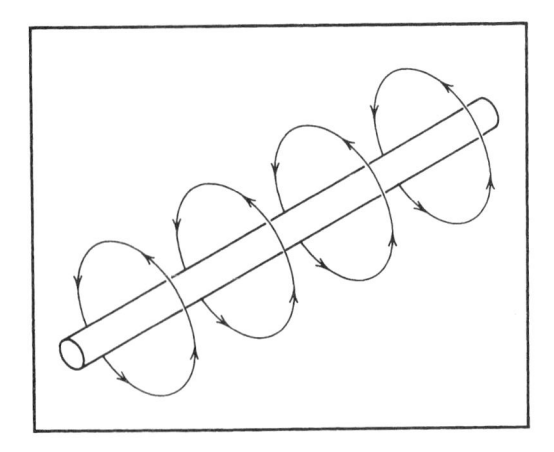

FIGURE 5–10 Magnetic lines of force around a current-carrying conductor

EFFECTS OF MAGNETIC FIELDS 53

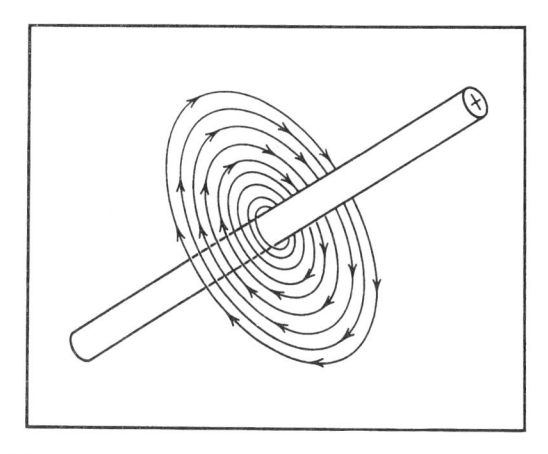

FIGURE 5–11 Flux density is greatest at the surface of a conductor

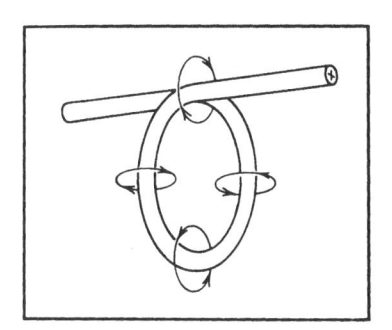

FIGURE 5–12 Lines of force within a loop

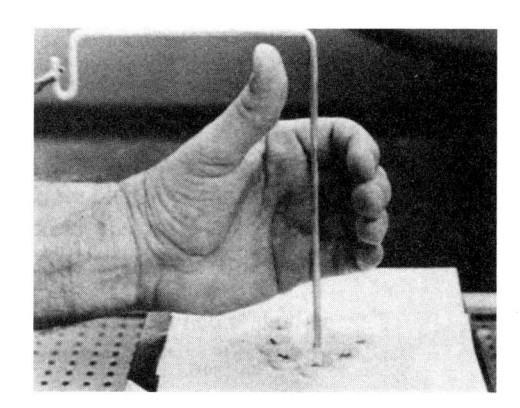

FIGURE 5–13 Illustration of the left-hand rule

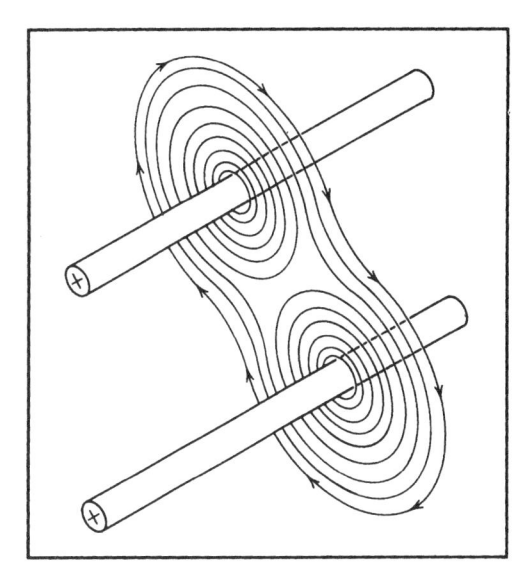

FIGURE 5–14 Merging lines of force in parallel conductors

the middle of the loop. Since lines of force will not cross each other, the flux density would increase if the middle of the loop were made smaller. Increasing the flux density produces a greater magnetic effect with the same amount of current.

The direction of a magnetic field can be determined if the direction of electron flow is known. A simple method known as the *left-hand rule* (Figure 5–13) is used. With the thumb of the left hand extended in the direction of electron flow, simply grasp the conductor. The fingers then point in the same direction as the magnetic lines of force. However, if the direction of electron flow is yet to be determined, place a compass atop the current-carrying conductor. The compass needle will align itself with the magnetic field. Place the left hand around the conductor with fingers toward compass north and extend the thumb—the thumb points in the direction of electron flow.

When current flow is in the same direction in two or more parallel conductors, each conductor creates a field of force running in the same direction (Figure 5–14). If these conductors are close enough to each other, a magnetic attraction between them will exist. If they are placed very close to each other, and each conductor is carrying about the same amount of current, the lines of force between them will merge.

The insulation between parallel wires must be sufficient to resist the magnetic force pressing them together. In many cases this insulation is merely a heavy coat of enamel; in others it is specially treated paper, cloth, or plastic. The thickness and type of insulation material is determined by the amount of current, which in turn determines the strength of magnetic attraction. The insulation does not reduce the magnetic lines of force; it only serves as a cushion to prevent damage to the conductors.

In many cases the type of insulation used on wires is determined by the voltage to be applied as well as the environment where the wire is used. The temperature, as well as the presence of chemicals, has a lot to do with insulation requirements.

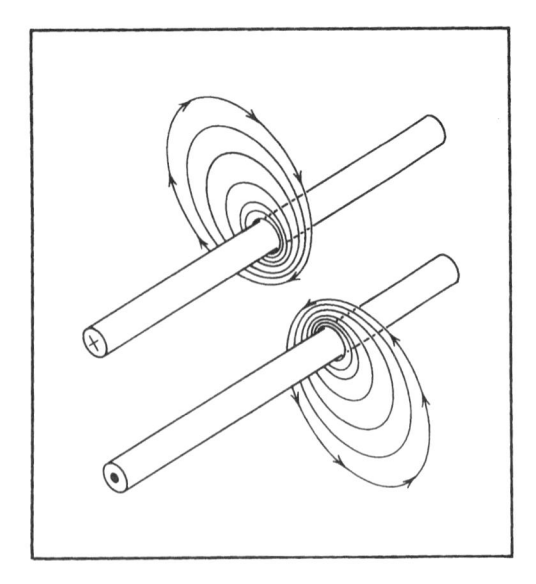

FIGURE 5–15 Opposing lines of force in parallel conductors

If current flow is in the opposite direction in two or more parallel conductors (Figure 5–15), the lines of force are counteractive and tend to repel each other. The closer the conductors are brought together, the more dense are the lines of force between them. The lines of force do not merge and will not cross each other. Although the two wires tend to repel each other, they also must be insulated to prevent damage.

It is now obvious that by looping (coiling) a conductor, the magnetic lines of force are merged and greatly concentrated. The greater the number of loops, the greater the magnetic lines of force. A coil of wire so constructed has a north and south pole only during the time of current flow; therefore, it is known as an electromagnet.

The magnetic polarity of a coil may be determined by the left-hand rule. To apply the rule, grasp the coil with the fingers pointing in the direction of current flow (negative to positive). The extended thumb will point toward the north pole of the coil.

Coil Strength

The strength of an electromagnetic coil is determined by the number of turns of wire in the coil and the amount of current flow, in amperes, passing through the wire. This is known as the *ampere-turn rating* of the coil. For example, if a current of one ampere were passed through a coil of 100 turns, the ampere-turn rating would be 100:

$$1 \times 100 = 100$$

The same rating would result if 0.5 amperes were passed through 200 turns or if two amperes were passed through 50 turns:

$$0.5 \times 200 = 100$$

or

$$2 \times 50 = 100$$

There are four methods of increasing or determining the strength of an electromagnet. They are as follows:

1. Increase the current through the coil.
2. Increase the number of turns in the coil.
3. Increase the current and the number of turns in the coil.
4. Insert a magnetic core into the center of the coil.

The following are examples of the first method, increasing the current through the coil:

- 1 ampere × 100 turns = 100 ampere-turns
- 3 amperes × 100 turns = 300 ampere-turns
- 6 amperes × 100 turns = 600 ampere-turns

The following are examples of the second method, increasing the number of turns in the coil:

- 1 ampere × 100 turns = 100 ampere-turns
- 1 ampere × 300 turns = 300 ampere-turns
- 1 ampere × 600 turns = 600 ampere-turns

The following are examples of the third method, increasing both current and number of turns in the coil:

- 1 ampere × 100 turns = 100 ampere-turns
- 1.2 amperes × 250 turns = 300 ampere-turns
- 2 amperes × 300 turns = 600 ampere-turns

The fourth method, inserting a magnetic core into the center of the coil, increases the magnetic lines of force without increasing either the current or the number of turns. The magnetic lines of force permeate the magnetic core much better than the air core. Depending on the original ampere-turn rating of a coil and the type

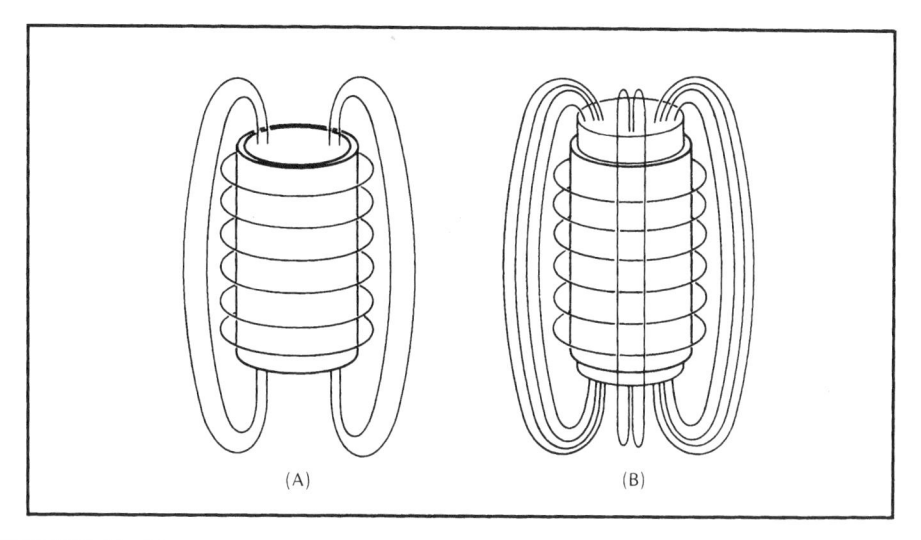

FIGURE 5–16 Relationship of permeability of coil (a) without core and (b) with core

of iron used as a core, the hypothetical relationships are shown in Figure 5–16. The illustration shows 100 ampere-turns in both coils. The coil with the air core has 100 lines of force; the coil with the soft-iron core has 600 lines of force. Therefore, in this example, the iron core is six times more permeable than the air core.

When electric current passes through the coil of an electromagnet, the soft-iron core itself becomes a magnet. But it is a magnet only as long as current is applied to the coil. The number of lines of force that disappear (dissipate) from the core when current to the coil is removed depends on the type of iron used. In most cores, a slight amount of magnetic flux will always remain. This is known as residual magnetism and produces a weak permanent magnet that usually causes no lasting or undesirable effects. However, for any application where residual magnetism may be a problem, there are methods to overcome its weak attraction.

PRACTICAL EXERCISE

This exercise requires a vehicle and the use of a magnetic compass.

1. Raise the hood of the vehicle.
2. Locate the wire(s) leading to a headlamp (two wires for single beam; three wires for dual beam).
3. Separate the wires. DO NOT disconnect them. Take care not to damage the wires.
4. Place the compass close to each wire. What happened with:

    a. the first wire? _____

    b. the second wire? _____

    c. the third wire (if dual beam)? _____

5. Again, hold the compass close to one of the wires.
6. While observing the compass, ask an assistant to turn the headlamps ON.
7. Turn the headlamps OFF.
8. Did the compass needle align itself perpendicular with the wire? _____ How do you explain what happened? _____
9. Repeat Steps 5, 6, and 7 with the other wire(s).
10. Does the compass needle align itself with the wire each time the headlamps are turned ON? _____
11. Can the magnetic field about a wire carrying current be detected? _____ How? _____

12. Can you determine a difference in direction of the magnetic field between either (or any two) of the wires? _____

13. Why do you think this happens? _____
_____
_____

SUMMARY

- Magnets have two poles.
- Like magnetic poles repel each other.
- Unlike magnetic poles attract each other.
- A wire carrying current has a magnetic field about it.
- A ferrous metal core is used with an electromagnet to increase the strength of the magnetism.

REVIEW

1. With magnets, like poles (attract) (repel) each other.
2. The concentration of a magnetic field in a bar magnet is at the (sides) (ends) of the bar.
3. Two adjacent wires carrying current in the same direction will (attract) (repel) each other.
4. Two adjacent wires carrying current in opposite directions will (attract) (repel) each other.
5. If a wire carrying current is formed into a loop or coil, the magnetic field will (increase) (decrease).
6. If a magnetic material, such as soft iron (Fe), is used as the core of an electromagnet, the strength of the magnet will (increase) (decrease).
7. Soft iron (Fe) is used as the core of a (permanent) (temporary) magnet.
8. Hard steel is used as the core of (permanent) (temporary) magnets.
9. (Nothing) (Glass) will insulate magnetic lines of force.
10. The magnetic field about a current-carrying conductor is strongest (close to) (far from) the wire's surface.

◆ UNIT 6 ◆

ELECTRICAL TESTING

OBJECTIVES

On completion of this unit you will be able to:

- Explain how an analog meter is constructed.
- Understand how meter movements are connected in circuits to provide ampere, voltage, and resistance measurements.
- Read and interpret meter readings.
- Understand what a digital meter is.
- Construct and use a test lamp.
- Discuss the various testing methods using meters and test lamps.

ELECTRIC TEST EQUIPMENT AND TESTING

The modern automotive vehicle has greatly improved in performance over the past years. Much of this improvement is due to increased use of electrical and electronic devices. Maintenance of these devices requires the use of electrical test equipment. The modern automotive technician is likely to be found with a volt ohmmeter in one hand and a wrench in the other.

Electric Meters

An electric meter is used to determine the characteristics of an electric circuit or component. The meters commonly used by the technician include the ammeter, voltmeter, and ohmmeter. Occasionally, the technician will have the opportunity to use a watt meter. The meters used in the trade today include analog and digital types. To become familiar with the equipment, always read the instruction manual that comes with the test equipment before using it.

Analog Meters

An analog electric meter uses a meter movement that operates on the electric motor principle. In order to obtain meter pointer deflection, a current must pass through the meter-movement coil. This current sets up a magnetic field in the meter-movement coil, which is repelled by the fixed magnetic field provided by the permanent magnets. The stronger the current flow through the meter-movement coil, the greater the deflection of the coil and pointer, as shown in Figure 6–1.

In the meter movement shown in Figure 6–1, the permanent magnets are indicated by N for north pole and S for south pole. The movement-deflection coil pivots at the center and has a pointer attached to it. A coil spring is attached to the meter movement holding the movement and pointer at zero (0) deflection when there is no current flow through the movement coil.

Physical stops in the form of pins are located just beyond the zero and full-scale deflection points in order to limit the movement of the pointer. When current flows through the movement-deflection coil, a magnetic field is set up around this coil. The magnetic field of the deflection coil will be repelled by the field of the permanent magnets. The meter movement will rotate around the pivot point, causing the pointer to move up the scale.

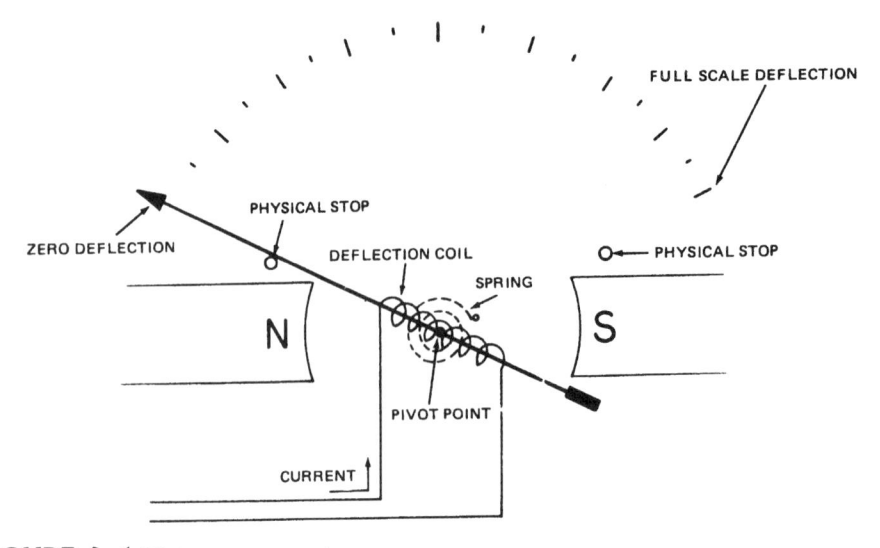

FIGURE 6-1 Meter movement

As the meter movement rotates up scale, pressure increases on the coil spring attached to the movement. The rotation will stop when the pressure of the repelling magnetic fields is balanced by the reverse pressure offered by the coil spring.

A different amount of current flow through the meter-movement coil will cause a different amount of meter movement and pointer deflection.

Ammeters

An *ammeter* is a meter used to measure current flow, the movement of electrons through the circuit. In order to measure current flow, the circuit must be opened and the ammeter inserted so that the current to be measured flows through the ammeter. The circuit connections are shown in Figure 6-2.

Normally, the current to be measured is much higher than the amount needed to deflect the meter movement to full scale. In order to make meters useful in line current-measuring systems, a shunt is installed within the meter. The *shunt* is a low-resistance current path that allows a proportional amount of current to flow through the meter movement. The remaining current bypasses the meter movement by going through the shunt. This is shown in Figure 6-3.

Suppose that a one-ampere full-scale meter were to be designed using a meter movement that required 0.001 ampere (1 mA) full-scale deflection. The meter-movement resistance is 100 Ω. According to Ohm's law, $I \times R = V$, there would be 0.001 A $\times$ 100 Ω = 0.1 V across the meter movement at full-scale deflection.

The scale of the meter would be changed to indicate 1 ampere instead of 1 mA, but it would still take only 1 mA through the meter movement to cause full-scale deflection. This arrangement is shown in Figure 6-4.

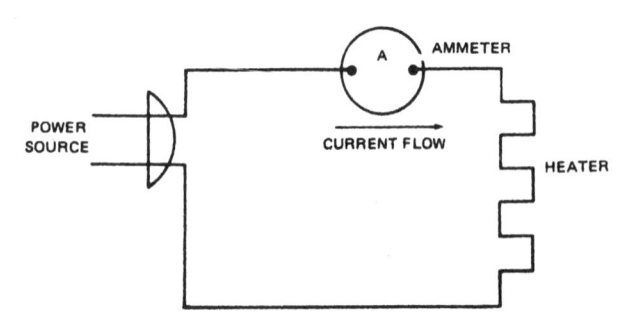

FIGURE 6-2 Ammeter connection

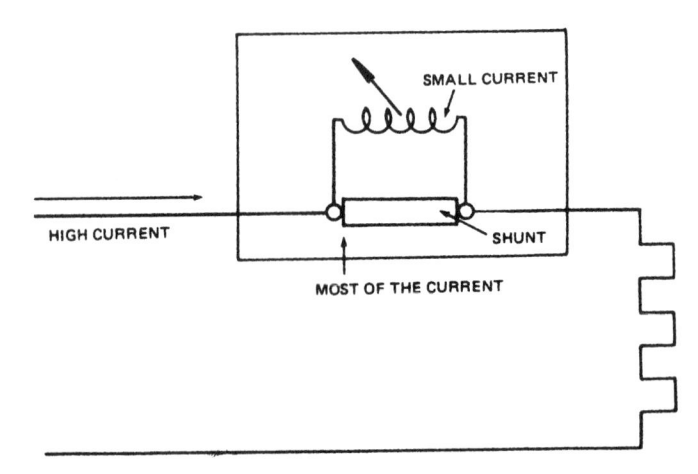

FIGURE 6-3 Shunt circuit

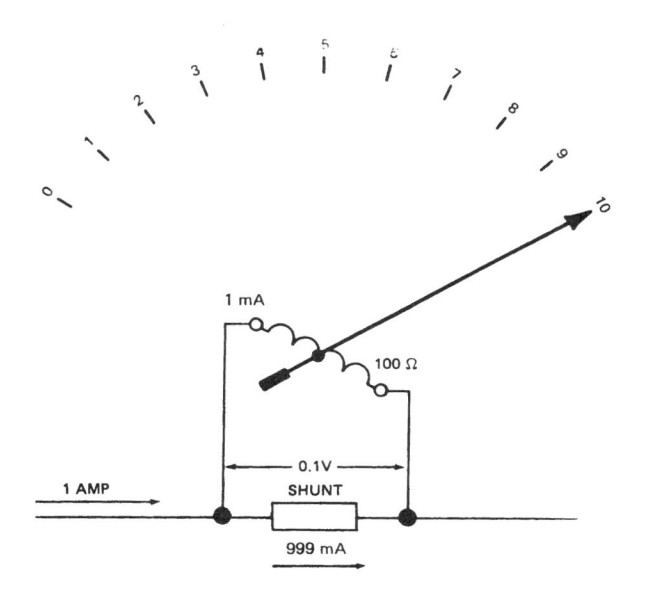

FIGURE 6–4 Shunt operation

One ampere is coming down the line. Only 1 mA goes through the meter movement, while 999 mA go through the shunt. The shunt would be in parallel with the meter movement. The resistance of the shunt would be:

$$R = V/I$$

$$0.1 \text{ V}/999 \text{ mA} = 0.1 \text{ V}/0.99 \text{ A} = 0.1001 \text{ }\Omega$$

The shunt in the ammeter is a low-resistance path for current flow. An ammeter is therefore a low-resistance device. An ammeter is never connected across the circuit. The low resistance of the ammeter would short the circuit, usually causing the line fuse to open. Ammeters are never connected in parallel with any component; they are always connected in series.

Many meters contain a switch system that provides more than one full-scale deflection sensitivity. A meter may have three or more full-scale readings, such as 0–1, 0–10, and 0–100 amperes.

Voltmeters

A voltmeter can be developed with the same meter movement used in the ammeter. That meter movement required 1 mA for full-scale deflection and had 100 Ω of resistance.

In Figure 6–5, a 300-volt meter is shown. The scale has been drawn using 300 V as the full-scale deflection. Because voltage is measured across a circuit, the voltmeter will have to be placed across the circuit in order to measure voltage. There will be 300 volts at the terminals of the voltmeter. According to Ohm's law, the total resistance will equal the voltage divided by the current:

$$R = V/I$$

$$R = 300/0.001 \text{ or } 300{,}000 \text{ }\Omega$$

The scaling resistor would be 299,900 Ω, with the meter-movement resistance of 100 Ω added to it for the 300,000 Ω total. The procedure used in making a voltmeter is to determine the required resistance by dividing the full-scale voltage wanted by the meter-movement full-scale current. The scaling resistor is then the total resistance minus the resistance of the meter movement.

Voltmeter Loading. A common problem occurs in using a voltmeter to measure voltages across resistances that are of the same order as the resistance of the meter. Remember that when a voltmeter is connected to a circuit, the internal resistance of the meter is connected across (in parallel with) that portion of the circuit.

Consider two circuits where voltage is to be measured. In Figure 6–6(a), the voltmeter of Figure 6–5 will indicate the correct voltage.

The meter in Figure 6–6(b) will indicate only 10 volts, which is the voltage across R_2 with the meter connected. Remember the internal resistance of the meter is 300 kΩ. When it is connected to the circuit, the circuit changes (Figure 6–7).

The meter indicates the voltage that is present after the meter is connected. Whenever a voltmeter is used to make measurements in a circuit, it is good practice to use a meter that has an internal resistance 10 times (10×) higher than the resistance of the components in the circuit.

Many meters now include amplifying components. The input resistance in some of these meters is 10 megohms or higher. This level of input resistance makes the meter particularly useful in high-resistance circuits.

Ohmmeters

The ohmmeter is developed using the same 1-mA full-scale meter movement. An ohmmeter requires the use of an internal power source. The standard source is a 1.5-volt cell. Observe the circuit shown in Figure 6–8. If

60 Unit 6 ELECTRICAL TESTING

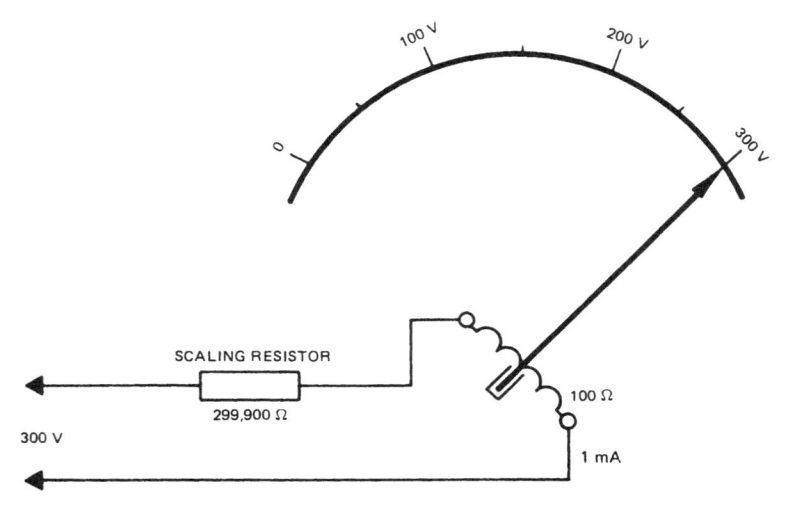

FIGURE 6–5 Voltmeter

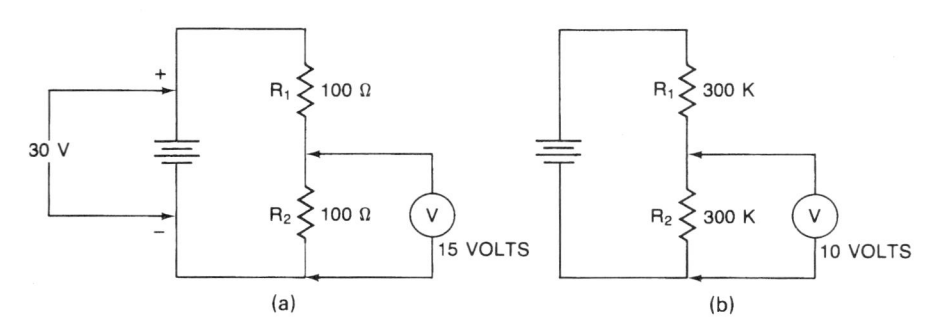

FIGURE 6–6 Voltmeter loading

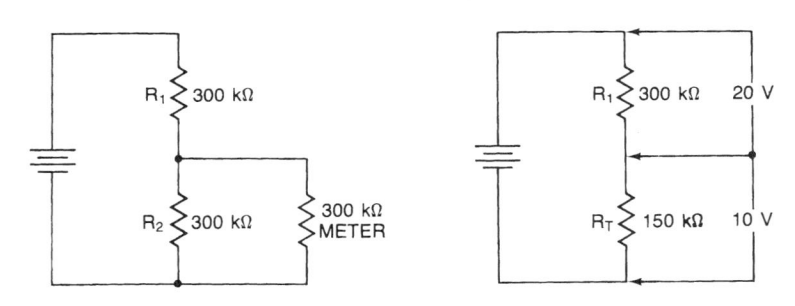

FIGURE 6–7 Effect of internal meter resistance

the terminals of the meters were shorted together, a complete series circuit would exist. The power source is 1.5 volts. The series resistances are the 100-Ω meter resistance, the 1300-Ω resistor, and the 300-Ω potentiometer. The potentiometer would be adjusted to provide a total resistance of 1500 Ω for the series circuit. Using Ohm's law, we would then find:

$$I = E/R = 1.50 \text{ V}/1500 \text{ }\Omega = 1 \text{ mA}$$

The current flow through the circuit would be 1 mA, the full-scale current of the meter movement. The meter movement would rotate to full scale where the scale is marked zero (0 Ω). The meter leads are shorted

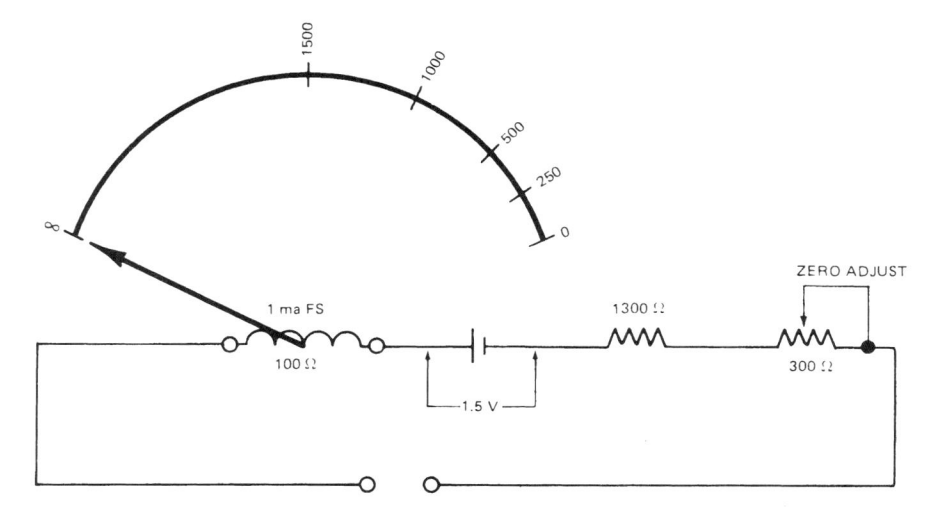

FIGURE 6–8 Ohmmeter

together. The resistance is zero (0 Ω). The meter indicates 0 Ω.

If the meter leads were connected across a 1500-Ω resistor, the series circuit would contain the 1500-Ω resistor, the 100-Ω meter movement, the 1300-Ω resistor, and the potentiometer set at 100 Ω. The total resistance would be 3000 Ω. According to Ohm's law:

$$I = E/R = 1.5/3000$$

$$I = \frac{1.5 \text{ v}}{3000 \text{ }\Omega}$$

$$I = 0.5 \text{ mA}$$

There would be 0.5 mA flowing through the circuit. The meter movement would move to approximately half scale, where the marking is 1500 Ω.

It is obvious that the value of the resistor connected between the terminals of the meter will determine the total current flow through the series circuit, and therefore the amount of current through the meter movement. The current determines how far up scale the meter will rotate. A high resistance will keep the current low and provide little meter movement. A low resistance will provide a higher current, and the meter may rotate close to zero ohms (0 Ω). Many ohmmeters provide three to five multipliers for resistance measurement. Common multipliers include $R \times 1$, $R \times 10$, $R \times 100$, $R \times 1000$, and $R \times 10,000$.

Digital Meters

Many new meters are becoming available as new digital electronic circuitry is developed. Digital meters such as the multimeter shown in Figure 6–9 are easier to read

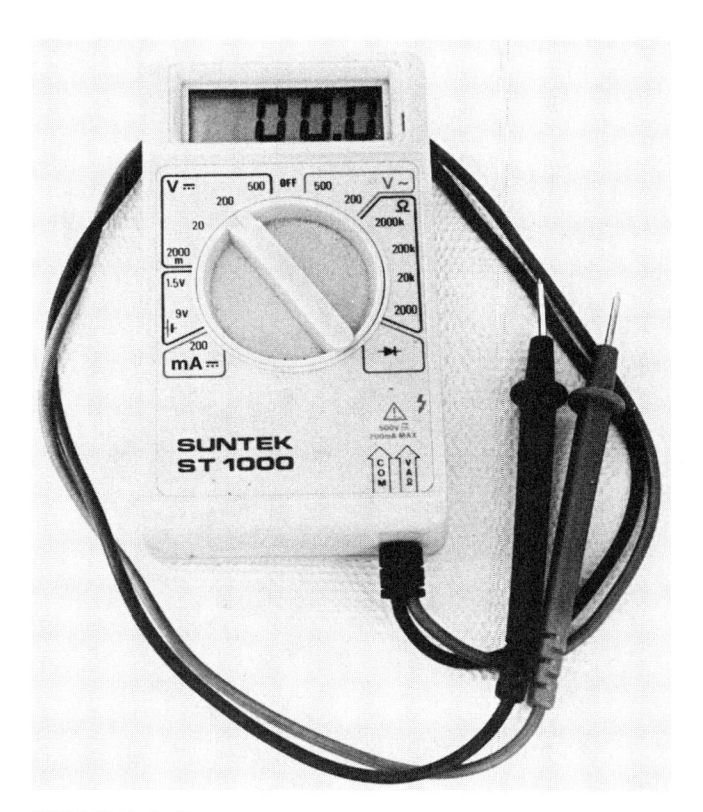

FIGURE 6–9 Digital volt-ohm-milliammeter (Courtesy of BET, Inc.)

than standard meters because the meter reader has only to read the digits. With the older analog meters, the meter reader must interpolate, that is, decide what the meter is indicating. This process takes practice if correct answers are to be obtained.

ELECTRICAL TESTING

For proper electrical testing it is essential to use accurate and reliable electrical test instruments, including the voltmeter, ammeter, and ohmmeter. These instruments are supplemented with variable resistors (piles), jumper wires, and test lamps. Although some meters are multipurpose, such as the volt-ohmmeter shown in Figure 6–10, each meter will be discussed separately.

Voltmeter Use

For automotive testing service, a voltmeter with a minimum range of 0–15 volts (dc) is required. If the meter is to be used for battery or alternator testing, a 0–20 volt range is desirable. Voltmeters are always connected into a circuit to be tested in parallel, *never* in series, with the circuit. Voltmeters are always connected across the circuit to be tested as shown in Figure 6–11.

In an automotive electrical circuit, voltmeters (Figure 6–12) are used for electrical testing to ensure that battery voltage (state of charge) is sufficient and that this same voltage is reaching the electrical components of the circuit.

Consider the simple circuit in Figure 6–13. If there is 12 volts available at the battery (meter A), and the

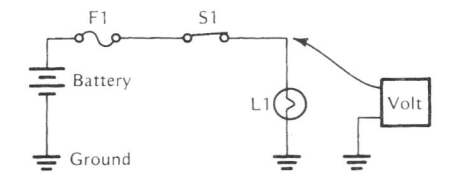

FIGURE 6–11 Voltmeters are always connected into a circuit in parallel

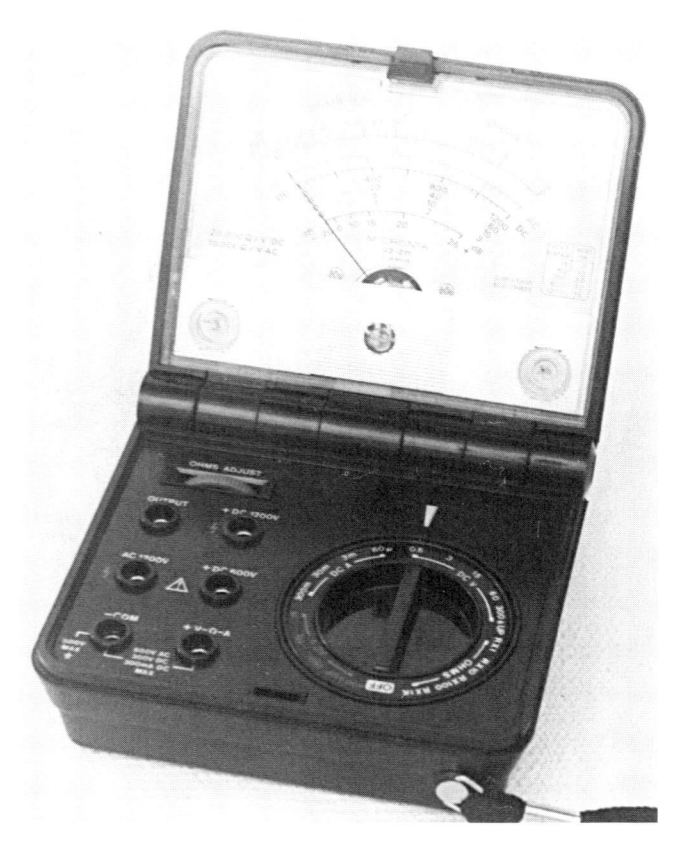

FIGURE 6–10 Analog volt-ohm-milliammeter (Courtesy of BET, Inc.)

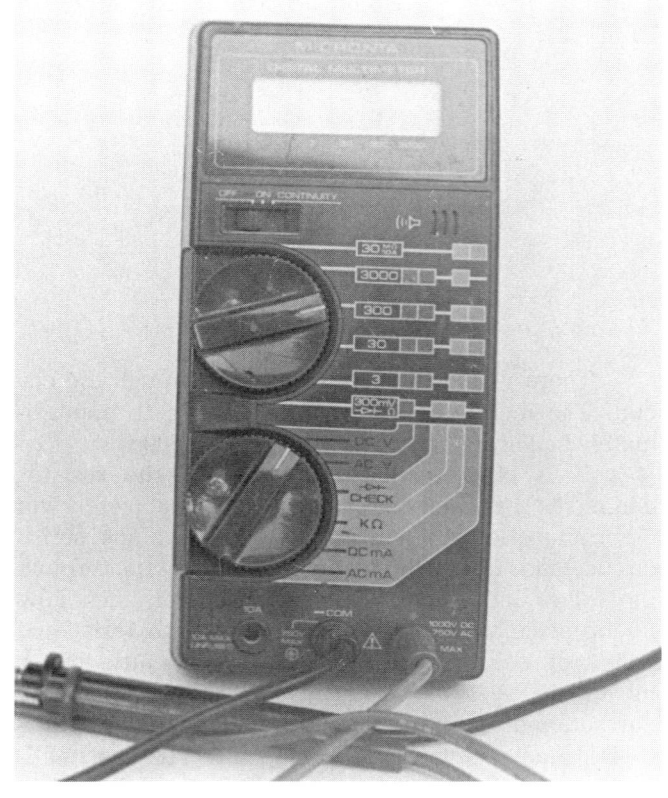

FIGURE 6–12 Typical voltmeter used to check battery load (Courtesy of BET, Inc.)

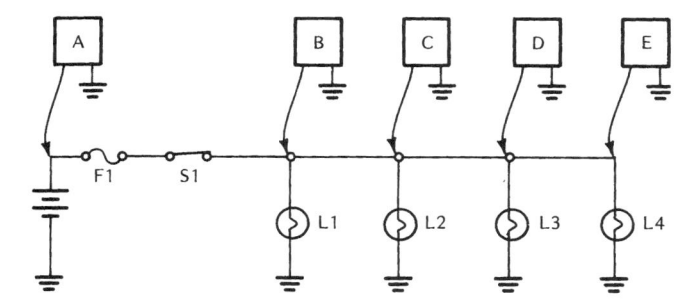

FIGURE 6–13 Checking a circuit using a voltmeter

switch (S1) is closed, there should also be 12 volts available at each lamp (meters B through E). If, for example, meters D and E indicated less than 12 volts, there would be a high resistance between lamps L2 and L3. In this case, lamps L3 and L4 would probably burn, but not at full brilliance.

Assume that the four lamps represent a tail-lamp circuit and that L3 and L4 do not light. If, as previously checked, meters D and E read 0 volts, the wire would probably be broken or disconnected between L2 and L3. If, on the other hand, L3 and L4 did not light but meter readings D and E indicated 12 volts available, there would be three possible causes:

1. Both bulbs (L3 and L4) are burned out.
2. The lamp sockets are defective.
3. The ground connection is defective.

If, after testing the lamps with an ohmmeter or substituting new lamps, they still did not light, proceed as shown in Figure 6–14. If the lamp sockets were defective, there would be no circuit through the lamps, and meters F and G would indicate 0 volts. If, on the other hand, meters F and G read 12 volts, a defective ground would be indicated.

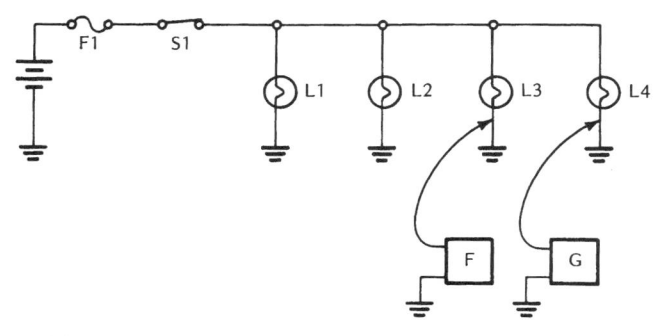

FIGURE 6–14 Using the voltmeter to check for "open" grounds

Ammeter Use

Ammeters for general use should have a high-range rating of 30 to 50 amperes (dc). Special-purpose meters (Figure 6–15), used for checking starter cranking current, should have a range of 300 amperes or more. Ammeters are always connected into a circuit to be tested in series—never in parallel—with the circuit (Figure 6–16). This means that a wire must be disconnected in order to connect the ammeter.

In the tail-lamp circuit previously tested with the voltmeter, all lamps now burn, but the customer's complaint is, "The fuse blows often." For illustration, assume that the circuit normally draws 5 amperes and is protected by a 6-ampere fuse. The customer's complaint indicates that the tail-lamp circuit draws in excess of 6 amperes. Refer to the section on fuses and circuit breakers in Unit 15. A high-resistance short is indicated somewhere in the circuit. Mathematically, each lamp should draw 1.25 amperes.

$$5 \div 4 = 1.25$$

Connect the ammeter as shown in Figure 6–17. Disconnect the circuit at D1, close switch S1, and read the ammeter. With the load (lamps) disconnected, the meter should read 0 amperes. If there is any reading, the wire between the fuse block and disconnect D1 is shorted (high resistance) to ground. If there is no reading, disconnect D2, D3, and D4.

Each lamp reconnected, starting with L1, should read 1.25 amperes. Connect D1; the reading should be 1.25 amperes. Now, connect D2; the reading should be 2.5 amperes.

$$1.25 + 1.25 = 2.5$$

Next, connect D3; the reading should be 3.75 amperes.

$$2.5 + 1.25 = 3.75$$

Finally, connect D4; the reading should be 5 amperes.

$$3.75 + 1.25 = 5$$

If, when making any connection, the reading is higher than expected, the problem is in that part of the lamp circuit. For example, when D3 is connected, the reading is expected to be 3.75 amperes. If it is 4.5

FIGURE 6–15 A starter current meter (Courtesy of BET, Inc.)

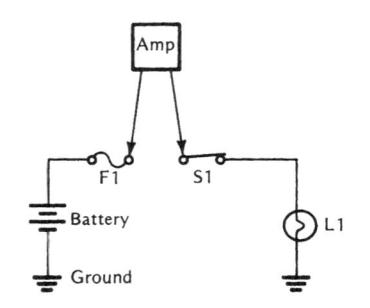

FIGURE 6–16 Ammeters are always connected into a circuit in series

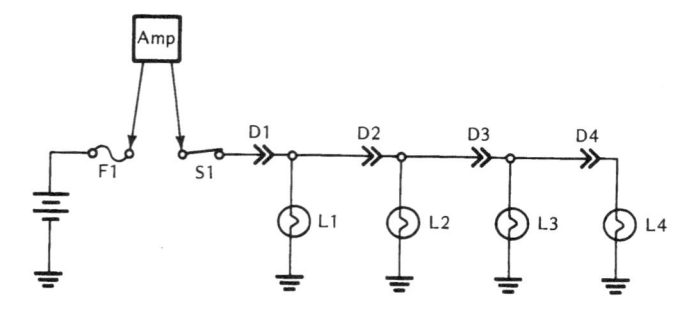

FIGURE 6–17 Checking a circuit using an ammeter

or 5 amperes, the problem is isolated to the circuit of lamp L3.

When testing, always use a fuse. The fuse should be rated at no more than 50 percent higher capacity than specifications. This offers circuit protection and provides enough amperage for testing. The 6-ampere tail-lamp fuse may be replaced with a 9-ampere fuse for testing only. After the problem is found and corrected, be sure to replace the original specification fuse for maximum circuit protection.

Ohmmeter Use

Ohmmeters are used to test circuit continuity and resistance with no power applied. In other words, the circuit or component to be tested must first be electrically disconnected. Connecting an ohmmeter into a "live" circuit will usually result in a damaged test instrument.

Ohmmeters, shown in Figure 6–18, usually have a multirange scale for reading from zero to several thousand ohms (often several hundred thousand ohms). The range-select switch is usually identified as R × 1, R × 10, R × 100, R × 1000, and so on. This identification simply means "scale reading times multiplier." For example, when the meter needle points to 5 on the scale, the reading is 5 Ω only if the range-select switch is at R × 1. When the select switch is at R × 100, the reading is 500 Ω.

ELECTRICAL TESTING 65

FIGURE 6–18 A typical ohmmeter (Courtesy of BET, Inc.)

$$5 \times 100 = 500$$

Ohmmeters are battery operated and have zero-adjust provisions to compensate for battery voltage drop caused by age and use. To adjust the zero setting, set the range-select switch to the scale to be used and short the two test probes together. Then turn the adjustment knob until the needle indicates zero.

Infinity is a term used with ohmmeters, often designated as ∞. *Infinity*, by definition, is "a hypothetical amount which is larger than any assignable amount." For example, assume that an ohmmeter has a high range of 100,000 Ω, and that a circuit being tested with the meter has a resistance of 500,000 Ω. The meter needle would "peg" at 100,000 Ω and would be noted as infinite (∞) ohms.

Ohmmeters are also used to trace and check wires or cables. Assume that one wire of a four-wire cable (Figure 6–19) is to be found. Connect one probe of the ohmmeter to the known wire at one end of the cable and touch the other probe to each wire at the other end of the cable. A meter needle deflection of 0 Ω indicates the correct wire. Using this same method, you can check a suspected defective wire by connecting the ohmmeter as shown in Figure 6–20. If the meter needle deflects, the wire is sound. If it does not, the wire is defective (open). If the wire is sound, continue checking by connecting probe P2 to leads 2, 3, and 4. Any deflection of the meter needle indicates that wire 1 is shorted to one of the other wires and that the harness is defective.

Other Test Instruments

Other test instruments used for testing specific automotive components include the variable resistor (pile) for battery testing (Figure 6–21) and the alternator output tester (Figure 6–22). Two popular, and often homemade, testing tools are the jumper wire and the test lamp. The following section describes the construction and use of these handy tools.

Jumper Wire. Jumper wires (Figure 6–23) are used to temporarily bypass circuits or circuit components for electrical testing. The use of jumper wires will be discussed throughout this text when applicable. A *jumper* consists of two alligator clips and a piece of wire; it may be constructed in a few minutes. Simply strip the insulation from each end of a predetermined length of wire and attach the alligator clips.

Heavy-gauge jumper wire sets (Figure 6–24), used for starting cars with discharged batteries, are usually purchased preassembled. The homemade jumper can be used to determine whether a component is defective. Consider the circuit in Figure 6–25. With switch S1 closed, the lamp should light. If it does not, the problem could be the fuse, the switch, a wire, the lamp, or the ground. To solve this problem, connect one end of a jumper to point A and the other end to points B through F in turn. If the lamp L1 lighted when the jumper was attached at point B, the wire (A to B) would be defective. If at point C the fuse were defective, at point D the wire (C to D) would be defective. If the lamp failed to light until the jumper was connected to point E, the switch would be defective; at point F the wire (E to F) would be defective. If, when the jumper was connected to point F, the lamp still did not light, the bulb could be burned out or the ground connection could be defective. To check for the defective ground, connect the jumper as shown in Figure 6–26.

Jumper Safety. When using jumper wires, do not deliberately short-circuit wire or cable. To do so may cause serious damage to the electrical wiring system. If in doubt, do not use a jumper for electrical testing.

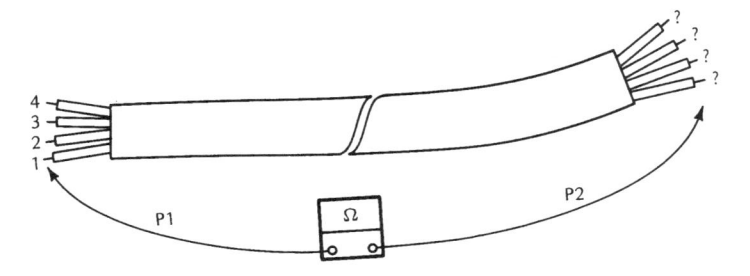

FIGURE 6–19 Tracing a circuit using an ohmmeter

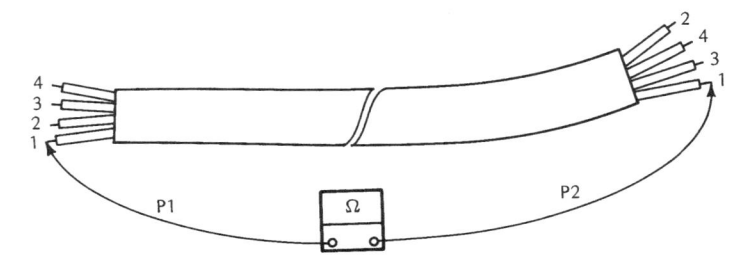

FIGURE 6–20 Using an ohmmeter to check for opens and shorts

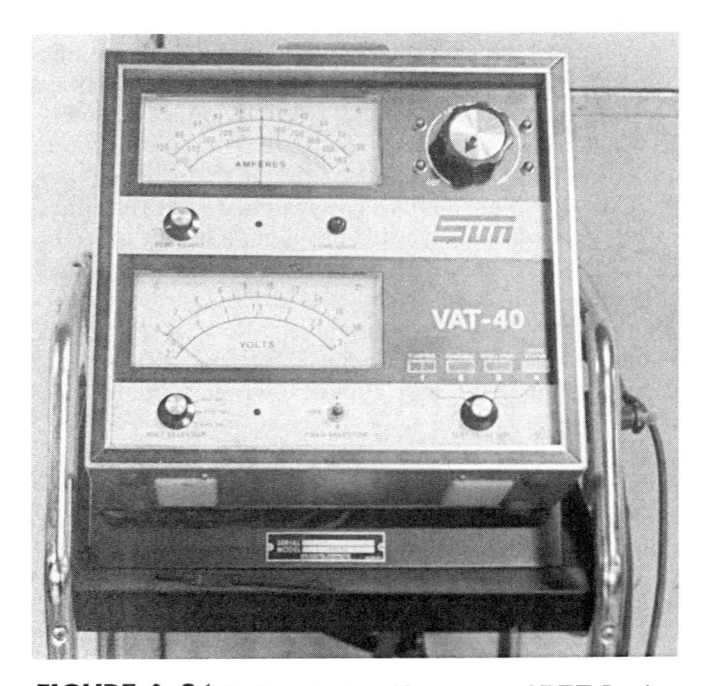

FIGURE 6–21 Battery tester (Courtesy of BET, Inc.)

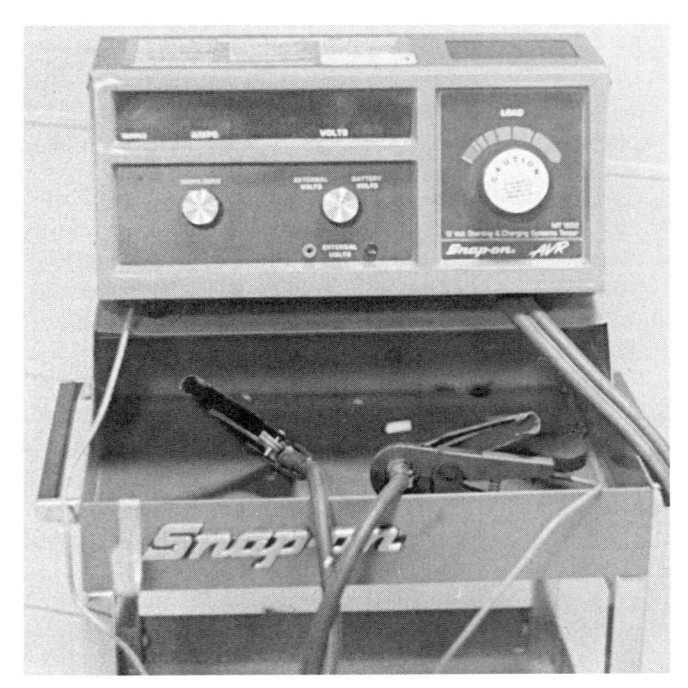

FIGURE 6–22 Alternator tester (Courtesy of BET, Inc.)

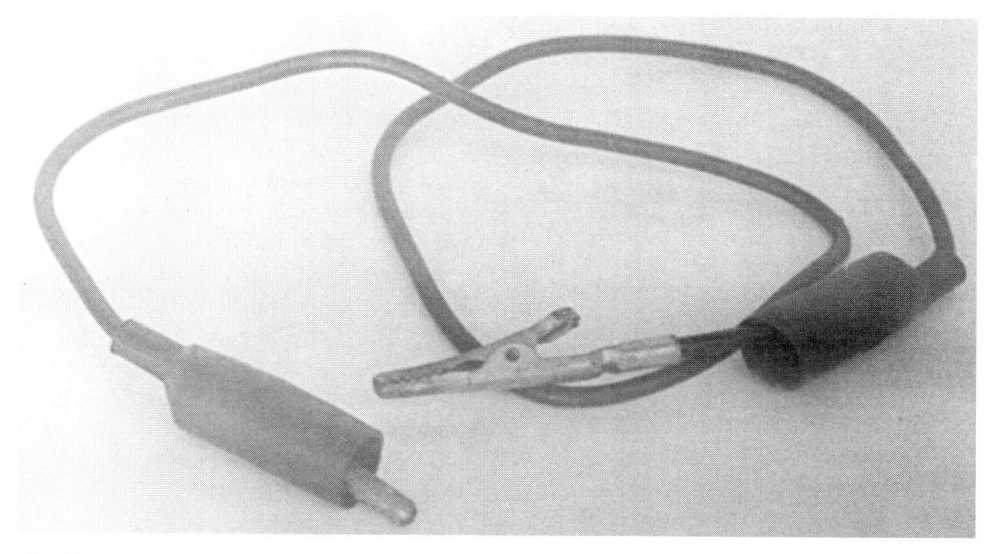

FIGURE 6–23 Typical jumper wire (Courtesy of BET, Inc.)

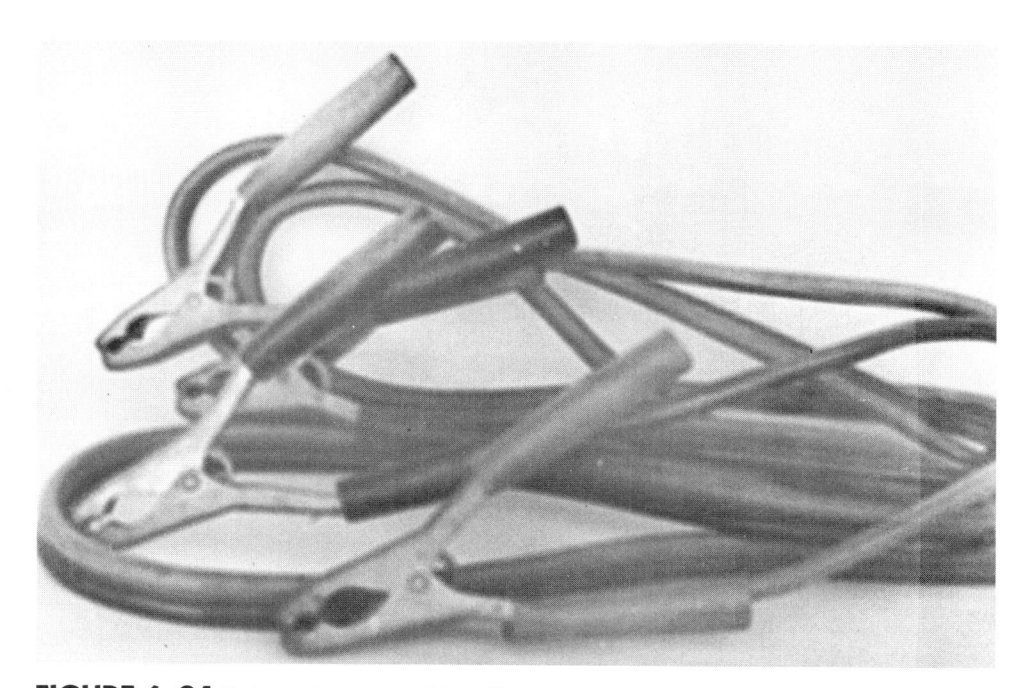

FIGURE 6–24 Battery jumper cables (Courtesy of BET, Inc.)

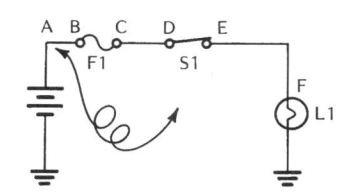

FIGURE 6–25 Testing circuit using a jumper wire

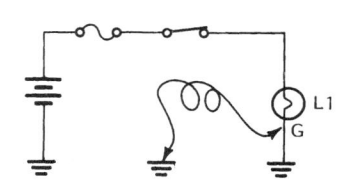

FIGURE 6–26 Using a jumper wire to test for an open ground

67

68 Unit 6 ELECTRICAL TESTING

When using battery jumper cables, it is important to use the following procedure to avoid damage or personal injury.

1. Connect one jumper cable from the charged battery positive (+) terminal to the discharged battery positive (+) terminal.
2. Connect the other jumper cable from the charged battery negative (−) terminal to a suitable engine ground point of the car with the discharged battery. Do not connect this cable to the discharged battery negative (−) terminal.
3. Do not connect positive (+) to negative (−) or negative (−) to positive (+). To do so may cause serious damage to the battery or the charging-circuit components.

Be certain to follow these instructions and those given in car owner's manuals. An improperly connected battery can explode, spewing sulfuric acid.

Test Lamps. There are two types of test lamps: external powered and internal powered. Externally powered test lamps are 12 volts to match the automotive battery voltage. Internally powered test lamps are usually 1.5 or 3 volts (one or two cells). An externally powered test lamp (Figure 6–27) can be constructed inexpensively with a lamp socket, a 12-volt lamp, two pieces of wire, and two alligator clips.

WARNING: Do not attempt to use a test lamp in anything other than power circuits. *Never* use a test lamp in a computer circuit.

The circuit in Figure 6–25 could be tested using a test lamp, as shown in Figure 6–28. One side of the test lamp would be connected to a good ground. The other side would be connected to points B through G. If the test lamp did not light at point B, the wire (A to B) would be defective; at point C the fuse would be defective; at point D the wire (C to D) would be defective; at point E the switch would be defective. If the test lamp did not light at point F, the wire (E to F) would be defective. Finally, the lamp L_1 and lamp socket should be checked.

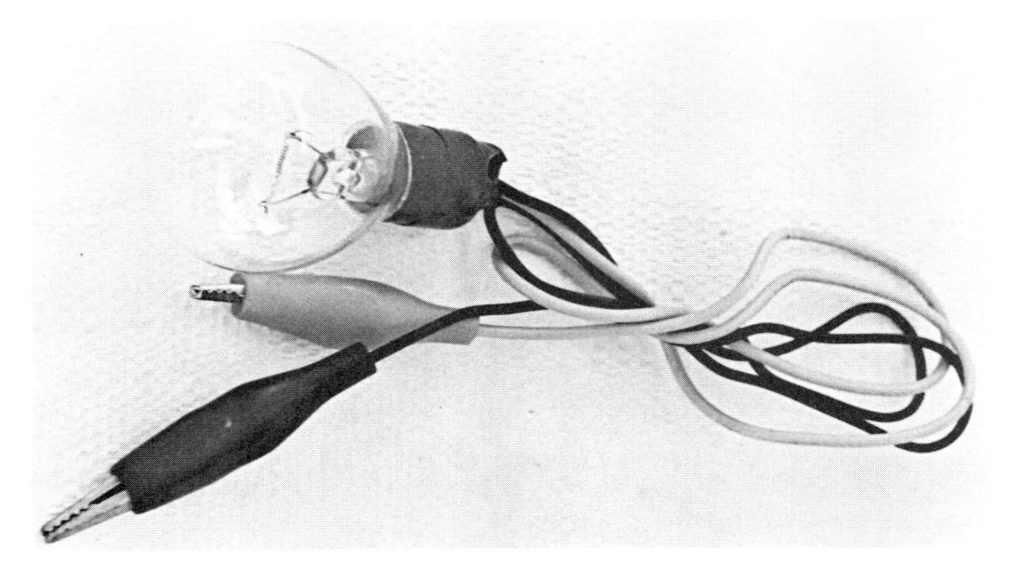

FIGURE 6–27 A test lamp made from a panel lamp assembly (Courtesy of BET, Inc.)

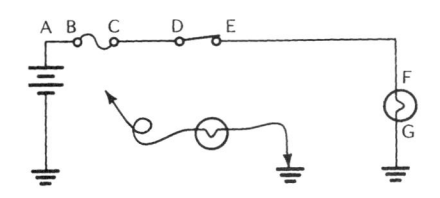

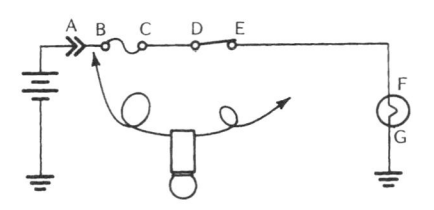

FIGURE 6–28 Testing a circuit with a test lamp

	First Band Color	Second Band Color	Third Band Color	Fourth Band (if present)	Measured Value
1					
2					
3					
4					
5					
6					
7					
8					
9					
10					

FIGURE 6-29 Resistance values

PRACTICAL EXERCISE

You will need a late-model vehicle, an ohmmeter, and assorted resistors, as used in the Unit 3 exercise, for this exercise.

EXERCISE 1

1. Using an ohmmeter, measure the resistance of 10 color-coded resistors.
2. Record your measurements on the chart in Figure 6–29.
3. Compare the meter readings with your findings in the exercise in Unit 3 (page 32).
4. Are the meter readings for Unit 6 close to the values determined by the color codes in Unit 3? _____

EXERCISE 2

1. Open the hood of the vehicle.
2. Select "Volts" on the meter and a value of 15 V or above.
3. Record the measurement of Steps 5 through 7, 9, and 10 in the chart (Figure 6–30).
4. Measure the voltage across the battery, positive (+) to negative (−).
5. Measure the voltage from the battery positive (+) terminal to the vehicle frame.
6. Measure the voltage from the battery negative (−) terminal to the vehicle frame.
7. Turn the headlamps ON.
8. Repeat the measurement of Step 5.
9. Measure the voltage from the lamp (+) socket/terminal to the vehicle frame.
10. Turn the headlamps OFF.

QUIZ

1. Was the voltage measurement in Step 8 the same as that measured in Step 5? _____ Why? _____
2. Was the voltage measurement of Step 9 the same as that measured in Step 5? _____ Why? _____
3. Was the voltage measurement of Step 7 the same as the measurement of Step 6? _____ Why? _____

MEASURED VOLTAGE

Step 5	Batt Voltage + to –	—— VOLTS
Step 6	+ Term to Frame	—— VOLTS
Step 7	– Term to Frame	—— VOLTS
Step 9	Batt Voltage + to –	—— VOLTS
Step 10	Lamp Socket + to Frame	—— VOLTS

FIGURE 6-30 Voltage measurements

SUMMARY

- Analog meters contain a meter movement (restrained electric motor).
- The same analog meter movement is often used to indicate current, voltage, and resistance.
- The connection of the meter movement to other electrical components, and the method of connection, determine whether current, voltage, or resistance will be measured.
- It requires experience to properly read an analog meter.
- A digital meter is not necessarily more accurate than an analog meter, although it is easier to read.
- A test lamp is often effective in checking for a complete circuit.
- Never use a test lamp to check digital circuits.

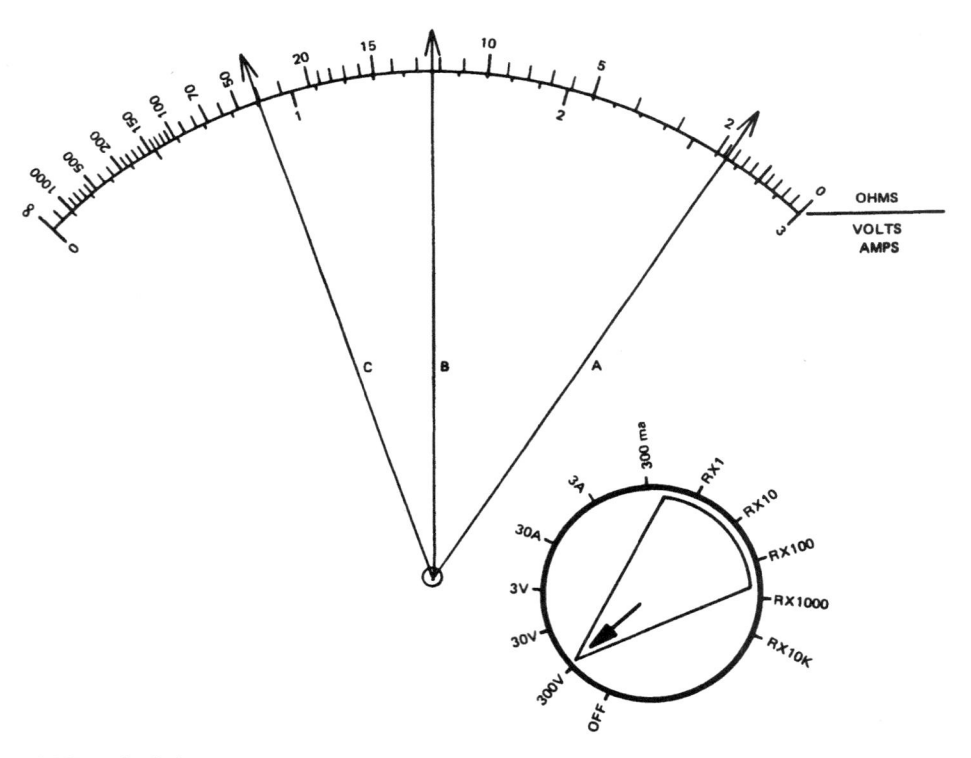

FIGURE 6-31 Test meter

REVIEW

The meter shown in Figure 6–31 is a volt-ohm-ammeter. The switch in the meter provides for the selection of volts, current, or resistance. The position of the switch selects the range of the meter. The meter is shown with a pointer placed at different positions. Read the statement of the question and indicate what value the meter is indicating.

1. The selector switch is in the 30 V position. Pointer B position indicates _____ volts.
2. The selector switch is in the 3 A position. Pointer A indicates _____ amperes.
3. The selector switch is in the 3 V position. Pointer C indicates _____ volts.
4. The selector switch is in the 300 V position. Pointer A indicates _____ volts.
5. The selector switch is in the R × 10 position. Pointer A indicates _____ ohms.
6. The selector switch is in the R × 1000 position. Pointer B indicates _____ ohms.
7. The selector switch is in the 30 A position. Pointer B indicates _____ amperes.
8. The selector switch is in the R × 100 position. Pointer C indicates _____ ohms.
9. The selector switch is in the R × I position. Pointer B indicates _____ ohms.
10. The selector switch is in the 300 mA position. Pointer A indicates _____ ma.
11. An ohmmeter is used to check circuit
    a. resistance   c. current
    b. voltage   d. wattage
12. When an ohmmeter range of R × 1000 is selected, what is the actual resistance if the meter needle points to 3.5?
    a. 3500 Ω   c. 35 Ω
    b. 350 Ω   d. 3.5 Ω
13. What is the symbol for infinity?
    a. Ω   c. π
    b. Δ   d. ∞
14. What is the desirable range of a voltmeter used for general testing in automotive circuits?
    a. 0–1.5 volts   c. 0–150 volts
    b. 0–15 volts   d. none of the above
15. Refer to Figure 6–4 and assume that lamp L2 is burned out. What voltage should be indicated on meter C?
16. When using battery jumper cables, the negative (−) cable from the charged battery is connected to
    a. the negative (−) terminal of the discharged battery.
    b. an engine ground of the car with the discharged battery.
    c. the positive (+) terminal of the discharged battery.
    d. the negative (−) cable is not connected to anything.
17. When battery jumper cables are properly connected, the two batteries are in (series) (parallel).
18. A battery is dangerous because it contains _____ acid.
    a. citric   c. sulfuric
    b. hydrochloric   d. hydrofluoric
19. Ammeters are never connected into a circuit in (series) (parallel).
20. Ammeters are used to check
    a. current   c. voltage
    b. resistance   d. temperature

◆ UNIT 7 ◆

BATTERIES

OBJECTIVES

On completion of this unit you will be able to:

- Describe what constitutes an electric cell.
- Understand the difference between a battery and a cell.
- Discuss the connection of cells for voltage and current.
- Define the term *internal resistance*.
- Discuss how batteries are rated.
- Test batteries.

The purpose of this unit is not to explain how chemical energy is stored, converted, and then released as electrical energy when needed, but rather to cover the connection of cells, internal resistance, and the terminal voltage of batteries.

As you will recall from Unit 1, Count Alessandro Volta (1745–1827) invented the battery in 1780. His battery, the voltaic pile, was the first source of "constant current electricity." For his discovery, the volt, a unit of electromotive force, was named in his honor in 1881.

Whenever two different metals are placed in an acid or salts solution in an insulated case, a voltage is produced between the metals by chemical action. This combination, the two metals and the solution, is known as a *cell* (Figure 7–1). One of the metal plates is positive (+), whereas the other is negative (−). All cells and batteries provide a source of direct current (dc).

The term *cell* refers to a single unit, the two metals in a solution. The term *battery* refers to cell combinations in series or in parallel that provide for higher current or voltage capabilities. A battery is used for storing and converting chemical energy into electrical energy. The term *battery* is now commonly used instead of cell; for example, a flashlight battery is actually a flashlight cell.

There are two main types of cells: primary and secondary. *Primary cells* are temporary (nonrechargeable) and use up the materials of which they are composed while providing electrical energy (Figure 7–2). *Secondary cells* change in chemical composition while providing electrical energy but may be recharged. Recharging converts the cell back to its original condition by passing current through it in the opposite direction. A flashlight battery is a good example of a primary cell, and a car battery is a good example of a secondary cell.

Chargers have become available that recharge primary cells, including some flashlight batteries. Most of the batteries are not returned to "original" condition, however. By design, some batteries are intended to be recharged, whereas others are intended to be discarded when discharged.

The symbol used for a cell is shown in Figure 7–3. The longer line is the positive (+) terminal of the cell; the shorter line is the negative (−) terminal.

The voltage output of a cell is determined by the materials used to make up the cell. The standard carbon-zinc (C-Zn) flashlight cell comes in many sizes. The small penlight cell has an output of 1.5 volts. The common D size flashlight cell also has an output of 1.5 volts; the D cell can, however, deliver higher current for a longer period of time than can the penlight cell.

74 Unit 7 BATTERIES

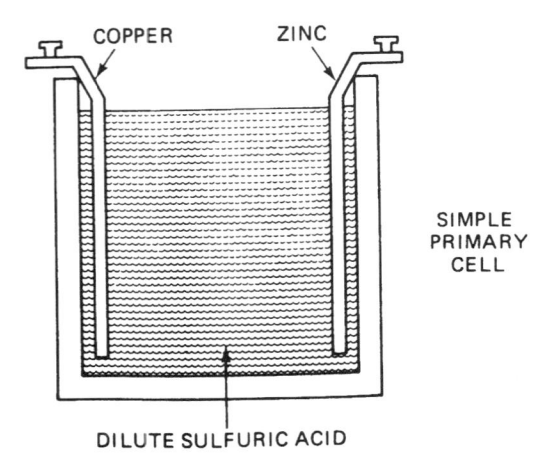

FIGURE 7–1 Simple primary cell

FIGURE 7–2 Primary cells (Courtesy of BET, Inc.)

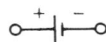

FIGURE 7–3 Symbol for cell/battery

INTERNAL RESISTANCE

Everything through which current flows has resistance. Batteries have an internal resistance. If the open circuit voltage of a cell is measured and current is then supplied (by that cell) to an external circuit, its voltage will decrease. This decrease in voltage is equal to the voltage developed across the internal resistance of the cell. Figure 7–4 illustrates a battery with the internal resistance shown before the output terminals.

A cell with a voltmeter in parallel, an ammeter in series, and a load resistor of 10 Ω in series controlled by a switch is shown in Figure 7–5. When the switch is open, the cell voltage is 1.5 volts. When the switch is closed, the ammeter indicates 0.14 ampere of current. The voltmeter indicates 1.4 volts. This means that 0.1 volt is being developed across the internal resistance of the battery.

$$1.5 - 1.4 = 0.1$$

The internal resistance of the cell may be determined by using Ohm's law:

$$R = E/I$$
$$R = 0.1/0.14$$
$$R = 0.714 \, \Omega$$

All batteries have different internal resistances. As a cell becomes discharged, its internal resistance increases. As the internal resistance of the cell increases, its ability to supply current decreases.

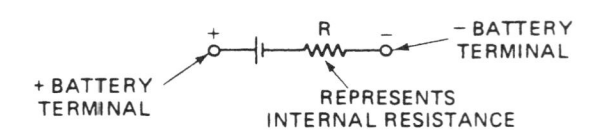

FIGURE 7–4 Battery or cell with internal resistance

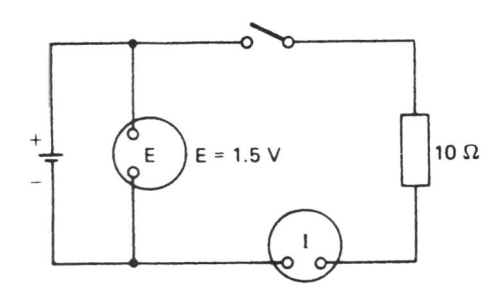

FIGURE 7–5 Determining internal resistance

BATTERIES/CELLS IN SERIES

Cells are connected in series whenever a voltage higher than that which can be supplied by a single cell is needed. Two flashlight cells are shown in series in Figure 7–6. The total voltage available is 1.5 plus 1.5, or 3 volts. This is standard for a two-cell flashlight. If three cells were connected in series, as in Figure 7–7, the available voltage would be 4.5 volts.

$$1.5 + 1.5 + 1.5 = 4.5$$

BATTERIES/CELLS IN PARALLEL

Whenever higher current capability is needed, cells are connected in parallel. Certain precautions must be taken when connecting cells in parallel. Most important, the cells must have the same terminal voltage. For example, a 1.2-volt cell should not be connected in parallel with a 1.5-volt cell.

A load device drawing a constant current of 3 amperes requires a voltage of at least 1.3 volts (Figure 7–8). Cells of 1.5 volts are available. Each cell has an internal resistance of 0.1 Ω. With one cell connected to the load device, there is 0.3 volt developed across the internal resistance of the cell, leaving 1.2 volts to appear across the load. If two cells are connected in parallel, the situation is improved.

In Figure 7–9, the load device is still drawing 3 amperes. Each cell, however, is supplying only 1.5

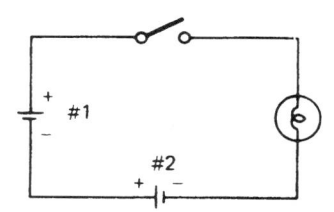

FIGURE 7–6 Two cells in series

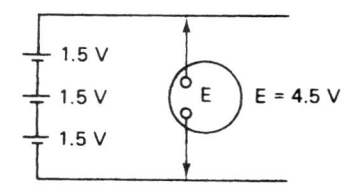

FIGURE 7–7 Three cells in series

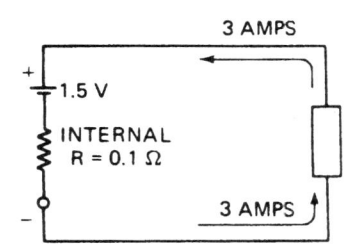

FIGURE 7–8 Internal voltage drop—single cell

amperes. The voltage developed across the internal resistance is:

$$E = I \times R$$
$$E = 1.5 \times 0.1$$
$$E = 0.15 \text{ volt}$$

The voltage at the terminals of the cells and across the load is:

$$E_{\text{out}} = 1.5 - 0.15$$
$$E_{\text{out}} = 1.35 \text{ volts}$$

This voltage meets the original requirement of at least 1.3 volts across the load.

Whether or not more cells should be connected in parallel depends on how long current is to be drawn. If three cells are connected in parallel, each has to supply only one ampere. The voltage developed across the internal resistance is only 0.1 volt. The terminal voltage of each cell and the voltage to the load is 1.4 volts (Figure 7–10). The three-cell combination will last longer than the two-cell combination.

Internal resistance is present in every device that produces electricity, whether it is a battery, generator, bimetal strip, or solar cell. The effect of internal resistance is the same on these devices as on batteries.

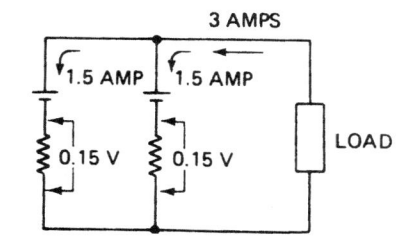

FIGURE 7–9 Lower internal voltage drop with parallel cells

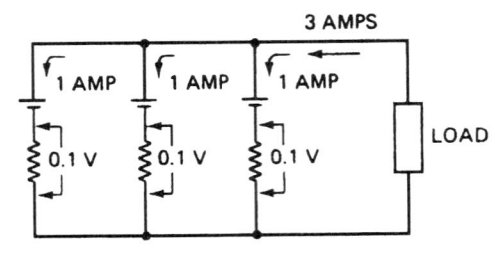

FIGURE 7-10 Further decrease in voltage drop with three-cell supply

MAXIMUM POWER TRANSFER

The importance of matching source resistance to load resistance will be emphasized. The facts regarding the matching requirement can be shown using a simple battery with its internal resistance and a number of load resistors.

In the example in Figure 7-11, a 6-volt battery will be considered. For the sake of simplicity, the battery will be considered to have an internal resistance of 3 Ω. Resistors ranging from 20 Ω to 1 Ω will be considered as the load on the battery. The chart in Figure 7-12 shows the relation of resistor value, current, load voltage, and power in the load resistor.

When the 20-Ω resistor is used as a load, the power in the load is 1.36 watts. The power in the load increases as the load resistor decreases in value. The maximum is 3 watts (W) with the 3-Ω load. The power in the load immediately starts to decrease as the load resistor is reduced in value below 3 Ω.

Maximum power is developed in the load when the resistance of the load matches the resistance of the source; in this case, it is the 3-Ω internal battery resistance.

Ohms RL	Ohms RT	Amps 6 V RT	Volts R Int.	Volts RL I × RL	Power Load I × ERL
20	23	0.26	0.78	5.22	1.36
15	18	0.33	1.0	5.0	1.66
10	13	0.46	1.4	4.6	2.12
8	11	0.55	1.6	4.4	2.4
6	9	0.66	2.0	4.0	2.64
5	8	0.75	2.25	3.75	2.81
4	7	0.86	2.58	3.42	2.95
3	6	1.0	3.0	3.0	3.0***
2.99	5.99	1.0+	3.01	2.99	2.99
2	5	1.2	3.6	2.4	2.88
1	4	1.5	4.5	1.5	2.25

FIGURE 7-12 Maximum power transfer chart

THE AUTOMOTIVE BATTERY

All the electrical energy required to start the engine of a car is found, in chemical form, in a "black box" under the hood. This box, the battery, is one of the most neglected and least understood components of the automotive electrical system. If a technician were to check 100 batteries at random, 5 percent would be defective and should be replaced; 25 percent would be undercharged; 25 percent would have defective battery cables that should be replaced. Only 45 percent would be good—but 10 percent of these would need to be replaced or serviced within the next 12 months.

Although there are many types of storage batteries, the lead-acid battery is most commonly used for automotive service (Figure 7-13). This battery can be charged and discharged continuously throughout its life.

BATTERY SAFETY

Extreme care must be taken when handling an automotive battery. Its inert solution, called the *electrolyte*, is a strong solution of sulfuric (or sulphuric) acid (H_2SO_4) and water (H_2O). Sulfuric acid is a highly corrosive, oily liquid that is colorless in its pure state. When mixed with water it has a yellow-tan color. A sulfuric acid (H_2SO_4) container must be clearly identified to avoid misuse.

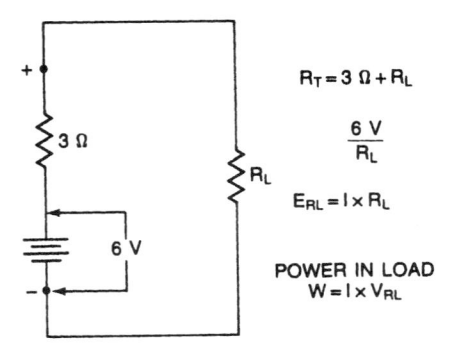

FIGURE 7-11 Checking maximum power transfer

FIGURE 7–13 (a) A typical 12-V top-post connected battery; (b) A typical 12-V side- (screw) connected battery (Courtesy of BET, Inc.)

When mixing sulfuric acid (H_2SO_4) and water (H_2O), never add the water to the acid because even a small amount of water will boil violently when added to the acid. Always add the acid to the water—very slowly—because this process also generates heat. Pure sulfuric acid (H_2SO_4), with a specific gravity of 1.835, should not be diluted by an inexperienced person because of the dangers involved. The correct electrolyte mix is available wherever automotive batteries are sold.

A battery that is being charged produces hydrogen (H) and oxygen (O) vapors that are potentially explosive. This vapor escapes from the battery through the metered vent caps and can ignite very easily. If a flame enters the battery it will explode, which could result in serious personal injury as well as damage to the vehicle. To avoid mishaps when handling batteries, the following guidelines are recommended.

1. Wear eye and body protection (safety glasses and garments).
2. Do not smoke. Do not work near open flames or sparks created by drill motors, welding equipment, and the like.
3. Do not connect or disconnect a battery charger cable while the charger is turned on.
4. Only work in well-ventilated areas. Do not inhale fumes.
5. When connecting jumper cables, always connect the positive (+) cable first.
6. Do not connect the negative (−) cable to a disabled battery—connect it to an engine ground only.
7. Keep an acid neutralizer handy and ready for use.

Neutralizing Sulfuric Acid

Sulfuric acid can be neutralized with household baking soda by mixing one pound (0.45 kg) of soda to one gallon (3.78 l) of water (Figure 7–14). If sulfuric acid gets on clothing, it should be washed immediately to avoid damage to the fabric. Remove the garment to avoid getting acid on the skin and wash it with clear running tap water. If acid does get on the skin, flush immediately with running water. Remove clothing, if it is also affected. Summon immediate medical attention if exposure is severe.

Sulfuric acid (H_2SO_4) in the eyes may result in blindness. Do not rub the eye. Force the eye open and flush it with running water at room temperature. Summon medical attention immediately, regardless of severity.

BATTERY COMPONENTS

The active materials of a battery react chemically to produce a flow of current whenever a demand is made.

FIGURE 7–14 A battery acid neutralizer can be made by mixing one pound (0.45 kg) of baking soda with one gallon (3.78 l) of water (Courtesy of BET, Inc.)

This assumes that the battery is in a state of "charge." A battery consists of cells containing both positive (+) and negative (−) plates. The plates consist of grids of lead-antimony (Pb-Sb) alloy and contain the "active" materials. These plates, shown in Figure 7–15, are flat rectangular castings with solid borders and grids of horizontal and vertical wires. The charged negative plates contain sponge lead (Pb), which is gray in color.

The charged positive (+) plates contain lead peroxide (PbO_2), which is chocolate brown in color. The electrolyte consists of a solution of sulfuric acid (H_2SO_4), which is essential for an interaction between the positive (+) and negative (−) plates.

BATTERY CHEMICAL ACTION

When a fully charged cell (Figure 7–16) is discharged (Figure 7–17), the following chemical action takes place. Oxygen (O) from the positive (+) plates combines with the hydrogen (H) in the electrolyte to form water (H_2O). Also, sulfate (SO) from the electrolyte combines with the sponge lead (Pb) of the positive (+) plates to form lead sulfate ($PbSO_4$). When a fully discharged cell (Figure 7–18) is charged (Figure 7–19), the chemical action is in reverse. The chemical changes occur, to some degree, at all times while the car is in use. For example, the battery is discharged slightly when the car is started. After the engine is running, the battery is returned to full charge by the alternator.

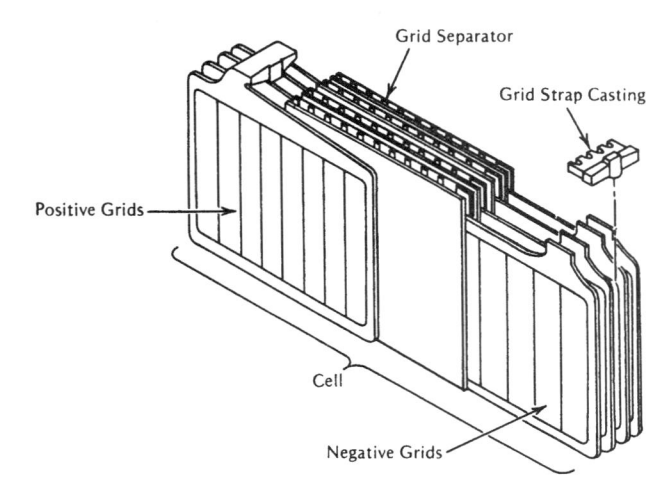

FIGURE 7–15 A cell consists of positive (+) and negative (−) grid plates

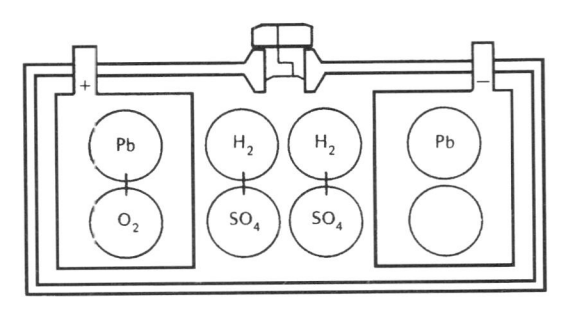

FIGURE 7–16 The chemical condition of a fully charged battery

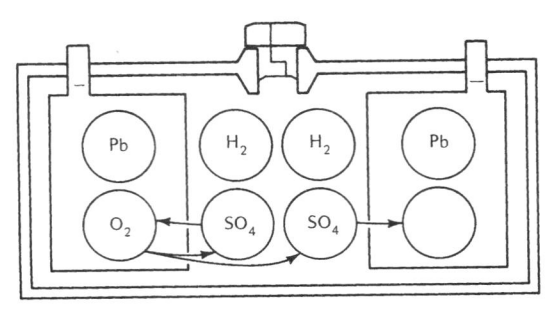

FIGURE 7–17 Chemical action of a battery being discharged

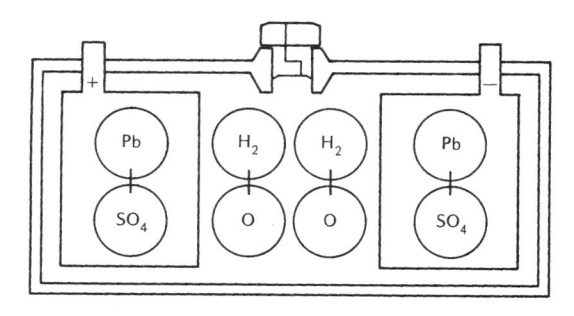

FIGURE 7–18 Chemical condition of a fully discharged battery

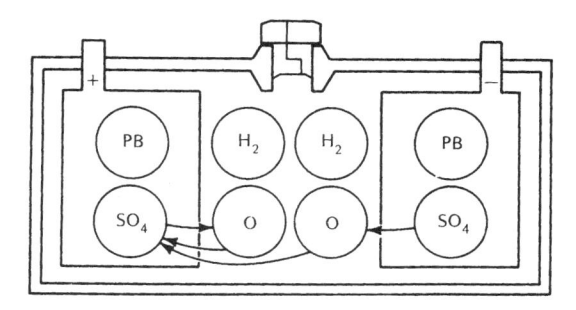

FIGURE 7–19 Chemical action of a battery being charged

THE ELECTROLYTE

The specific gravity of the sulfuric acid (H_2SO_4) and water (H_2O) solution—the electrolyte—changes as battery conditions change.

For example, the specific gravity of the electrolyte in a fully charged battery is 1.26. In a fully discharged battery it is 1.11. As a standard, these readings are valid only at a temperature of 80°F (26.6°C).

A calibrated glass float within a glass tube, known as a *hydrometer* (Figure 7–20), is used to take samples of the electrolyte in a serviceable battery to determine its specific gravity. As the sample is pulled into the tube, the float seeks its equilibrium level, and the specific gravity reading is taken from the scale at the fluid surface level. This test is not possible for a maintenance-free battery.

For accurate readings at temperatures above or below 80°F (26.6°C), a temperature compensation chart (Figure 7–21) is used. For every 10°F (5.6°C) above 80°F, four points (0.004) must be added. For every 10°F below 80°F, four points (0.004) must be subtracted from the reading. This compensation is necessary because of expansion and contraction of the electrolyte at various temperatures.

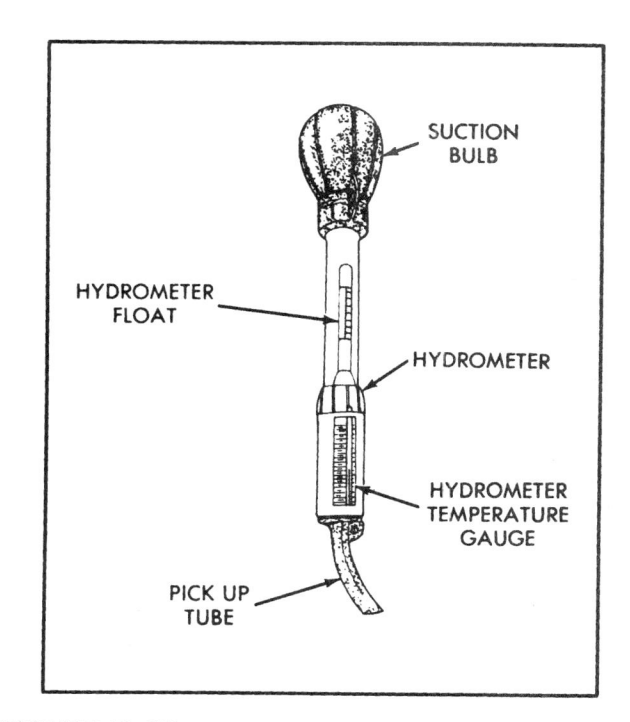

FIGURE 7–20 A typical glass float-type hydrometer

An all-glass thermometer may also be used to determine the electrolyte temperature of serviceable batteries. The thermometer (Figure 7–22) should have a scale that reads to at least 150°F (65.5°C) and should be

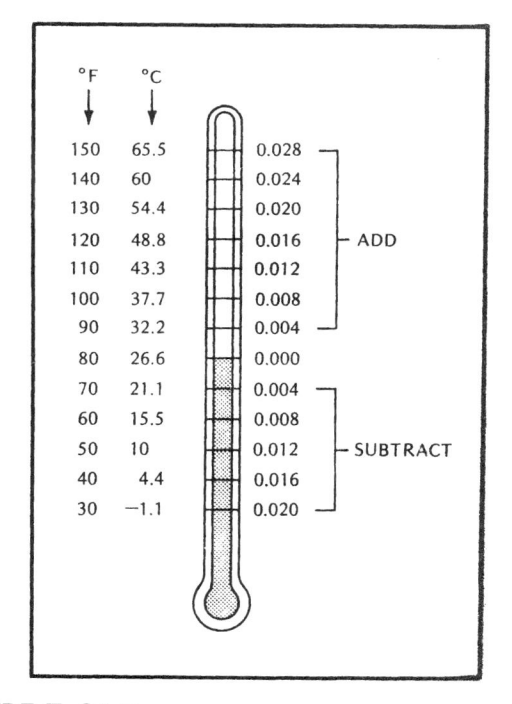

FIGURE 7–21 Temperature compensation chart

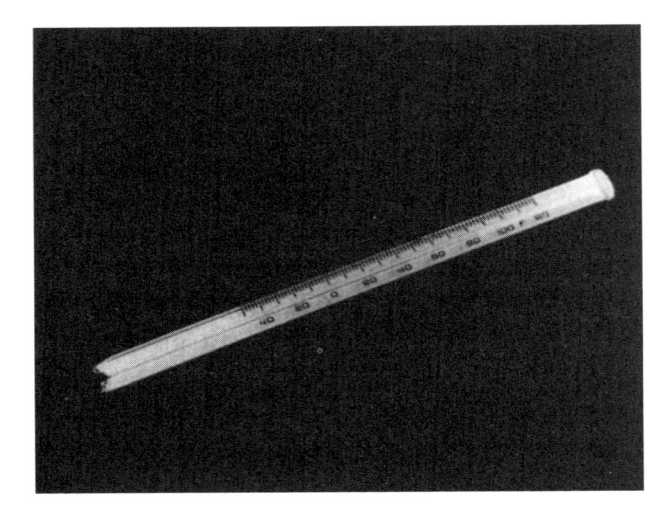

FIGURE 7–22 A typical "spirit" glass thermometer in °F scale. Thermometers are also available in °C scales.

designed for about 1-in. (25.4-mm) immersion. For obvious reasons, such as corrosion and electrical conductivity, metal or metal and glass thermometers must not be used. Also, this test is not possible for maintenance-free batteries.

BATTERY POWER RATINGS

The most popular battery power rating standard is the Association of Automobile Battery Manufacturers (AABM) twenty-hour rating, which is defined as the discharge rate (in amperes) multiplied by the time (in hours) required to reduce a fully charged 12-volt (actually 12.6-volt) battery to a discharged condition of 10.5 volts. This rating is determined by laboratory testing under controlled conditions. For example, if a battery delivered 4 amperes for 20 hours, its rating would be 80 ampere-hours.

$$4 \times 20 = 80$$

This also means, in effect, that the battery would deliver 80 amperes for one hour, or one ampere for 80 hours, and so on. Some battery manufacturers give their power ratings in watts, which is an extension of the ampere-hour rating. Watts are determined by multiplying the volts times the amperes times the time (hours). Using the preceding example, if a 12-volt battery delivered 4 amperes for 20 hours, what would the power rating be in watt-hours?

The solution is as follows:

$$12 \text{ (volts)} \times 4 \text{ (amperes)} \times 20 \text{ (hours)}$$
$$= 960 \text{ (watt-hours)}$$

Actually, the formula is somewhat more complicated because the voltage would constantly decrease over the 20-hour period from 12.6 to 10.5 volts. Mathematically, the decrease would be about 0.105 volts per hour. The average voltage would be about 11.55 volts; therefore, the watt-hour rating would be closer to 924, as follows:

$$11.55 \times 4 \times 20 = 924$$

At 10.5 volts the battery would be about 83 percent charged. If the electrical and ignition system were in good repair, and the engine were sound and in reasonably good "tune," it should start and run at this reduced voltage.

BATTERY CARE

The average automotive battery life is only a few years. But some batteries have been known to provide many years of service. Taking good care of a battery will greatly increase its life. The following points will help to ensure long battery service.

1. Be sure that a battery is of sufficient power rating for the application.
2. Keep the outside case (top and sides) clean.
3. Keep terminals and connections clean and tight.
4. Keep the electrolyte at full level by adding distilled water (H_2O). (Some maintenance-free batteries are sealed, so this attention is not possible).
5. Keep the battery fully charged.
6. Keep the charging system properly adjusted to prevent overcharging.

BATTERY TESTING

Specific instructions for testing are included with the various battery testers. However, the following tests can be used as a guide for quick checks of battery condition. Note, however, that the specific gravity test cannot be performed on maintenance-free batteries.

Specific Gravity Test

The specific gravity test uses a hydrometer. It indicates the state of charge of a battery based on the specific gravity of the electrolytic solution. The specific gravity of water (H_2O) is 1.000, and the specific gravity of sulfuric acid (H_2SO_4) is 1.835. When properly mixed, the water to acid ratio should be between 3.1:1 and 2.8:1. The specific gravity of the solution in a fully charged battery will be 1.260 to 1.280 at 80°F (26.6°C) (Figure 7–23). To check the specific gravity of a battery, pull a sample of fluid from a cell into the hydrometer until the inner float rises. Read the specific gravity at eye level at the surface of the fluid (Figure 7–24). After reading, return the fluid sample to the cell from which it was removed. Repeat this procedure with the other cells.

Drain Load Test

The drain load test is conducted whenever it is determined that there is an unwanted load on the battery. When this occurs, the battery will eventually become discharged. There are, however, several types of load, some of them intentional. Following is a description of the types of load that may be found in today's vehicle.

LOADS

A *load* refers to any device that requires electrical current for operation. This would be the case when

Specific Gravity	State of Charge	Open Circuit Cell Voltage
1.260	100%	*2.10
1.230	75%	2.07
1.200	50%	2.04
1.170	25%	2.01
1.110	0	1.95

FIGURE 7–23 Specific gravity vs. state of charge at 80°F (26.6°C)

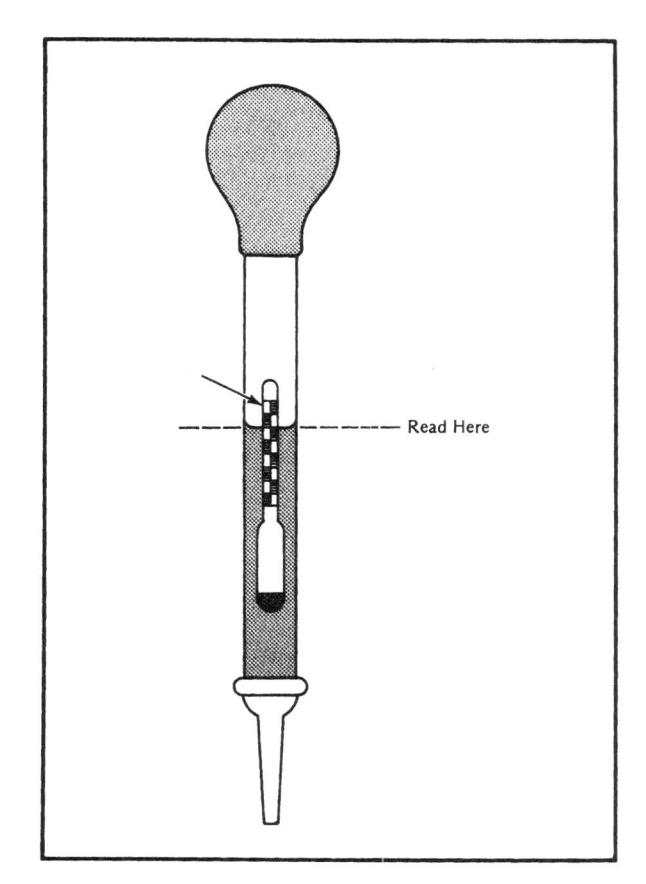

FIGURE 7–24 Read liquid at eye level; disregard curvature at liquid surface

turning on the headlamps, wiper motor, air conditioner, and so on. These may be thought of as intentional loads. There are two other types of load, however, that the automotive electrical technician must be aware of, the parasitic load and the sneak or phantom load.

Parasitic Load

A *parasitic load* is one that constantly draws a small amount of current from the battery even though the ignition switch may be OFF. This draw is continuous, 24 hours per day. A parasitic load varies from vehicle to vehicle depending on electrical/electronic equipment. Table 7–1 shows examples of parasitic loads, in milliamps (mA), for various devices.

Note that the parasitic draw is in milliamps (mA), which is one thousandths (1/1000) of an ampere. From the example given, all parasitic loads combined, at maximum, amount to less than one quarter (¼) ampere.

82 Unit 7 BATTERIES

TABLE 7-1 Parasitic loads for various accessory devices

Device	Load (mA)
Analog clock	6.5–7.5
Digital clock	3.0–4.5
Electronic control module (ECM)	6.5–8.5
Electronic radio/clock	6.5–7.5
Load leveler	3.5–4.5
Memory seat	2.5–3.5
Twilight sentinel	2.5–3.5
Voltage regulator	0.7–1.3

Sneak or Phantom Loads

A *sneak load*, also known as a *phantom load*, occurs whenever a switch device does not turn a load OFF when expected. This may be caused by a misadjustment or malfunction of the switching device. This would be the case, for example, with a defective trunk-light switch that allowed the trunk lamp to remain ON when the trunk is closed. Other phantom loads result, for example, from a "pinched" wire to ground when adding an accessory such as a tape deck. Often a sneak load may be difficult to locate because it is an intermittent condition.

Previous practice for testing for a sneak load was to connect an ammeter in series with the battery ground cable. The ammeter was connected to the battery ground cable and negative (−) battery post (Figure 7–25). However, because parasitic loads are expected in today's vehicle, this is no longer a valid test. An ammeter, capable of measuring current in milliamps (mA), must be used to check the total battery drain. The automotive technician must now be able to distinguish between normal parasitic loads and unwanted loads.

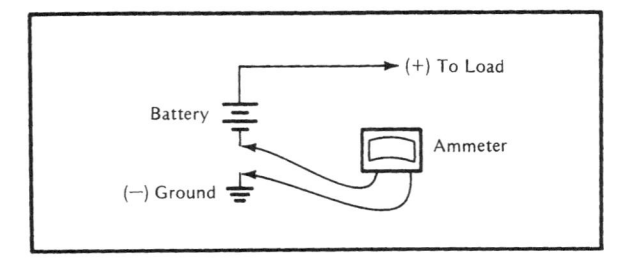

FIGURE 7–25 An ammeter connected in series with the battery ground cable

Self-Discharge

Automotive batteries also have a tendency to self-discharge. If a vehicle is to be out of service or stored for 30 days or longer, the battery ground cable should be disconnected. If this is not practical, the battery should be charged every 30 to 60 days. Often a totally discharged battery cannot be restored to original condition if allowed to remain in that condition for an extended period of time.

CAPACITY TEST

A quick capacity test can be made of a serviceable battery only if the specific gravity reading is 1.22 or more. With battery cables connected, connect a 0–15-volt voltmeter positive (+) to positive (+) and negative (−) to negative (−) across the battery. Disable the ignition system so that the engine will not start while cranking. Follow manufacturers' prescribed procedures for disabling the ignition system, particularly for late-model solid-state ignition systems. Crank the engine for 15 seconds while observing the voltmeter. Assuming a 12-volt system, the reading should not drop below 9.5 volts.

NOTE: A defective starter, a tight engine, defective cable connections, or an undersized battery will affect the capacity test accuracy.

TESTING MAINTENANCE-FREE BATTERIES

The green dot in the built-in hydrometer indicates that the battery is at least 65 percent charged. This condition is sufficient to hold the load test. If the dot is not visible, stratification may have occurred as a result of the battery's being deeply discharged. *Stratification* is a condition whereby the sulphuric acid separates from the electrolyte and settles at the bottom of the battery case, which can result from cranking for a long period of time. This condition can usually be corrected by gently shaking the battery. Take care, however, not to splash electrolyte out the vent provisions.

If the hydrometer is green the battery is ready for testing. If it is black, however, the battery must be charged before testing. If it is clear to light yellow, the battery is defective and must be replaced.

To test the battery follow this procedure carefully:

TABLE 7-2 Temperature vs. voltage comparisons

Temperature °F	°C	Minimum Voltage
70*	22.2	9.6
60	15.6	9.5
50	10.0	9.4
40	4.4	9.3
30	−1.1	9.1
20	−6.7	8.9
10	−12.2	8.7
0**	−17.8	8.5
		8.0

*and above
**and below

1. Connect a carbon pile to the battery.
   a. If the battery is in the vehicle, connect the pile to the battery terminals.
   b. If you have removed the battery from the vehicle, use an adaptor charging tool to attach the load clamps of the pile.
2. Apply a 300-ampere load for 15 seconds to remove the surface charge.
3. Remove the load and wait 15 seconds for the battery to recover.
4. Apply the specific load
   a. As shown on the battery, or
   b. One-half (50 percent) of cold cranking amps (CCA) rating.
5. Read the voltage after 15 seconds.
6. Remove the load and compare the reading to specifications based on temperature, as shown in Table 7–2.

CHARGING RATE

For all practical purposes the charging rate of a battery can be determined by checking the open circuit voltage (OCV) rating generally found on the battery. The charging time, then, relates to the reserve capacity rating (RCR) of the battery. If a constant potential battery charger is not available, refer to the chart in Table 7–3 for charging information.

JUMP STARTING

When it is necessary to jump start a vehicle because of an undercharged battery, certain procedures should be followed step by step to be safe. Actually, hooking up jumper cables from one vehicle to another is a relatively simple matter. If not done correctly, however, serious damage to one of the vehicles or, worse yet, serious injury to the operator may occur. The following, then, is the recommended procedure:

1. Position vehicles so the jumper cables will reach. Be certain, however, that the vehicles do not touch each other.
2. Turn all switches of both vehicles to OFF. Turn blower motor switch of disabled vehicle ON. This provides a load to prevent spikes in the electrical system that could cause damage to sensitive electronic components.
3. Connect positive (+) jumper cable:
   a. To positive (+) battery terminal of service vehicle.
   b. To positive (+) battery terminal of disabled vehicle.

Do not let the positive and negative cables touch one another.

TABLE 7-3 Charging rate vs. battery capacity

| Reserve Capacity Rating | 80 or less ||| 81–125 ||| 126–180 |||
Charge Rate (Amperes)	5	10	20	5	10	20	5	10	20
Open Circuit Voltage				Charge Time (Hours)					
12.25–12.39	4	2	1	6	3	1.5	8	4	2
12.10–12.24	6	3	1.5	8	4	2	12	6	3
11.95–12.09	8	4	2	12	6	3	16	8	4
<11.95	10	5	2.5	14	7	3.5	10	10	5

4. Connect negative (−) jumper cable:
   a. To negative (−) battery terminal of service vehicle.
   b. To good metallic point of the disabled vehicle. DO NOT connect to disabled vehicle battery. Some late-model vehicles are equipped with a remote jump start terminal for the negative (−) connection.
5. Ensure that the cables are clear of any moving parts of both engines.
6. Start the engine of the service vehicle. Run at moderate speed.
7. Start the engine of the disabled vehicle. Leave all switches OFF except the blower (Step 2).
8. Reduce engine speed of both vehicles to normal idle.
9. Disconnect negative (−) cable:
   a. From formerly disabled vehicle first.
   b. Then, from service vehicle.
10. Disconnect positive (+) cable: Again, be sure that the cables do not touch before all are disconnected.
    a. From formerly disabled vehicle.
    b. Then, from service vehicle.

WARNING: Hydrogen (H) and Oxygen (O) vapors are produced during normal battery operation. These vapors may cause an explosion if open flames or sparks are produced in the area. It is important to follow the following safety precautions:

1. Work in a well-ventilated area.
2. Never charge a battery in an enclosed space.
3. Wear protective clothing when handling batteries.
4. Shield face and eyes when servicing a battery.

Batteries are dangerous; they contain sulfuric acid. Avoid contact with eyes, skin, or clothing. If a mishap occurs, immediately flush the area with large quantities of water for a minimum of 15 minutes. If ingested, drink large quantities of milk followed by milk

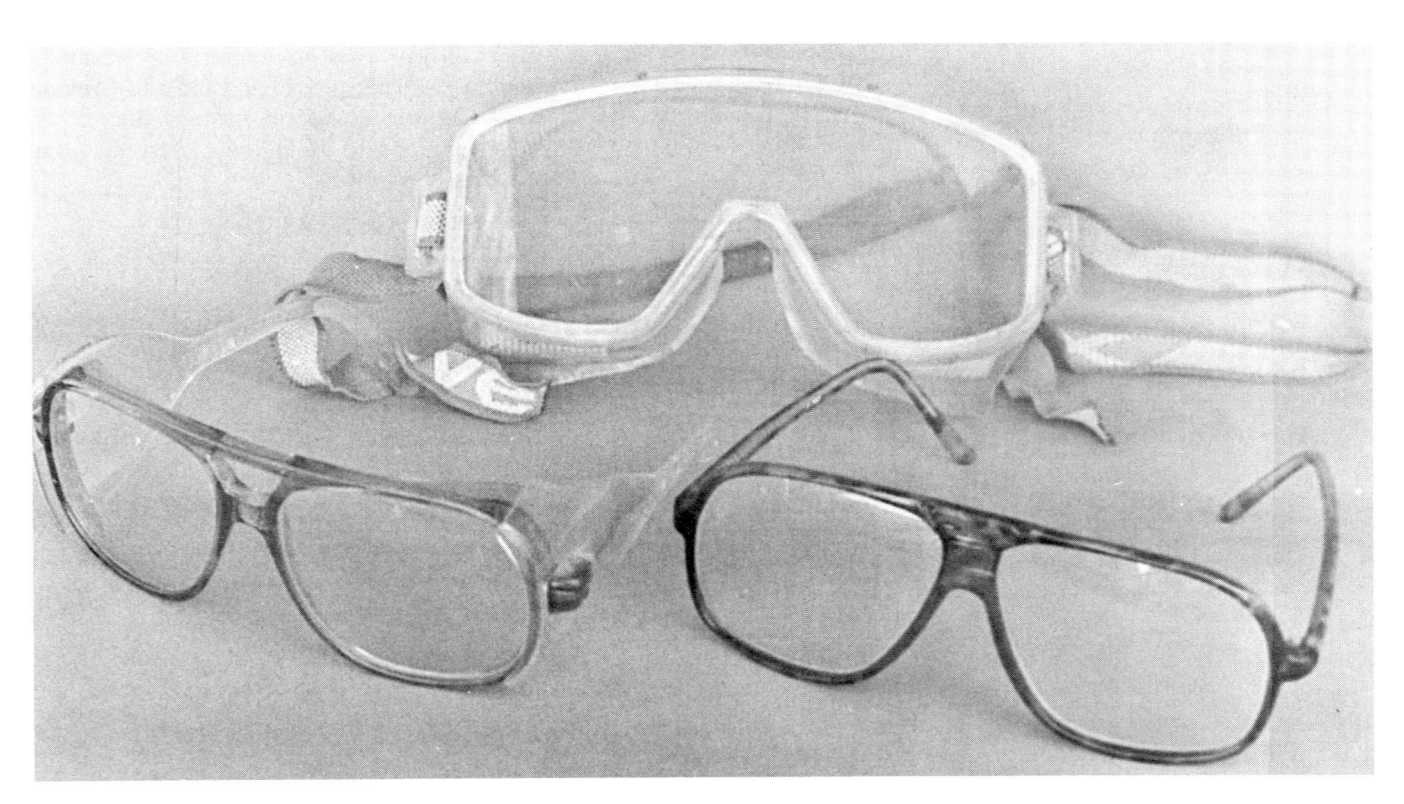

FIGURE 7–26 Eye protection (Courtesy of BET, Inc.)

of magnesia or a beaten raw egg. In either case, call a physician or poison control center immediately and follow instructions given.

When lifting plastic-case batteries, avoid excessive pressure on the sides. Compressing the soft plastic sides could cause sulfuric acid to spurt out of the vent caps. Always use a battery carrier. If a carrier is not available, carefully handle by grasping opposite corners.

Batteries contain hazardous materials. Dispose of them in a manner consistent with local rules and regulations. Disposal of hazardous materials is now governed by federal and state law. Improper disposal of hazardous material can result in severe penalties. Local parts stores are generally aware of local regulations and, in most cases, will assist in disposal.

CONCLUSION

Batteries, well serviced and maintained, often provide years of dependable service even after the warranty has expired. But batteries must be considered dangerous, and adequate safeguards must be taken when servicing them. This includes, above all, adequate eye protection (Figure 7–26). Remember that the acid–water solution of a battery is highly corrosive and its gas (vapor) is highly explosive.

SUMMARY

- A battery is made up of cells connected in series or parallel.
- As a battery becomes discharged its internal resistance increases, lowering its ability to deliver current.
- Automotive batteries can be dangerous. A technician must take care when working with them.

PRACTICAL EXERCISE

You will need two vehicles, one able and one disabled, and a set of jumper cables for this exercise. Suitable dress is also required. This includes long-sleeve cotton shirt and adequate face protection (safety glasses or face shield).

This exercise is to be carried out only under the direct supervision of a qualified technician, such as your instructor.

1. Position the able vehicle so the jumper cables will reach. Be certain that the vehicles do not touch each other.
2. Turn OFF all switches and controls of both vehicles.
3. Turn ON the blower motor switch of the disabled vehicle.
4. Put on safety glasses or other suitable eye protection.
5. Carefully connect the positive (+) jumper cable:
   a. First, to the positive (+) battery terminal of the service (able) vehicle.
   b. Next, to the positive (+) battery terminal of the disabled vehicle. Be sure the positive (+) and negative (−) cables do not touch.
6. Connect the negative (−) jumper cable:
   a. First, to the negative (−) battery terminal of the service (able) vehicle.
   b. Next, to a good metallic point of the disabled vehicle. DO NOT connect this cable to the disabled vehicle battery.
7. Ensure that the cables connected in Steps 3 and 4 are clear of all moving parts of both engines.
8. Start the engine of the service (able) vehicle. Run at moderate speed.
9. Start the engine of the disabled vehicle. Leave all switches OFF except the blower (Step 3).
10. Reduce the engine speed of both vehicles to normal idle.
11. Disconnect negative (−) cable:
    a. First, from the formerly disabled vehicle.
    b. Then, from service (able) vehicle.
12. Disconnect positive (+) cable: Keep the cables from touching before all are disconnected from the vehicles.
    a. First, from the formerly disabled vehicle.
    b. Then, from service (able) vehicle.

Unit 7 BATTERIES

QUIZ

The following questions are for open classroom discussion.

1. What are the dangers involved in "jump starting" a vehicle?
2. Why do you think the disabled vehicle's battery ground (−) terminal is not used?
3. Why do you think that the cables are connected to the able vehicle before the disabled vehicle?

REVIEW

1. The electrolyte of a battery is made up of a strong solution of _____ acid and water (H_2O).
2. The fumes (vapor) from a battery are highly _____.
3. An acid neutralizer consists of one pound of _____ to _____ of water (H_2O).
4. The specific gravity of a fully charged battery is _____.
5. Specific gravity readings are taken with an instrument known as a(n) _____.
6. Whenever a higher current capability is needed, batteries are connected in (series) (parallel).
7. When a higher voltage is needed, batteries are connected in (series) (parallel).
8. Current flow through the internal resistance of a battery causes the output voltage to _____.
9. A battery has a no-load terminal voltage of 13 volts. When 1 ampere of current is drawn from the battery, the terminal voltage is 12 volts. What is the internal resistance of the battery? _____.
10. As a battery becomes discharged, its internal resistance (increases) (decreases).
11. A 12-volt battery has an internal resistance of 0.5 ohm. How much current can be drawn from the battery before the terminal voltage decreases to 10 volts? _____.
12. How many 1.5-volt batteries must be connected in series in order to obtain 13.5 volts? _____.
13. A battery's terminal voltage is measured as 13.5 volts. When 11 amperes is drawn from the battery the terminal voltage decreases to 12.9 volts. What is the internal resistance of the battery?
14. A battery's terminal voltage is measured at 13.6 volts. When a load resistor of 0.52 Ω is placed across the battery, 24.8 amperes of current flows. What is the internal resistance of the battery?
15. A battery's terminal voltage is measured at 13.5 volts. When a load resistor of 0.6 Ω is placed across the battery, the terminal voltage decreases to 13.1 volts. What is the internal resistance of the battery?

◆ UNIT 8 ◆

CAPACITANCE AND INDUCTANCE IN DC

OBJECTIVES

On completion of this unit you will be able to:

- Explain how capacitors are constructed.
- Understand the effect of capacitance in a circuit.
- Understand that a capacitor in a circuit opposes a change in voltage.
- Explain how capacitors charge and discharge.
- Discuss the time required to charge a capacitor.
- Discuss the effects of inductance.
- Understand that inductance opposes a change in current.
- Explain the time required for current to change in a circuit containing inductance.

In early motor vehicles capacitance and inductance were of little significance to the auto mechanic. There was a capacitor in the distributor that was changed when needed, and the "spark" came from an inductor, the coil. Otherwise, capacitors and coils were in the radio, not something the general auto mechanic worked with.

In the modern motor vehicle electronics is involved with almost every system. Capacitors and inductors are part of most electronic systems, and relays and solenoids are common items. The modern automotive electrical/electronic technician must have a good basic understanding of capacitors and inductors and their effects in electronic circuits.

CAPACITANCE

Capacitance is the property of an electric circuit that opposes a change in voltage. Electrons must move before the voltage across a capacitor will change. It takes time for electrons to move; therefore, it takes time for the voltage to change. The first hint of capacitive action was introduced in Unit 2. A generator was shown pulling electrons from a wire connected to the positive (+) terminal of the generator and pushing them out on the wire connected to the negative (−) terminal. A similar situation is shown in Figure 8–1; the generator is running and wires are connected to the positive (+) and negative (−) terminals. In order for a voltage to appear between positive (+) and negative (−) wires, electrons have to be pulled from the positive (+) side and forced out on the negative (−) side.

If a circuit constructed with a sensitive meter were connected as shown in Figure 8–2, the meter would indicate electron movement or current flow during the charging of the wires. The meter needle would move up as the switch is closed and immediately drop back to zero as the wires became charged. The number of free electrons that move would control the meter needle movement.

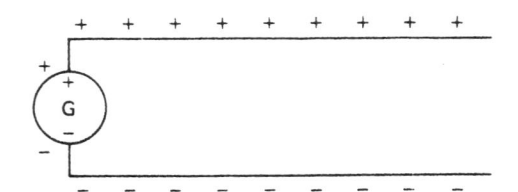

FIGURE 8–1 Generator moving electrons, which charge wires

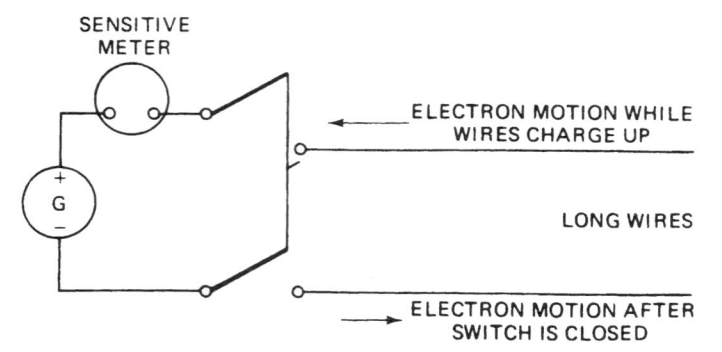

FIGURE 8–2 Meter indicating that wires are charging

The number of electrons that have to move in order to "charge" the wires could be increased by replacing the external wires with large, flat plates. Increasing the surface area enlarges the "charge" capability.

Metal plates with a large surface area separated by some high-resistance insulating material make up a capacitor. In Figure 8–3, air is the insulating material.

CURRENT FLOW IN CAPACITIVE CIRCUITS (DC)

It was indicated in the explanation for Figure 8–2 that, after the switch is closed, there is a momentary movement of electrons. With the dc generator connected to the wires, some of the free electrons move from the top wire through the generator and out on the bottom wire. After the wires become charged, voltage appears between all points on the top and bottom wires. Electron movement then stops.

A similar situation exists if a capacitor is connected across a dc generator or any other dc power source. As the capacitor is connected across the generator, the current jumps up. As the capacitor charges, the current flow drops to zero and the source voltage appears across the capacitor terminals. A graph of voltage and current as related to time is shown in Figure 8–4.

Current

1. As the switch is closed at time zero, the current rises immediately to maximum.
2. The current decreases as the capacitor charges.
3. The capacitor is fully charged; the current drops off to 0 amperes.

Voltage

1. As the switch is closed, the voltage across the capacitor rises rapidly from 0 volts.
2. The voltage rise starts to taper off as the voltage across the capacitor approaches the source (generator) voltage.
3. The voltage across the capacitor reaches the source voltage; the capacitor is fully charged. The voltage across the capacitor remains constant at the source voltage level.

The number of electrons that flow into and charge a capacitor is related to the size of the capacitor in farads (F) and the applied voltage. The charge in

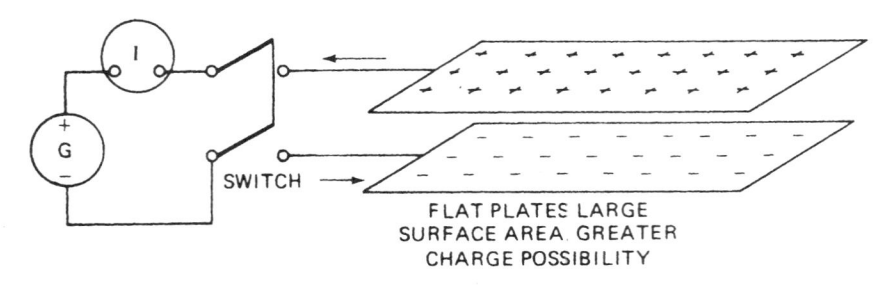

FIGURE 8–3 Increased meter current with increase in surface area

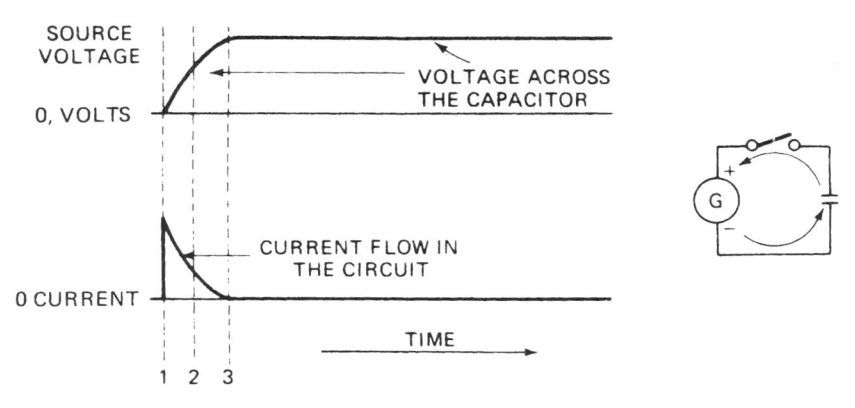

FIGURE 8–4 It takes time to charge a capacitor

coulombs (Q) is equal to the capacitance (C) in farads (F) times the voltage (V).

$$Q = C \times V$$

EXAMPLE 1

Determine the charge on a 10 microfarad (µF) capacitor that has 30 volts applied to it.

Solution

$$Q = C \times V$$
$$Q = 10 \times 10^{-6} \times 30$$
$$Q = 300 \times 10^{-6}$$

How many electrons are stored in the capacitor?

$$e = 3 \times 10^{-4} \times 6.25 \times 10^{18}$$
$$e = 1.875 \times 10^{15}$$

There are 1.875×10^{15} electrons stored on the negative (−) plate of the capacitor. There is an absence of the same number of electrons from the positive (+) plate.

This simple formula, $Q = C \times V$, will prove helpful in the investigation and understanding of integrated amplifier circuits covered in later units.

RC TIME CONSTANT

In electric circuits containing resistance and capacitance there is a relationship called the *time constant* (TC) of the circuit. Any dc *resistance-capacitance* (RC) circuit will exhibit characteristics similar to those shown in Figure 8–5. The time constant, $R \times C$, provides an indicator of how fast the capacitor will charge.

EXAMPLE 2

In Figure 8–5, a 1,000,000-Ω (1-MΩ) resistor is connected in series with a 2-microfarad (µF) capacitor. What is the time constant of the circuit?

Solution

$$TC = RC$$
$$TC = I \times 10^6 \times 2 \times 10^{-6}$$
$$TC = 2 \text{ s}$$

The capacitor in Figure 8–5 will become fully charged in five time constants, or in this case 10 seconds. As S1 is closed, the rate at which the capacitor

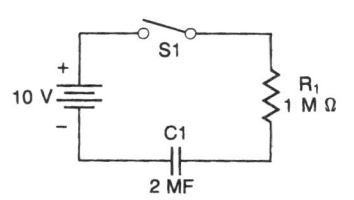

Solution

$$T = RC$$
$$T = 1 \times 10^6 \times 2 \times 10^{-6}$$
$$T = 2 \text{ sec}$$

FIGURE 8–5 RC circuit, time constant 2 seconds.

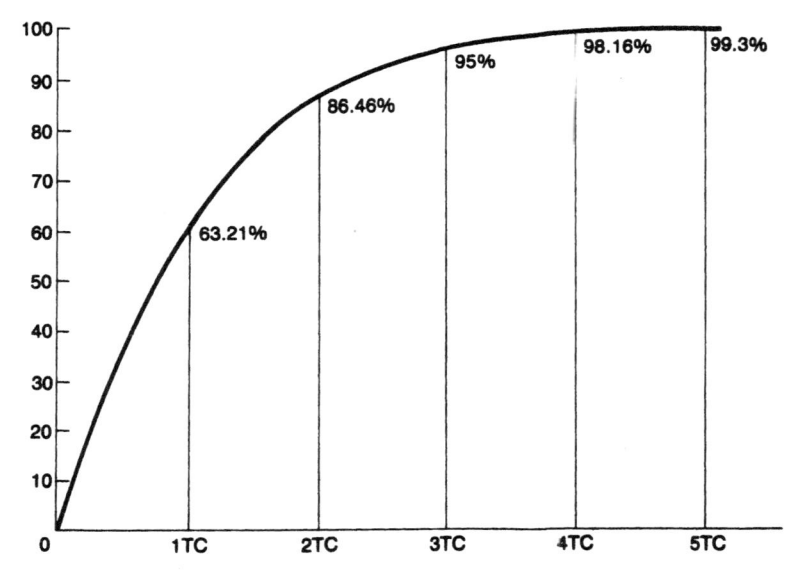

FIGURE 8–6 Universal time constant curve

charges changes. The capacitor charge is shown as a curve in Figure 8–6. This is known as a *universal time-constant curve*.

The capacitor charges 63.21 percent of the voltage left to charge in each time constant. In Figure 8–6, with 10 volts applied, the capacitor will charge to 6.321 volts in the first time constant (2 seconds). This leaves 3.679 volts left to charge. The capacitor will charge 2.325 volts in the next time constant, bringing the voltage across the capacitor to 8.646 volts.

This process could go on forever, with 36.79 percent of the voltage always being available to charge in the next time constant (see Table 8-1).

For practical reasons, the capacitor is said to be fully charged at the end of five time constants, when the capacitor will be charged to 99.3 percent of the available voltage. Normally, an approximation of the voltage across the capacitor in an RC circuit is all that is necessary. The universal time constant curve can be used to make this determination. The voltage across the resistor equals the supply voltage minus the voltage across the capacitor.

In special cases where the exact voltage across the capacitor is needed, use the formula

$$V_c = V_s(1 - e^{-\frac{t}{TC}})$$

where V_c = voltage across the capacitor
V_s = supply voltage
e = natural logarithm base 2.718
t = time in seconds the capacitor charges
TC = RC time constant

TABLE 8–1 Capacitor Voltage by Time Constant

By Time Constant	
0.632 × 10 V = 6.32 V	Charge after the 1st TC
10 V − 6.32 V = 3.679 V	Volts left to charge
0.632 × 3.679 V = 2.325 V	Change in voltage
6.32 V + 2.325 V = 8.646 V	Charge after the 2nd TC
10 V − 8.646 V = 1.354 V	Volts left to charge
0.632 × 1.354 V = 0.8558 V	Change in voltage
8.646 V + 0.8558 V = 9.50 V	Charge after the 3rd TC
10 V − 9.5 V = 0.5 V	Volts left to charge
0.632 × 0.5 V = 0.316 V	Change in voltage
9.5 V + 0.316 V = 9.816 V	Charge after the 4th TC
10 V − 9.816 V = 0.814 V	Volts left to charge
0.631 × 0.814 V = 0.116 V	Change in voltage
9.816 V + 0.116 V = 9.932 V	Charge after the 5th TC
10 V − 9.932 V = .068 V	Volts left to charge

EXAMPLE 3

Consider the circuit of Figure 8–5. What is the voltage across the capacitor 1.8 seconds after S1 is closed?

Solution

By formula:

$$V_c = V_s(1 - e^{-\frac{t}{RC}})$$
$$V_c = 10(1 - 2.718^{-\frac{1.8}{2}})$$

$$V_c = 10(1 - 2.718^{-0.9})$$
$$V_c = 10(1 - 0.407)$$
$$V_c = 10 \times 0.593$$
$$V_c = 5.93 \text{ V}$$

By formula using a scientific calculator:

$$V_c = V_s \left(1 - e^{-\frac{t}{RC}}\right)$$

STEP	ENTER	PRESS	DISPLAY
1	2.71828	y^x	2.71828
2	0.9	+/−	−0.9
3		=	0.40657
4		+/−	−0.40657
5	1	+	0.59343
6		×	0.59343
7	10	=	5.9343

The voltage across the capacitor will be 5.9343 volts, 1.8 seconds after switch S1 is closed.

One point worth noting concerns the relationship of current to voltage in the capacitor. As shown in Figure 8–4 at position 1, the current is maximum when the voltage is zero; at position 3 the voltage is maximum when the current is zero. The current is ahead of the voltage.

CAPACITOR CONSTRUCTION

For practical reasons, capacitors are constructed with thin metal sheets, usually aluminum (Al), as the plates. Ceramic, plastic (Mylar), or a chemical electrolyte, with paper, is used as the high-resistance material between the plates. The three common types of capacitors are ceramic, plastic, and electrolytic.

Electrolytic capacitors are shown in Figure 8–7. Electrolytic capacitors are used where large capacitance values, in a small package, are required. They are standard where filtering is needed. A *filter* is a device used to maintain a voltage. Capacitance opposes a change in voltage. Large capacitance values offer strong opposition to a change in voltage.

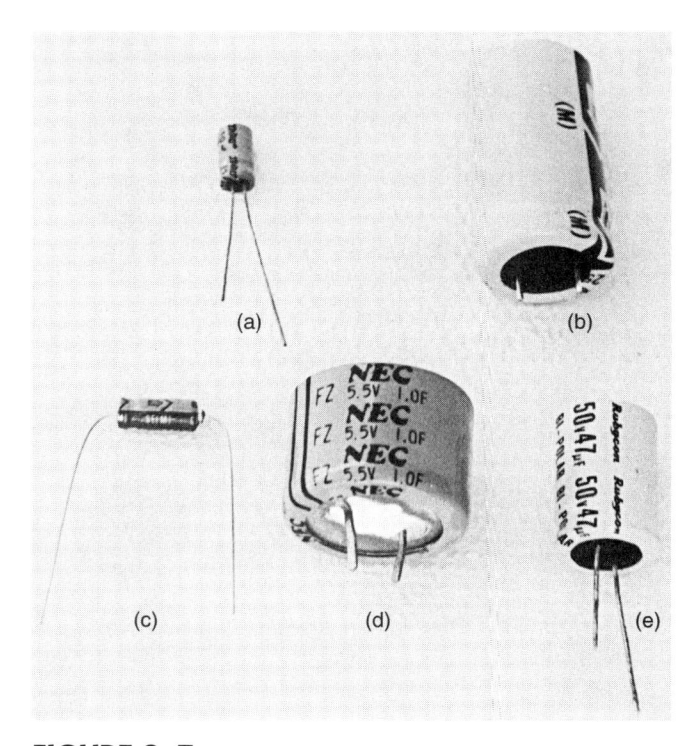

FIGURE 8–7 Electrolytic capacitors

Most electrolytic capacitors are polarized. They must be installed in a circuit with the capacitor positive (+) terminal connected to positive (+) voltage and the negative (−) terminal to negative (−) voltage.

In Figure 8–7 capacitors a and b are regular, radial lead, aluminum (Al) electrolytic capacitors. Capacitor a is 10 µF/35 V, and capacitor b is 3300 µF/25 V. Capacitor c is an axial lead 10-µF/25 V capacitor. Capacitor d is 1 farad at 5.5 volts (1 F/5.5 V), and e is a 47-µF/50 V nonpolarized electrolytic capacitor.

In Figure 8–8 ceramic, plastic, tantalum, and surface-mount capacitors are shown. Capacitors a and b are dipped Mylar, 0.068 µF/200 V and 0.47 µF/630 V. Capacitors c, d, and e are dipped ceramic capacitors, 0.05 µF/50 V, 0.1 µF/50 V, and 100 pF/500 V. Capacitors f and g are tantalum capacitors, 15 µF/30 V and 150 µF/20 V. Capacitor h is a surface-mount capacitor, 0.022 µF/250 V. Tantalum capacitors are polarized.

CAPACITOR RATINGS

Capacitors are usually rated in capacity and maximum voltage rating. This capacity unit is the farad (F), which is a very large unit. Capacitors are usually rated in microfarads (µF, one millionth of a farad), which is a

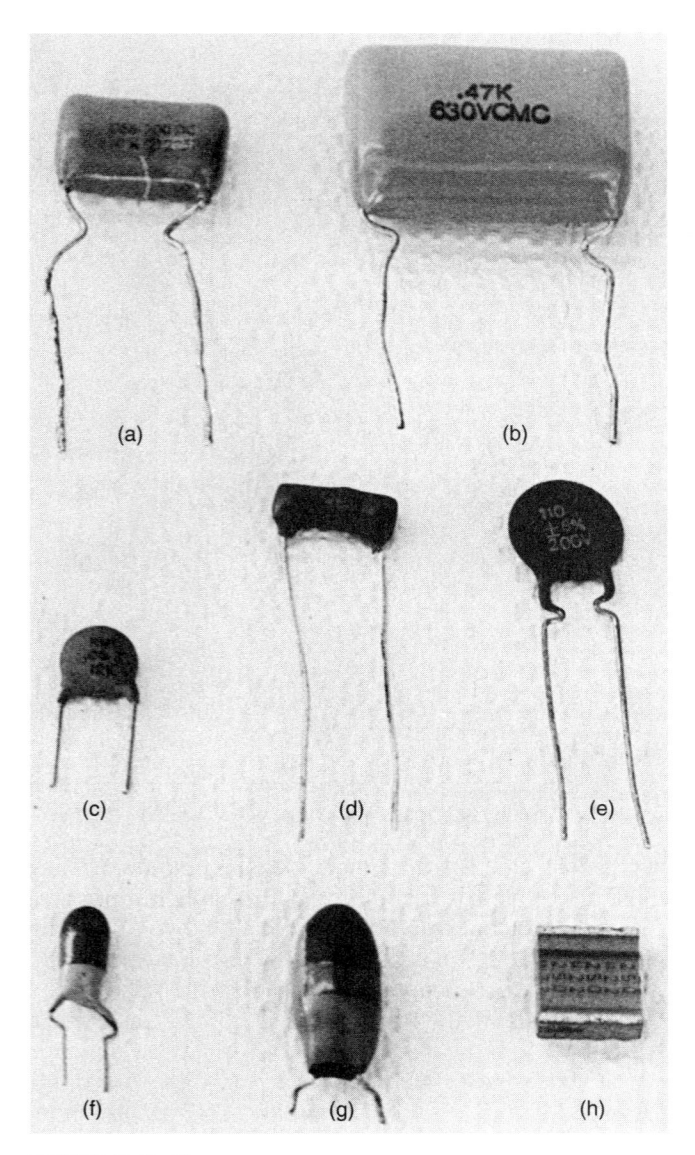

FIGURE 8-8 Regular capacitors

practical unit for most applications. A capacitor may be marked 10 MFD, 10 mfd, 10 MF, 10 mf, or 10 µF, any of which means 10 microfarads. The common abbreviation for microfarad is µF.

INDUCTANCE

Inductance is the property of an electric circuit that opposes changes in current.

In Figure 8–9, a coil of wire is shown connected to a resistor, a switch, and a power source. This is known as an LR circuit. The L is for the inductance of the coil, whereas the R is for the resistor. When the switch S1 is closed, current starts to flow in the circuit. The current flowing through the coil will cause a magnetic field to build up around the coil. The magnetic field cutting through the coil will induce a voltage in the coil. The induced voltage is in direct opposition to the applied voltage from the battery.

INDUCTOR RATINGS

Inductors are usually rated in inductance and maximum current-carrying capability. The unit of inductance is the henry (H).

A 1-henry (H) coil will have one volt induced in it when the current is changing at a rate of one ampere per second. Practical sizes of coils used in electronics include coils rated in millihenrys (mH) and microhenrys (µH).

CURRENT AND VOLTAGE RELATIONSHIP

The relationship between current and voltage in an inductor is as shown in Figure 8–10. At the instant S1 is closed, the voltage across the coil is maximum while the current is zero. As the voltage across the coil decreases, the current increases (2). When the voltage across the coil is minimum, the current is maximum (3). This is exactly opposite the effects that were noted with the resistor capacitor (RC) circuit. In the *inductor-resistor* (LR) circuit, the current lags behind the voltage.

As a timed operation at point 1 (Figure 8–10), the current (I) is rising most rapidly. The magnetic field is changing most rapidly and the back electromotive force (bemf) is greatest, as is the voltage across the coil.

At point 2, the rate of change of current has decreased. The bemf across the coil decreases, as does the voltage across the coil.

At point 3, the current has reached its maximum (E applied/R) steady-state value. The magnetic field is not changing. There is no bemf across the coil. The voltage across the coil is minimum.

LR TIME CONSTANT

As with the R circuit, there is a direct relation between the values of the inductor and resistor in an LR circuit

LR TIME CONSTANT   93

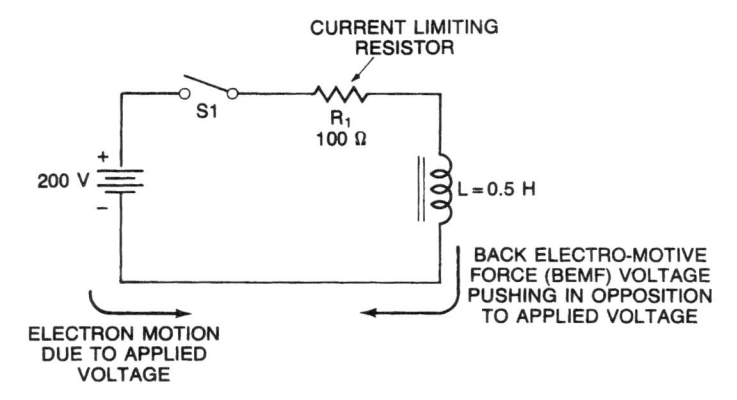

FIGURE 8–9 LR current, time constant 0.005 seconds.

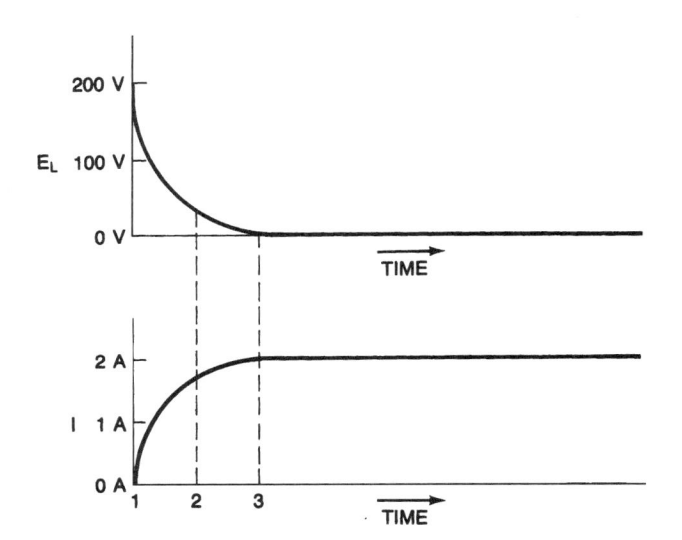

FIGURE 8–10 Current and voltage relationship in LR circuits

and the time required for the current to reach its steady-state value. The time constant (TC) in an LR circuit is equal to the inductance divided by the resistance in the circuit.

$$TC = L/R$$

The universal time constant curve may be used to determine the current level at any point up to 5 time constants (TC). At 5 TC, the current for all practical purposes equals V/R.

The current at any instant may also be found using the formula

$$TC = \frac{L}{R}$$

$$i = \frac{V_S}{R}(1 - e^{-\frac{t}{TC}})$$

where i = the instantaneous value of current
V_S = the supply voltage
R = the circuit resistance
e = natural logarithm 2.718
t = the time the circuit is connected
TC = the time constant of the circuit; TC = L/R
 = time in seconds

As an example, consider Figure 8–9, the LR circuit. What is the value of the current and the voltage across each component 2 milliseconds after S1 is closed?

$$i = \frac{V_S}{R}(1 - e^{-\frac{t}{RC}})$$

$$i = \frac{200}{100}(1 - 2.718^{-\frac{0.002}{0.005}})$$

$$i = 2(1 - 2.718^{-0.4})$$

$$i = 2(1 - 0.67)$$

$$i = 2 \times 0.33$$

$$i = 0.66 \text{ A}$$

$$V_R = i \times R$$

$$V_R = 0.66 \text{ A} \times 100$$

$$V_R = 66 \text{ V}$$

$$V_L = V_S - V_R$$

$$V_L = 200 \text{ V} - 66 \text{ V}$$

$$V_L = 134 \text{ V}$$

Unit 8 CAPACITANCE AND INDUCTANCE IN DC

$$i = \frac{V_s}{R}(1 - e^{-\frac{t}{TC}})$$

where $e = 2.718$

$$\frac{t}{TC} = \frac{0.002}{0.005} = 0.4$$

STEP	ENTER	PRESS	DISPLAY
1	2.718	y^x	2.718
2	0.4	+/−	−0.4
3		=	0.67034
4		+/−	−0.67034
5		+	−0.67034
6	1	=	0.32966
7		×	0.32966
8		(	0
9	200	÷	200
10	100	)	2
11		=	0.6593

For a scientific-calculator solution of the problem using the formula, examine the following:

The current, 0.002 seconds after S1 is closed, will be 0.6593 amperes.

SUMMARY

- Capacitors used in automotive circuits are of many sizes and shapes.
- The formula for RC time constant is T = RC.
- It takes five time constants for a capacitor to charge or discharge.
- Capacitance opposes a change in voltage, whereas inductance opposes a change in current.

PRACTICAL EXERCISE

An analog ohmmeter and several capacitors are required for this exercise.

1. Select a high-resistance scale on the ohmmeter (R × 1000 or higher).
2. While observing the meter needle, connect the meter leads to a capacitor.
3. Carefully, disconnect the meter from the capacitor. Take care not to short the capacitor terminals or leads.
4. Record your observation of Step 2 in the chart of Figure 8–11.
5. Repeat Steps 2 through 4.
6. Carefully, short the capacitor leads together.
7. Repeat Steps 2 through 4.
8. Repeat Steps 2 through 7 with each of the other capacitors.

QUIZ

For open classroom discussion.

- Why was the meter needle movement different between the first and second try?
- Why was the third try result like the first try result?

Capacitor #1
What did the Meter Pointer Do?

Step 4	
Step 4 (Of Step 5)	
Step 4 (Of Step 7)	

Capacitor #2
What did the Meter Pointer Do?

Step 4	
Step 4 (Of Step 5)	
Step 4 (Of Step 7)	

Capacitor #3
What did the Meter Pointer Do?

Step 4	
Step 4 (Of Step 5)	
Step 4 (Of Step 7)	

FIGURE 8–11 Capacitor study

1. In Step 2, did the needle "jump" toward zero, then rise to a high resistance? _____ Why? _____
2. In Step 5, did the needle jump toward zero, then rise to a high resistance as in Step 2? _____ Why? _____
3. In Step 7, was the indication more like that of Step 2 or Step 5? _____ Why? _____
4. What would be the effect, in Step 2, of an open capacitor? _____
5. What would be the effect, in Step 2, of a shorted capacitor? _____

REVIEW

Questions 1 through 10 refer to Figure 8–5.

1. If the resistor R_1 were reduced to 500 kΩ, capacitor C1 would charge (more quickly, more slowly).
2. What is the voltage across capacitor C1 four seconds after switch S1 is closed?
3. What is the voltage across resistor R_1 six seconds after switch S1 is closed?
4. What is the voltage across capacitor C1 one minute after switch S1 is closed?
5. What is the charge, in coulombs, on capacitor C1 one minute after switch S1 is closed?
6. How many excess electrons are stored on the negative (−) plate of the capacitor 2 time constants after S1 is closed?
7. What is the voltage across C1 2.72 seconds after switch S1 is closed?
8. What is the voltage across R_1 3.16 seconds after switch S1 is closed?
9. What is the value of the current in the circuit 2.86 seconds after switch S1 is closed?
10. How many time constants must pass after S1 is closed before the current is reduced below 50 percent of maximum? (less than 1, 2, 3, 4, or 5 time constants)

Questions 11 through 20 refer to Figure 8–9.

11. Increasing the value of the inductance in the circuit will (increase) (decrease) the time required for the current to reach maximum.

12. Increasing the value of the resistance in the circuit will (increase) (decrease) the time required for the current to reach maximum.
13. What is the value of voltage across R_1 0.10 milliseconds after S1 is closed?
14. What is the value of the voltage across the coil 0.005 seconds after S1 is closed?
15. What is the value of the current in the circuit 3 time constants after switch S1 is closed?
16. What is the value of the voltage across the coil 0.0092 seconds after S1 is closed?
17. What is the value of the voltage across R_1 0.012 seconds after S1 is closed?
18. What is the value of the current in the circuit 1.373 time constants after S1 is closed?
19. Inductance in an electric circuit opposes a change in (current) (voltage).
20. Capacitance in an electric circuit opposes a change in (current) (voltage).

UNIT 9

RELAYS, SOLENOIDS, AND MOTORS

OBJECTIVES

On completion of this unit you will be able to:

- Understand the construction of relays, solenoids, and motors.
- Discuss the use of diodes to eliminate the effects of back electromotive force (bemf).
- Explain motor speed and torque.
- Understand starting motors and starter control.
- Apply a working knowledge of speed control with pulse width modulation.

Relays, solenoids, and motors are used in many applications in the modern motor vehicle. Relays control the turning ON and OFF of devices throughout the vehicle, a process that is being taken over by the field effect transistor (FET). Solenoids are used to provide a linear action. The fuel injector, for example, is a form of solenoid.

Electric motors are used in many applications in present-day motor vehicles. Some examples are the windshield wiper, seat positioner, fuel pump, air conditioner air mover, radiator fan, and starter motor. Some electric motors (for example, the starter motor) are of extreme importance in motor vehicle operation.

This unit covers relays, solenoids, motor theory, motor speed control, and the starter motor.

RELAYS AND SOLENOIDS

Simple relays were discussed in Unit 5, in the section on electromagnets. A *relay* is simply a stationary electromagnet with provision for one or more sets of electrical contacts or point sets, which make or break as the electromagnet is energized. A *solenoid* is similar in function but has a movable core that slides in or out as the electromagnet is energized. The solenoid may be with or without provision for electrical contacts.

Relays

The *coil* of a relay consists of many turns of copper wire wound around a piece of soft iron. The coil is connected to the battery circuit by a switch. When the switch is open, there are no magnetic lines of force around the coil (Figure 9–1). When the switch is closed, magnetic lines of force are set up around the coil, as illustrated in Figure 9–2.

Whenever a relay or solenoid is used in a modern automobile, a diode is placed across the device coil. The purpose of the diode is to provide a discharge patch for the back electromotive force (back voltage or bemf). This bemf is generated when the magnetic field around the coil collapses as the circuit to the coil is interrupted. Diodes and bemf will be covered in greater detail in later units. The diode is shown in Figure 9–3.

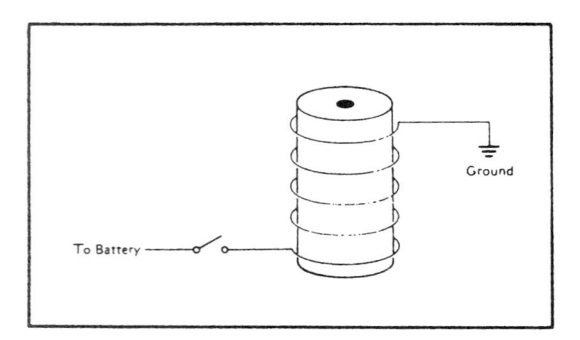

FIGURE 9-1 There are no magnetic lines of force when the switch is open

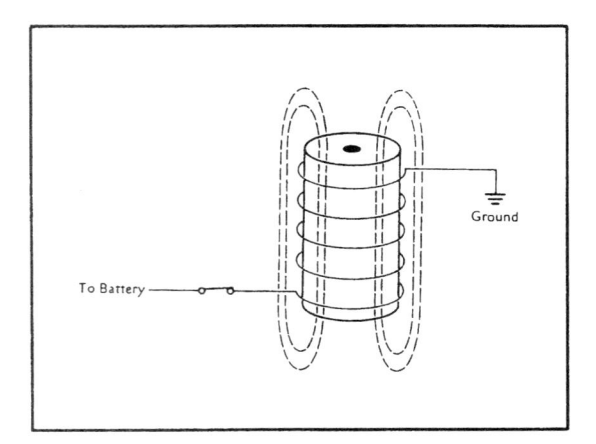

FIGURE 9-2 When the switch is closed, magnetic lines of force are set up around the coil

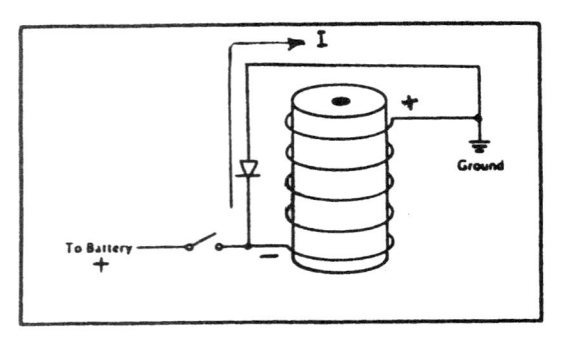

FIGURE 9-3 A diode is placed across the coil to reduce bemf when the switch is opened

bemf Polarity

In Figure 9-3 a diode is shown across the coil. When the switch is opened, the magnetic field about the coil collapses, developing the back emf with polarity as shown. The diode will conduct, limiting the voltage developed to a low value, approximately 0.6 volt. If the diode were not connected, high-voltage spikes would be present. These high-voltage spikes could damage sensitive computer components included in modern automobiles.

Consider a simple relay with one set of points. When there is no magnetic field, the points are held open by a spring attached to a pivot arm, as shown in Figure 9-4. When there is a magnetic field, the spring tension is overcome, and the pivot arm moves toward the electromagnet, closing the relay contact points, as shown in Figure 9-5.

The relay spring is of a calibrated length and tension so that it will perform properly in a specific application. Its purpose is to open and hold open the electrical contact points when electrical power is removed from the coil. Therefore, it is recommended that the spring calibration not be altered. Too little tension may result in an inoperative relay, in that the points may not open. Too much tension may result in chattering, a rapid opening and closing of the points when power is applied to the coil.

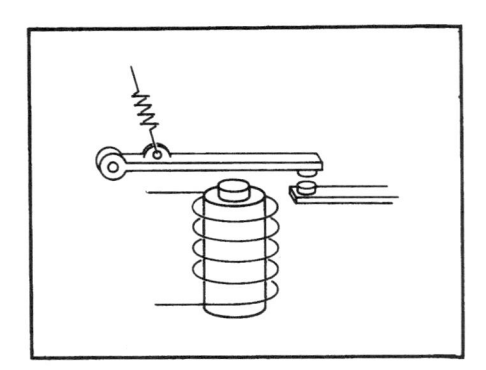

FIGURE 9-4 With no power applied to the coil spring, tension holds the points open

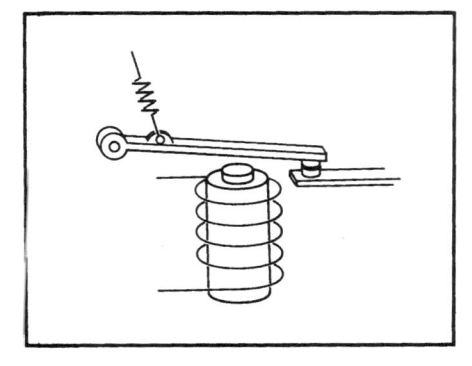

FIGURE 9-5 When power is applied to the coil, magnetic attraction overcomes spring tension to close the points

Solenoids

A *solenoid* is an electromagnet used to produce a mechanical action. It consists of a movable soft-iron core inside a coil of many turns of copper wire. The coil is connected to the battery circuit by a switch. When the switch is open, there are no magnetic lines of force around the coil. A compression spring holds the armature (core) in position, as shown in Figure 9–6(a). When the switch is closed, magnetic lines of force are set up around the coil with sufficient strength to overcome the spring action, and the armature is pulled into the core cavity, as shown in Figure 9–6(b). If the armature were connected by mechanical linkage to a device such as a door lock, the device would be activated whenever the solenoid coil was energized (Figure 9–7). To provide for simultaneous electrical switching action, there could be a set of contacts actuated by the mechanical linkage or armature action, as shown in Figure 9–8.

The solenoid spring is of a calibrated length and compression so that it moves the armature back to its original position when no power is applied to the coil. Decreasing tension will defeat this purpose; increasing tension may result in no or insufficient armature movement when the coil is energized. No attempt should be made to change its calibration.

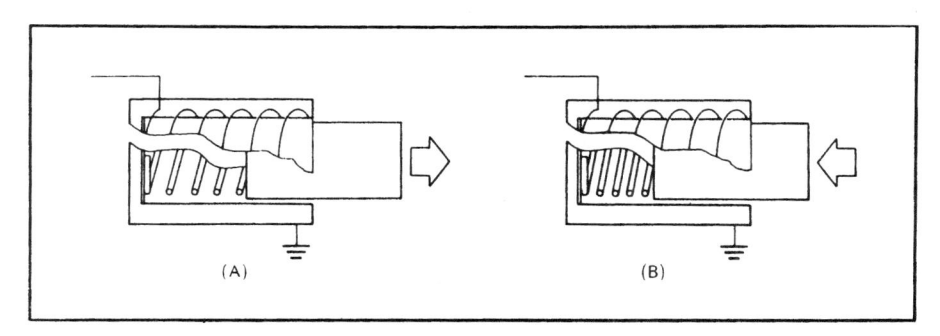

FIGURE 9–6 (a) A compression spring holds the core in position; (b) Magnetic attraction overcomes spring pressure

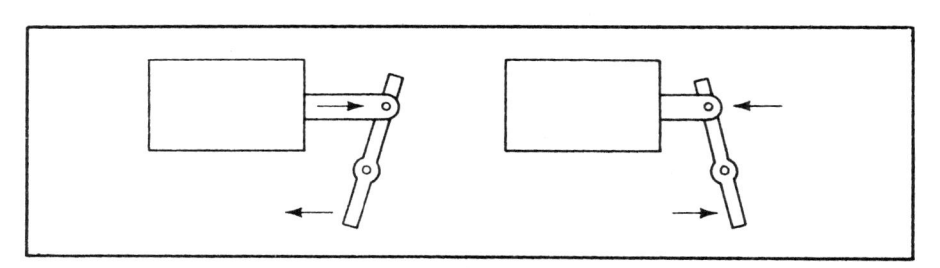

FIGURE 9–7 Mechanical linkage actuated by a solenoid

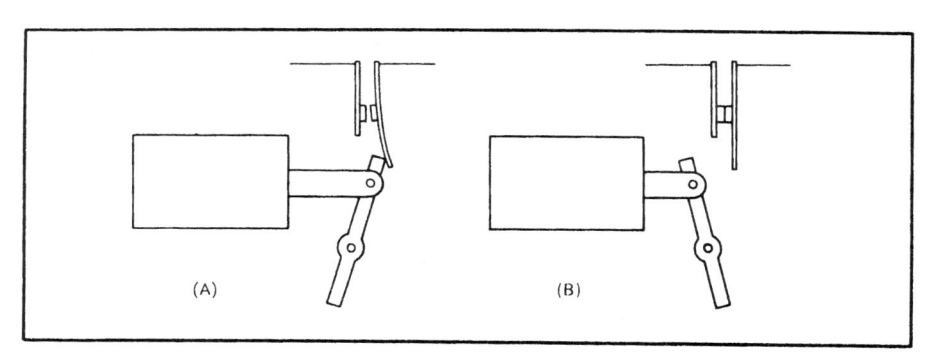

FIGURE 9–8 Solenoid with mechanical linkage and electrical contacts

Unit 9 RELAYS, SOLENOIDS, AND MOTORS

MOTOR PRINCIPLES

We have established that current-carrying conductors have a magnetic field surrounding them and that magnets have both attracting and opposing poles. It has also been established that magnetic fields are produced by both permanent and temporary magnets. To this point, attraction and repulsion of magnetic poles has been in a linear (straight) motion (Figure 9–9). However, if one magnet were attached to an immovable housing, and another magnet attached to a movable pivot, the mechanical motion of attraction and repulsion would be changed to an orbital (circular) motion (Figure 9–10). If both magnets were permanent magnets, the rotating shaft would come to rest at a point, in balance, limited by the fixed location of the magnetic poles. To obtain a full rotating motion, either the stationary magnet poles or the rotating magnet poles must change polarity. This is accomplished in most automotive motors by the use of electromagnets on the rotating shaft (Figure 9–11).

Remember that when a current is passed through a wire, a magnetic field is generated around that wire. Therefore, reversing the direction of current through the wire reverses the magnetic field. This effect produces an electromagnet capable of reversing its polarity.

To reverse the direction of current flow in a rotating wire (part of an armature), a set of brushes in contact with conductive strips of copper (Cu) called a *commutator* is used. For illustration, only one winding and one set of commutators are shown in Figure 9–12.

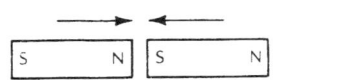

FIGURE 9–9 Linear attraction and repulsion of magnets

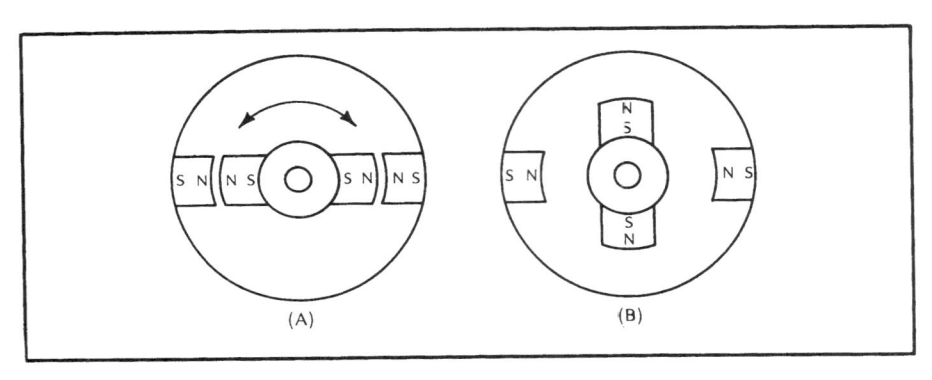

FIGURE 9–10 The fixed poles of the rotating magnets (a) come to rest at a balance between the fixed poles of the stationary magnets (b)

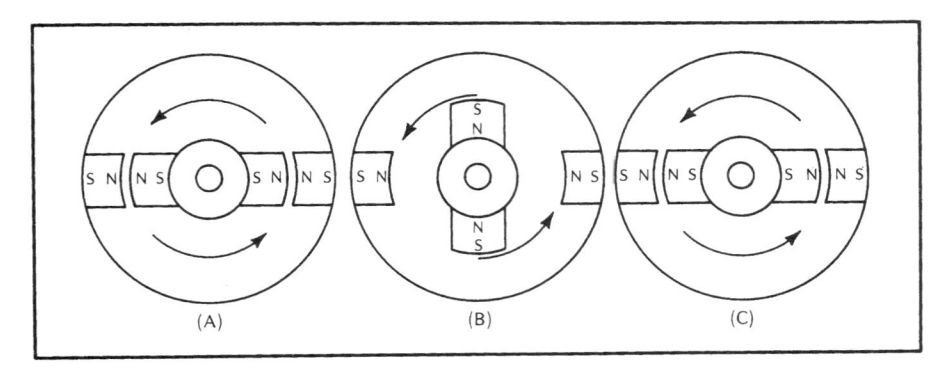

FIGURE 9–11 The changing poles of the rotating magnets by repulsion (a), attraction (b), and repulsion (c) provide a motor action

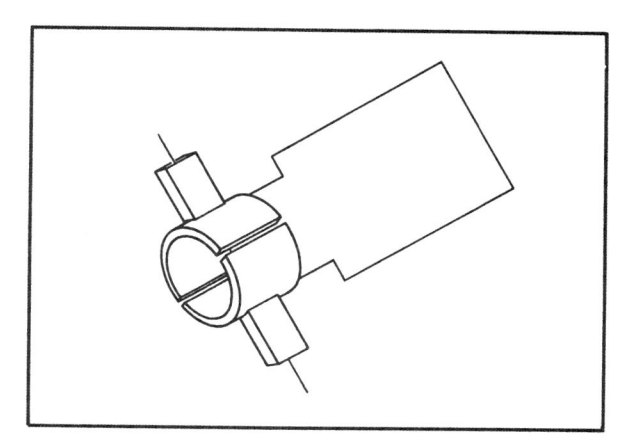

FIGURE 9–12 Brushes "ride" against commutator bars to provide current to rotating windings

Actually, there could be 8, 10, or 12 windings and sets of commutator bars. Regardless of the number of windings, however, there would be only one set of brushes.

In the simple one-winding motor, the armature would rotate because of the magnetic field of the stationary magnets, called pole shoes, and the magnetic field of the armature wire. The wire loop would rotate only one-quarter turn (90°) and stop. However, the commutator-brush arrangement would reverse the polarity of the loop, and the opposing loop segment would become attracted. Because of inertia and magnetic attraction, rotation would continue.

A one-loop motor would be very inefficient. It lacks torque characteristics essential for good motor action. The torque, or power, of a motor can be increased in three ways:

1. By increasing the number of segments in the armature.
2. By increasing the number of pole shoe pieces.
3. By adding field coils to the pole shoes.

Field coils are windings of many turns of fine copper (Cu) wire around the pole shoe pieces. Passing current through these windings produces electromagnets, thereby increasing motor torque and efficiency.

TORQUE

Torque is defined as a turning or twisting pressure exerted by a rotating shaft. Torque is measured in either foot-pounds (ft-lb) or the metric measure, newton meters (N · m). The torque, which causes an armature to turn, is proportional to the following:

1. The current in the coil.
2. The density of the magnetic flux.
3. The length of the conductor within the magnetic field.

It should not be necessary for an automotive technician to determine torque by calculation. The following information is given only for a basic understanding of the definition of the term, and is given in English measures only. The formula to determine the torque exerted by an armature shaft of a motor is:

$$\text{Torque (ft-lb)} = \frac{\Phi \times Z \times IA}{425 \times 10^6 \times m}$$

Where Φ = the flux passing through the armature (flux per in.^{2} × area per in.^{2})
Z = the number of wires in the armature
IA = the total armature current
m = the number of (parallel) paths through the armature

For example, consider a typical starter motor with an armature 5 inches long with 200 turns of wire and a magnetic flux of 28,000 lines per square inch (in.^{2}). Assume that total armature current is 175 amperes and that the pole pieces (Figure 9–13) are 20 square inches (20 in.^{2}). Using the formula,

Φ = 560,000 (28,000 × 20)
Z = 200
IA = 175
m = 2

becomes:

$$\frac{560,000 \times 200 \times 175}{425,000,000 \times 2} = \frac{19,600,000,000}{850,000,000}$$
$$= \frac{1960}{85} = 23,058$$

Some motors used for moving heavy loads have a high starting torque. Other motors used for light loads have a low starting torque. An example of a high-starting torque motor is the starter. A good example of a low-starting torque motor is the heater and air conditioner blower motor.

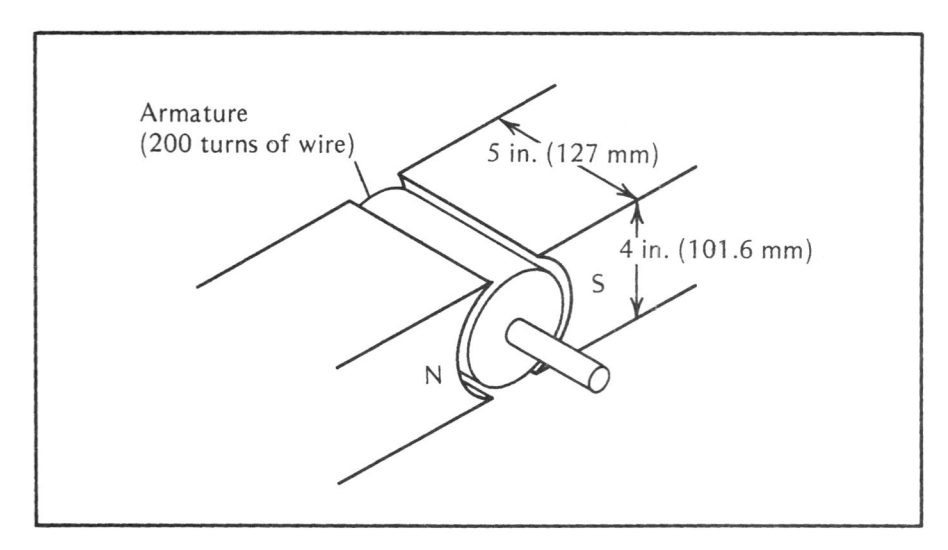

FIGURE 9–13 Armature between two pole pieces

STARTER MOTOR

The automotive starter is a motor with a special clutch-gear arrangement designed to engage the engine flywheel only during the time the starter motor is running. It is a dc motor of special design that turns at about 2000 revolutions per minute (rpm or r/min) under load (Figure 9–14). Because of its work load of turning a high-compression engine, it draws heavy current (100 to 500 amperes) and is designed for intermittent use only. Consequently, a starter motor should not be operated for more than 20 to 30 seconds at any one time. Because of its high current draw, heat buildup is rapid. Therefore, to allow time for this heat to dissipate, a

FIGURE 9–14 A typical (Ford) starter assembly (Courtesy of BET, Inc.)

pause of 1 to 2 minutes between operations is recommended. Also, a starter motor is designed to operate only under load. If it is allowed to operate with no load, its speed would be much greater than its design speed, which could damage the armature or other internal parts.

When the starter drive is engaged into the engine flywheel (Figure 9–15), the gear ratio is usually about 20 to 1 (20:1). This means that for every 20 revolutions of the starter, the engine makes 1 revolution. Since the engine, if properly tuned, starts in 2 to 3 seconds, the starter turns fewer than 100 revolutions during each starting period.

Properly cared for, a starter motor requires little service. But, of course, many drivers do not maintain their cars in tip-top shape; therefore starters, like other components, require service from time to time.

The starter motor operates on the same principle as any other dc motor. Along with our coverage of motor principles in this unit, the following section on starter motor circuits should be most helpful.

Starter Motor Circuits

The current entering most starters passes through both the field winding and the armature winding (Figure 9–16), both of which are made of heavy copper (Cu) ribbon, providing low resistance and high current flow. As discussed in Unit 2, the greater the current flow, the higher the torque that can be developed in a motor. Actually, the copper ribbon is only heavy-gauge wire that has been flattened until it is somewhat oval shaped.

The field coils of a starter motor are secured to the starter housing or case by pole shoes. Most starters have four field coils held in place with four pole shoes. Some starters, however, have four pole shoes with only two field coils. Regardless of the number of fields, the

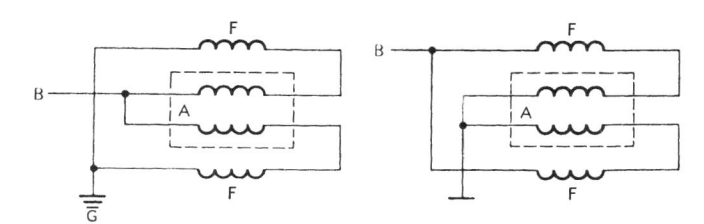

FIGURE 9–16 Battery current (B) passes through both the field (F) and armature (A) windings to ground (G)

pole shoes are arranged in a N-S-N-S sequence, as shown in Figure 9–17. Most starters have four brushes that ride against the commutator. Two of the brushes supply current to the armature, and the other two brushes provide the ground circuit. The series-parallel circuit starter also has four brushes, but the ground path is through the fields. Figure 9–18 illustrates some common field and armature circuits used in starters. Note that most pole shoes have a long tip on one side, as shown in Figure 9–19. This long tip should point in the direction of the armature rotation for the best magnetic field (flux). This is important when replacing the pole shoes after field coil repairs.

Checking field coils for opens and shorts can be accomplished using an ohmmeter (Figure 9–20). A low-resistance reading from the field coil to ground usually indicates a shorted field. A high-resistance reading across the field coil indicates an open field. For proper resistance values, manufacturers' specifications should be followed.

The armature of the starter motor (Figure 9–14) has a few turns of heavy-gauge copper ribbon attached

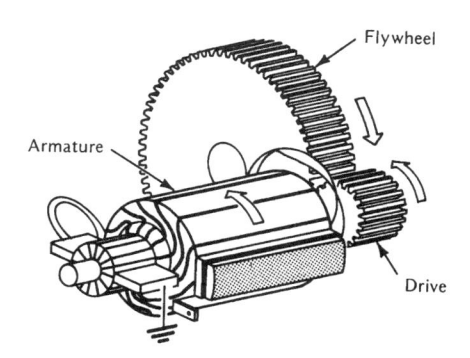

FIGURE 9–15 Illustration of gear ratio—starter drive to engine flywheel

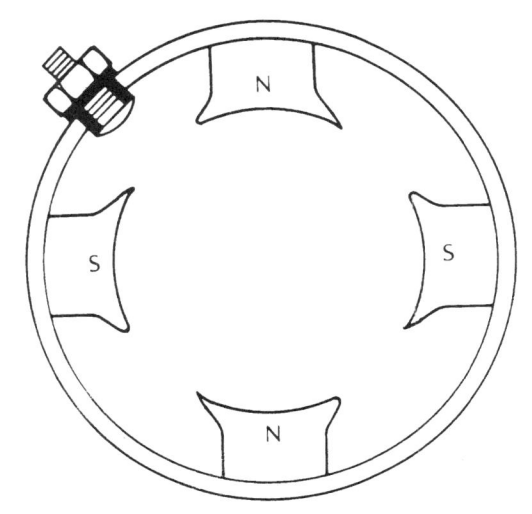

FIGURE 9–17 Starter housing with pole shoes showing a N-S-N-S arrangement

104 Unit 9 RELAYS, SOLENOIDS, AND MOTORS

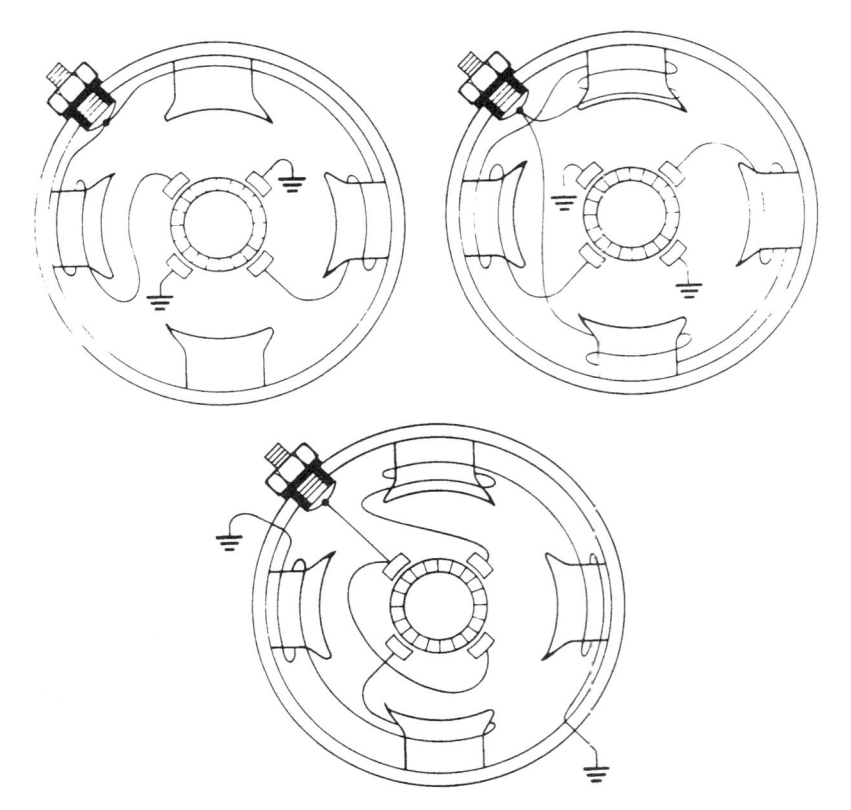

FIGURE 9-18 Some of the field and armature circuits found in a starter

to commutator bars. One end of the armature shaft is much longer than the other end to accommodate the drive assembly. Common problems of the armature include the following:

1. Out-of-round commutator.
2. Flat spots on commutator.
3. Shorted commutator segments (shorted to each other).
4. Open conductor at connection to commutator segment.
5. Physical damage caused by high-speed operation.

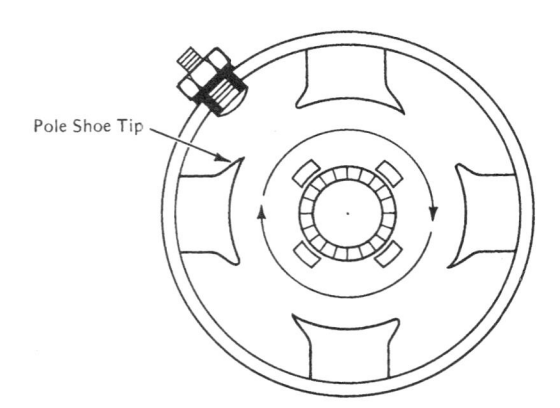

FIGURE 9-19 Extended pole shoe tip increases magnetic flux

FIGURE 9-20 Checking field coils for opens or shorts (Courtesy of BET, Inc.)

Shorted commutator segments can be cleaned with a special cutter called an undercutter. However, the undercutter is usually employed only after the commutator has been "trued" on a lathe or similar device. A growler can be used to check for opens or shorts, but an armature with an open or short usually must be replaced.

Starter Control Circuits

Because of the high current required by the starter, a magnetic switch, called a solenoid or solenoid switch, is used. This low-current circuit, when energized, closes a set of contacts to complete the starter motor high-current circuit (Figure 9–21).

Basically, there are two types of starter solenoid switches. Typical circuits for both types of starter solenoid switches are shown in Figure 9–22. One type (Figure 9–22(a)) mounts on the starter motor and provides the additional function of shifting the starter drive into the engine flywheel. The other type (Figure 9–22(b)) usually mounts on the fender well, remote from the starter motor. With this type, shifting the starter drive into the engine flywheel is accomplished by the inertia of the turning starter motor. The neutral safety switch is used on cars equipped with automatic transmissions. It prevents engine starting unless the transmission is in either neutral (N) or park (P) position.

Starter Drives

The starter drive is on the starter motor armature shaft end opposite the commutator end. It engages the starter motor into the engine flywheel only when the starter motor is turning. The starter drive, being a mechanical device, will not be covered in detail in this text.

Starter Motor Update

A new kind of starter motor is being introduced in the automotive industry. Although it is similar in shape and size to present starters, it is much different internally. The new motor uses permanent magnets in place of field coils.

Recent developments in the production of magnetic materials have provided magnetic pole pieces strong enough to function in starter motors. This is the current trend in automotive starter motor technology.

MOTOR CONTROL

The use of direct current (dc) motor control in the modern automobile is increasing with every model year. Some of the motor use is straightforward; that is, the motor is designed to consider power and gearing for output speed and torque, such as a starter motor. The motor is installed and connected to 12-volt dc power through switches and operates well within its requirements. It almost always operates at full speed. Other examples include window motors, seat position motors, antenna motors, and the like.

Other motors, such as the air conditioning blower motor and fuel pump motor, are designed for different speed operations, depending on the design demands of the system. For example, for maximum cooling, the air conditioning blower motor should operate at high speed. For minimum cooling, a slower speed is preferred. In the past, air conditioning blower motor speed was usually controlled by the use of resistors in series with the motor winding. The resistors consume power that was otherwise wasted. Wasted electrical power is not acceptable in the modern automobile, because of the heavy power demands placed on the alternator by other essential load devices.

There are two methods of controlling the speed of dc motors in the modern automobile:

1. Resistive (series resistance)
2. Electronic (pulse width modulation)

Of the two methods, electronic pulse width modulation provides the more efficient speed control.

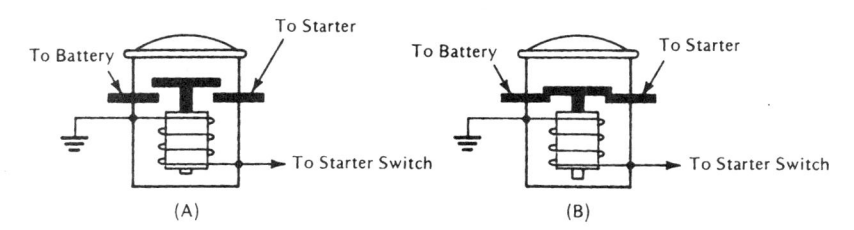

FIGURE 9–21 Starter switch: (a) open and (b) closed

Unit 9 RELAYS, SOLENOIDS, AND MOTORS

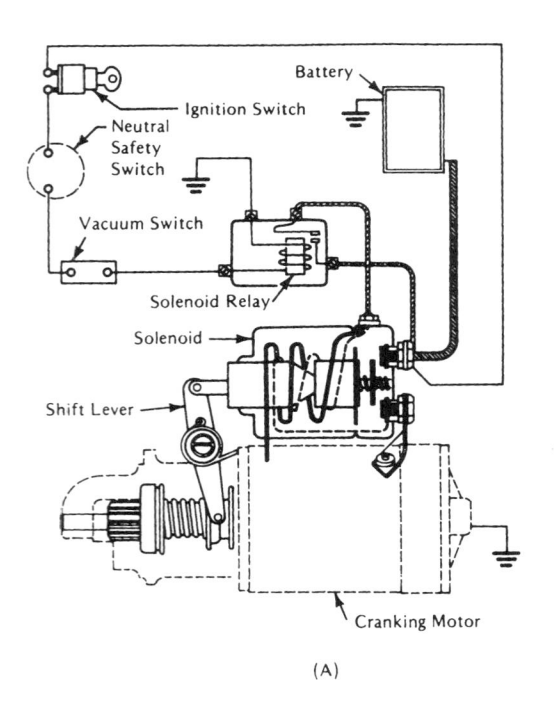

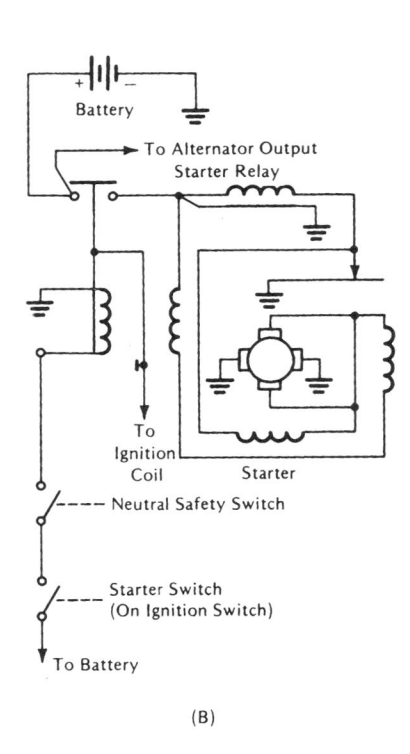

FIGURE 9-22 Typical starter switch circuits: (a) Starter mounted solenoid and (b) remote mounted solenoid

Series Resistive Method

With most dc motors the motor speed is directly related to the voltage applied to the motor terminals. In a 12-volt dc system, full battery voltage across the motor terminals provides maximum speed. If the motor draws, say, 3 amperes at 12 volts, a total of 36 watts provides power to the motor. At 6 volts the motor might draw 1.7 amperes, for a total power of 10.2 watts. When resistors are added in series with the motor, part of the supply voltage appears across the series resistor. The voltage across the motor and the current through the motor decrease. The motor slows down because less power is available to it.

The control shown in Figure 9-23(a) is a four-step motor speed control. The resistors might appear as in Figure 9-24.

Continuously variable speed control may be obtained using a rheostat, a variable resistor, as shown in Figure 9-23(b). A rheostat is shown in Figure 9-25. Rheostats, also called potentiometers or pots, were covered in Unit 3.

Pulse Width Modulation

Electronic speed control with pulse width modulation provides for continuously variable speed with very little power loss. Low electrical power loss, as indicated earlier, is a requirement in modern automobiles.

Effective speed control with pulse width modulation may be obtained at relatively low frequencies. The range from 200 Hz to 500 Hz operates well for this purpose. In this example 300 Hz is used as the operating frequency. It is not the frequency, but the width of the pulse, that determines motor speed. If the motor shown in Figure 9-26 is considered for speed control

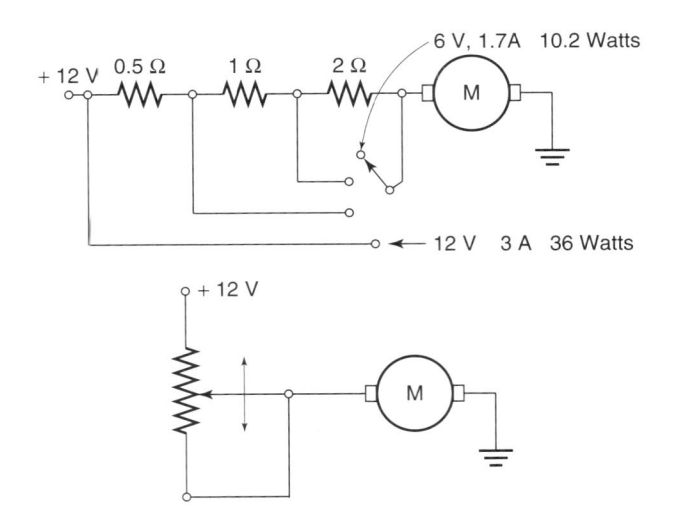

FIGURE 9-23 Motor speed control (resistive)

MOTOR CONTROL 107

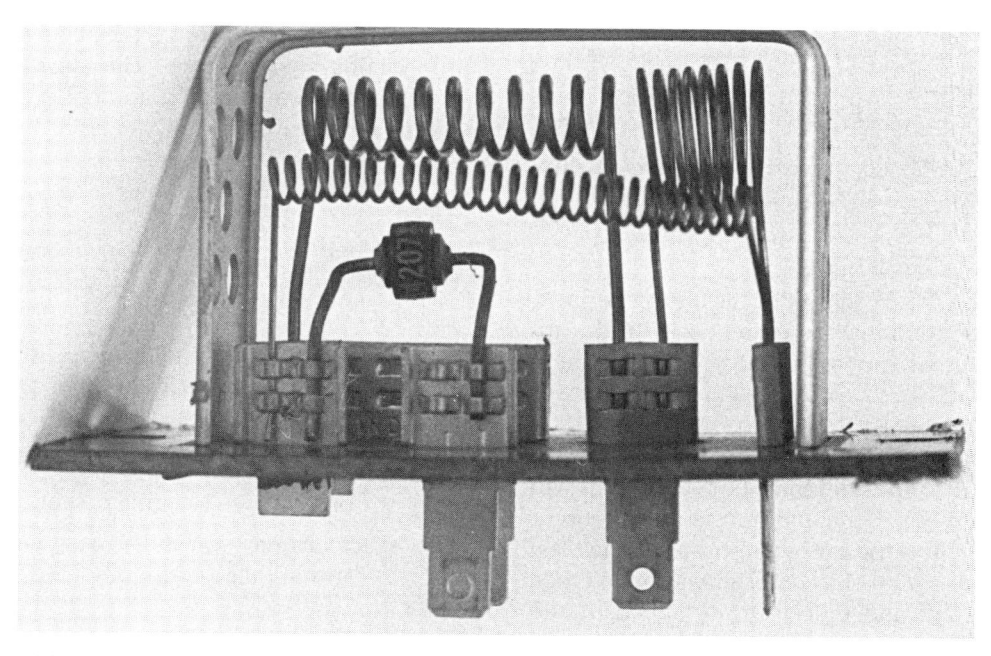

FIGURE 9–24 Step speed control (Courtesy of BET, Inc.)

with pulse width modulation, a comparison may be made.

First, consider the motor with 12 volts applied for one minute and the power off for one minute. The motor would run at full speed for one minute, then stop running for one minute. The power consumed for the first minute would be 36 watts. The power consumed during the second minute would be 0 watts. The average power for the two-minute period, then, would be 18 watts.

TOTAL POWER   36 + 0 = 36
AVERAGE POWER   36/2 = 18

The average speed of the motor over the two-minute period would theoretically be half the full speed. This is an example of pulse width modulation. A good example, but not very practical, since the motor would be running only half the time.

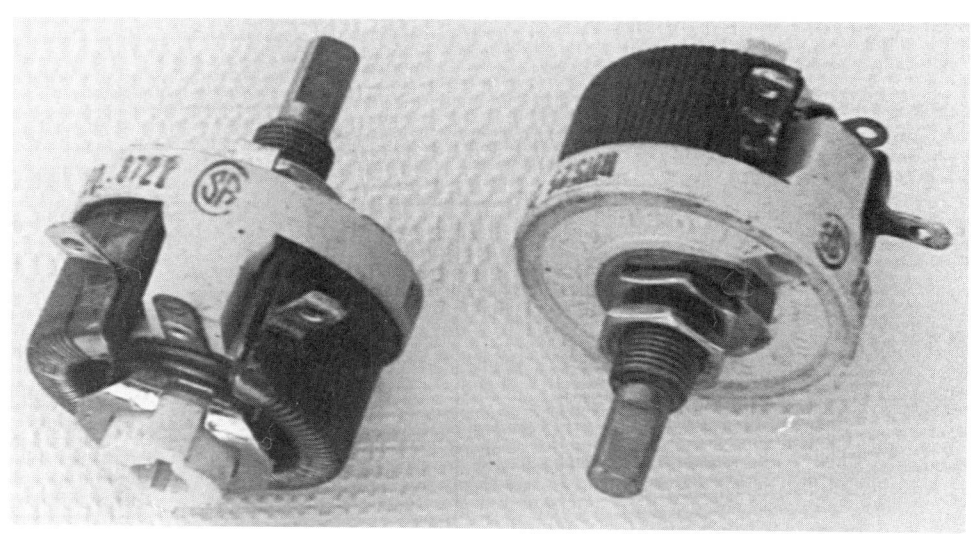

FIGURE 9–25 Rheostat speed control (Courtesy of BET, Inc.)

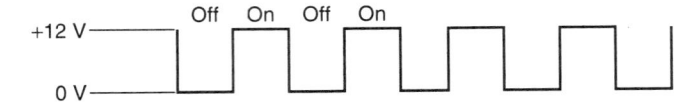

FIGURE 9–26 300-Hz square wave (50 percent duty cycle)

By increasing the frequency of the ON-OFF periods, a more stable operation may be obtained. With the application of voltage as shown in Figure 9–26 the same average power of 18 watts is obtained at a frequency of 300 Hz.

Since the motor is being turned ON and OFF 300 times each second, it does not have time to stop. It will run at a speed between 0 rpm and full speed—theoretically, at half speed. It is the pulse width versus the total wave period (duty cycle) that determines the average power to the motor, and therefore, its speed. If the width of the positive voltage pulse is changed, the motor speed will change. Variations in voltage pulse widths from very narrow to almost fully on may be obtained from a one-shot multivibrator, also referred to as a monostable multivibrator. In the automotive trade the one-shot multivibrator is also referred to as a voltage-to-duty-cycle converter. Multivibrators are covered in Unit 13, Semiconductor Circuits. An example of output voltage wave forms is shown in Figure 9–27.

If the voltage supplied to the motor is as in Figure 9–27(a), the motor will run very slowly. If the voltage supplied is as in Figure 9–27(b), the motor will operate at nearly full speed.

The output pulse width of a one-shot multivibrator is continuously variable from nearly zero to fully on. This ratio of ON time to total time is called the *duty cycle*. The duty cycle of the one-shot multivibrator or voltage-to-duty-cycle converter may be controlled with a dc analog input voltage. The speed of a motor can therefore be controlled by an analog dc voltage representing temperature or full demand, and so on.

SUMMARY

- Relays and solenoids are used in many applications in automotive systems.
- A diode in parallel with a coil provides a discharge path for back emf (bemf).
- Torque is the twisting power of a motor.
- Electronic speed control usually involves a form of pulse width modulation.

PRACTICAL EXERCISE

You will need a charged 12-volt battery, an ohmmeter, two jumper wires, and a DPDT relay for this exercise.

1. Record the readings of Steps 2, 4, 6, and 9 in the space provided in Figure 9–28.
2. Connect the ohmmeter to the relay coil and measure the resistance.
3. Disconnect the meter.
4. Connect the meter to the normally open (NO) contacts of the relay and measure the resistance.

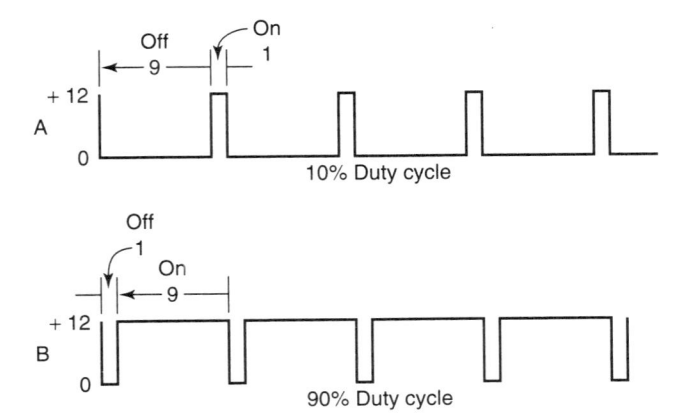

FIGURE 9–27 10 percent and 90 percent duty cycle waveforms

Step 2	Coil resistance _____ Ω
Step 4	N.O. Contacts Resistance _____ Ω
Step 6	N.O. Contacts Resistance _____ Ω Coil Energized
Repeat of Steps 4 and 6 N.C. Contacts	
Step 4	N.C. Contacts Resistance _____ Ω
Step 6	N.C. Contacts Resistance _____ Ω Coil Energized

FIGURE 9–28 Relay action

5. While the meter is connected to the NO relay contacts (Step 4), connect the relay coil to the battery.
6. Again, measure the resistance.
7. Carefully remove the jumper wires from the battery.
8. Disconnect the meter.
9. Repeat Steps 4 through 6 with the meter connected to the normally closed (NC) contacts of the relay.
10. Carefully disconnect the jumper wires from the battery.
11. Disconnect the meter.
12. Compare the readings of Steps 4 and 6 with the reading of Step 9.

REVIEW

1. The starter is a specially designed (ac) (dc) motor for (intermittent) (continuous) use.
2. The starter motor turns at about (100) (2000) rpm to turn the engine at about (100) (2000) rpm.
3. Most starters have (two) (four) pole pieces arranged in a (N-S) (N-S-N-S) sequence.
4. A(n) (growler) (undercutter) may be used to repair (opens) (shorts) of an armature commutator.
5. A series-parallel starter with (two) (four) fields is grounded through the (fields) (brushes).
6. The strips of copper (Cu) on which the motor brushes make contact are called _____.
7. The rotating portion of a motor is called the _____.
8. Torque is defined as _____.
9. When current is passed through a wire, a _____ is generated around that wire.
10. What is the torque, in ft-lb, of a motor with an armature 2 in. (50.8 mm) long with 100 turns of wire and a magnetic flux of 22,000 lines per square inch (6.45 cm^2)? Assume that the total armature current is 12 amperes and that the pole pieces are 3 in.^{2} (19.36 cm^2).

(T) (F) 11. Increasing spring tension of a relay may cause chattering.

(T) (F) 12. Many mechanical solenoids are being replaced in late-model cars with solid-state devices.

(T) (F) 13. A relay is considered an electromechanical switch.

(T) (F) 14. Solenoid cores are made of soft iron (Fe).

(T) (F) 15. Solenoids perform a mechanical action.

16. Increasing spring tension of a relay
    a. results in insufficient armature movement.
    b. results in no armature movement.
    c. Either a or b may be correct.
    d. Neither a nor b is correct; increasing spring tension improves armature movement.

17. A bar magnet is an example of
    a. a permanent magnet.
    b. a temporary magnet.
    c. an electromagnet.
    d. Any of the above may be correct.

18. The starter motor should be operated no longer than _____ at any one time.
    a. 10–20 seconds   c. 20–30 seconds
    b. 15–25 seconds   d. 25–35 seconds

19. What is the total power of a 12-volt motor that draws 3 amperes?
    a. 12/3 = 4 watts   c. 12 × 3 = 36 watts
    b. 12 + 3 = 15 watts   d. 12 − 3 = 9 watts

20. Motor speed control may be obtained by using
    a. a rheostat   c. resistors
    b. electronic speed control   d. Any of these is correct.

◆ UNIT 10 ◆

ELECTRIC POWER AND ENERGY

OBJECTIVES

On completion of this unit you will be able to:

- Understand that *power* is the rate (speed) of doing work, whereas *energy* relates to the amount of work done.
- Explain that electrical power, measured in watts, is given by the formula $W = I \times E$.
- Understand that electrical energy is determined by the additional factor of time; energy $= I \times E \times T$

ELECTRIC POWER

Power (P) is the measure of the rate of doing work. Mechanical power is measured in horsepower (hp). Electrical power is measured in watts (W). Mechanically, 550 ft-lb per second equals one horsepower (hp). Electrically, 746 watts equals one horsepower (hp). In the electrical system current (I) times voltage (E) equals power (P) in watts (W).

$$W = I \times E$$

A simple power-consuming device is a resistor with current flowing through it. Consider Figure 10–1, in which a 3-Ω resistor is connected between a 12-volt power source (battery) and ground.

According to Ohm's law, four amperes of current is flowing through the resistor. The power developed in the resistor is determined using the power formula as shown in Figure 10–1.

Another good example of a power-consuming device is an electric toaster. In Figure 10–2 an electric toaster is shown connected to a 120-volt electrical source. Ten amperes of current is being drawn from the power source. The power developed in the toaster may be determined by the formula

$$\text{watts} = \text{amperes} \times \text{volts}$$
$$W = I \times E$$
$$W = 10 \times 120$$
$$W = 1{,}200 \text{ watts or } 1.2 \text{ kW}$$

Another formula is sometimes used to determine power in electrical circuits:

$$W = I^2 \times R$$

In the form of a statement, the formula says that power (W) is equal to amperes squared (I^2) times resistance

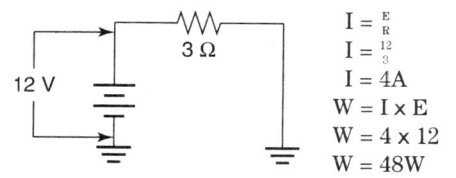

$$I = \frac{E}{R}$$
$$I = \frac{12}{3}$$
$$I = 4\text{A}$$
$$W = I \times E$$
$$W = 4 \times 12$$
$$W = 48\text{W}$$

FIGURE 10–1 Simple power-consuming circuit

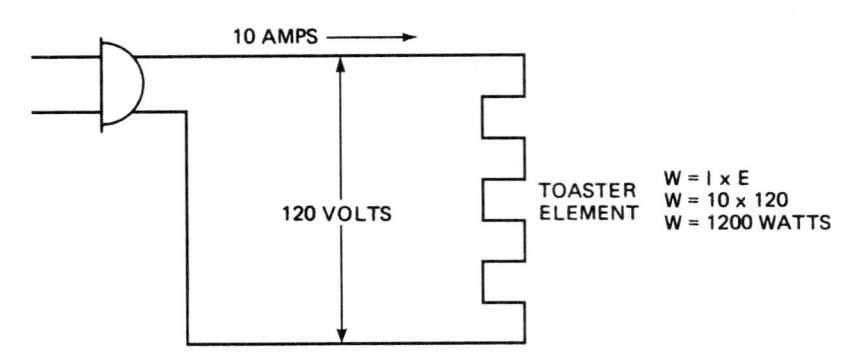

FIGURE 10-2 Toaster connected to a 120-volt source

(R). Another way of stating the formula is that power (W) is equal to amperes (I) times amperes (I) times resistance (R).

$$W = I \times I \times R$$

Since in Ohm's law $E = I \times R$, the factors $I \times R$ can be substituted for E in the original power formula:

$$W = I \times E$$

Substituting $(I \times R)$ for E:

$$W = I \times (I \times R)$$
$$W = I \times I \times R$$
$$W = I^2 \times R$$

In the toaster example, the resistance of the toaster heating element, according to Ohm's law, is

$$R = E/I$$
$$R = 120/10$$
$$R = 12 \, \Omega$$

When this factor is used in the formula $W = I^2 \times R$,

$$W = 10 \times 10 \times 12$$
$$W = 100 \times 12$$
$$W = 1{,}200 \text{ W}$$

This same answer was obtained using the formula $W = I \times E$.

It may be necessary at times to calculate power without information about the current. For example, consider a 30-Ω resistor with 90 volts applied to it. How much power will be developed in the resistor?

$$W = I \times E$$
$$I = E/R$$
$$W = 90/30 \times 90$$
$$W = 3 \times 90 = 270 \text{ W}$$

The problem could also have been solved using the following method:

$$W = I \times E$$
$$I = E/R$$
$$W = E/R \times E$$
$$W = E^2/R$$
$$W = 90 \times 90/30$$
$$W = 8{,}100/30$$
$$W = 270$$
$$W = E^2/R$$

This expression shows that the power developed in the circuit is equal to the voltage squared, divided by the resistance. The three power formulas are:

$$W = I \times E$$
$$W = I^2 \times R$$
$$W = E^2/R$$

HORSEPOWER

Electric power is required to operate electric motors. In a motor, the desired mechanical output is rotation of the motor shaft. Some electrical power is lost in the motor because of I^2R power loss in the resistance of the windings. An electric motor is not 100-percent efficient.

Consider the electric motor used to start an engine. How much power is needed to turn the engine over?

Starting current is measured at 312 amperes. Battery voltage is measured during start at 6.9 volts. Remember voltage drop across the battery's internal resistance, as covered in Unit 7 (Figure 10–3).

$$W = I \times E$$
$$W = 312 \times 6.9$$
$$W = 2152.8 \text{ W}$$

If the starter motor were 60-percent efficient, 861 watts would be converted into heat in the motor. Sixty percent efficiency means 40 percent loss:

$$2152.8 \times 40\% = 861 \text{ watts}$$

This is almost the same as the power developed in the toaster (Figure 10–2). This is a considerable amount of heat. That is why the starter should not be engaged for long periods of time, nor should it be engaged repeatedly without allowing time for cooling off.

With 861 watts lost in the starter, 1291 watts are available to turn the engine. That is equivalent to 1.73 hp.

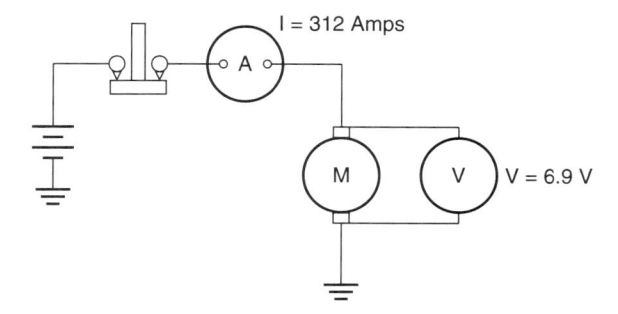

FIGURE 10–3 Battery voltage and current during engine start

$$\text{hp} = \text{Watts}/746$$
$$\text{hp} = 1291/746$$
$$\text{hp} = 1.73$$

As an aid in remembering the formula for watts, recall that Columbus discovered America in 1492; 1492 divided by 2 equals 746:

$$746 \text{ watts} = 1 \text{ hp}$$

If the no-load voltage of the battery is known the internal resistance of the battery can be determined. Assuming 12.8 volts no load:

$$12.8 - 6.9 = 5.9 \text{ volts (dropped across internal } R)$$
$$R = E/I$$
$$R = 5.9/312$$
$$R = 0.189 \text{ }\Omega$$

ELECTRIC ENERGY

Electric power was defined earlier as a measure of the rate of doing work. Electric *energy* refers to the total amount of work done. Stated a different way, the rate of doing work (power) is multiplied by time.

$$\text{Energy} = \text{rate} \times \text{time}$$

The power company charges its customers for the total amount of electrical energy it supplies. Since one watt-hour (Wh) is a very small unit of energy, the power company charges per kilowatt hour (kWh). One kilowatt hour, for example, is equal to:

$$1000 \text{ watts for one hour}$$
$$1000 \times 1 = 1000 \text{ Wh (1 kWh)}$$
or
$$500 \text{ watts for 2 hours}$$
$$500 \times 2 = 1000 \text{ Wh (1 kWh)}$$
or
$$100 \text{ watts for 10 hours}$$
$$100 \times 10 = 1000 \text{ Wh (1 kWh)}$$

"Kilo" (k), then, means thousand. The power company bases its charge for providing electrical energy on its cost of production plus a small profit. The cost varies by geographic location and is obtained only from your local company. The following example, based on $0.06 per kWh, illustrates how the cost of operation may be determined when power consumption is known.

EXAMPLE 1

An electric power company charges its customers 6 cents for one kilowatt hour (kWh) of energy. How much does it cost the customer to heat for 8 hours if the heat is obtained from a space heater of 22 Ω connected to a 220-volt source (Figure 10–4)?

Solution

Solve for current:

$$I = E/R$$
$$I = 220/22$$
$$I = 10 \text{ amps}$$

Solve for power:

$$W = I \times E$$
$$W = 10 \times 220$$
$$W = 2200 \text{ watts}$$

Solve for energy:

$$Wh = W \times time$$
$$Wh = 2200 \times 8$$
$$Wh = 17600 \text{ watt-hours or } 17.6 \text{ kWh}$$

Solve for cost at 6 cents per kWh:

$$Cost = kW \times 0.06$$
$$Cost = 17.6 \times 0.06$$
$$Cost = \$1.056 \, (\$1.06)$$

MEASURING ELECTRIC POWER

One method of determining electric power in a circuit is to measure the voltage across the circuit and the current through the circuit. The power developed is found by the formula:

$$W = I \times E$$

In Figure 10–5, the circuit of a lamp is shown. The voltage across the lamp is 13.5 volts as indicated by the voltmeter. The ammeter indicates 9.6 amperes.

$$W = I \times E$$
$$W = 9.6 \times 13.5$$
$$W = 129.6 \text{ watts}$$

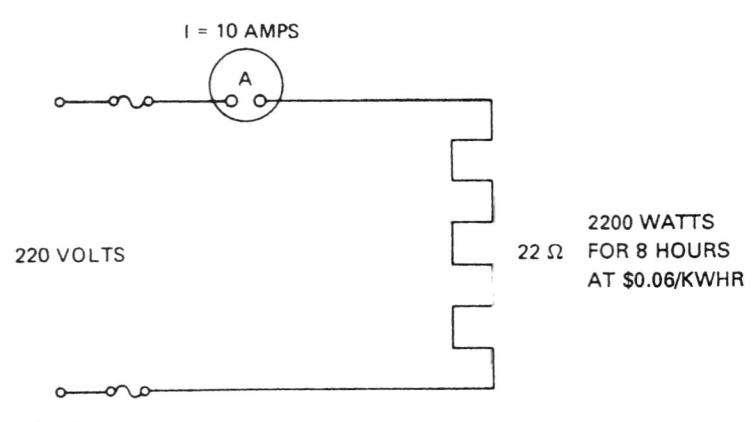

FIGURE 10–4 Typical space heater circuit

9.6 A

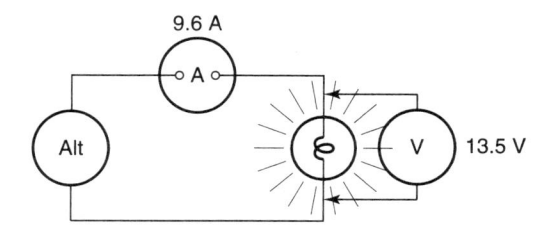

FIGURE 10–5 Typical automotive lamp circuit

SUMMARY

- Power in Watts = $I \times E$
- Energy = Power × Time
- Energy in Watt/Hour = $I \times E \times$ Time (hr)

PRACTICAL EXERCISE

A standard incandescent bulb, an automotive tail-lamp bulb, three jumper wires, and a multimeter are needed for this exercise.

1. Use the ohmmeter to determine the resistance of the incandescent bulb. The resistance is _____

2. Assuming the average household voltage of 115 volts, calculate the current that would be consumed by the bulb, considering the resistance found in Step 1. _____

    $I = V/R = 115/R =$ _____ ohms

3. Calculate the power that would be consumed by the bulb at this resistance, by multiplying the voltage times the current (Step 2).

    _____

    $I = I \times E =$ _____ $\times$ 115 V = _____ watts

4. Was the power calculated in Step 3 higher or lower than the wattage indicated on the bulb? _____ NOTE: The hot resistance of the bulb is much higher than the cold resistance when measured with the ohmmeter. Make no attempt to measure the bulb's hot resistance with power applied.

5. Measure and record the resistance of a tail-lamp bulb using an ohmmeter. The resistance is _____ ohms.

6. Using the 15-volt scale of the multimeter, measure the battery voltage. The battery voltage is _____ volts.

7. Calculate the current that the bulb will draw at battery voltage (Step 6).

8. Select an ampere scale on the multimeter that is higher than that calculated in Step 7.

9. Connect the ammeter in series with the battery and tail-lamp bulb. The measured current draw is _____ amperes.

10. Was the current measured in Step 9 higher or lower than the current calculated in Step 7? _____ NOTE: As the filament of the bulb becomes hot, its resistance rises rapidly. Do not attempt to measure resistance of the bulb with power applied.

REVIEW

1. The formula for power using voltage and current is W = _____.

2. The formula for power using current and resistance is W = _____.

3. A resistor draws 2.2 amperes of current at 13.7 volts. How much power developed in the resistor?

4. If a lamp of 86 watts is operated for 100 hours, how much electrical energy will be converted to light and heat?

5. A power company is selling electric energy at $0.08 per kW/h. How much will it cost to operate a 10,000-watt heater for 60 hours?

6. A certain starter motor draws 289 amperes with a motor terminal voltage of 7.2 volts. How much power is being fed to the motor?

7. If the no-load voltage of the battery in Problem 6 is 12.9 volts, how much voltage is dropped across the battery internal resistance?

8. Determine the internal resistance of the battery of Problems 6 and 7.

9. The voltage at the motor terminals will be (higher) (lower) if a smaller wire is used to connect the motor to the battery. Why? _____.

10. Why is it necessary to have good sound connections throughout the circuit from the battery to the starter? _____.

◆ UNIT 11 ◆

WIRING CIRCUITS

OBJECTIVES

On completion of this unit you will be able to:

- Trace automotive wiring circuits using a schematic.
- Understand the use of a schematic wiring diagram.
- Interpret schematic diagrams in relation to actual electrical-electronic systems and subsystems.
- Recognize and understand electrical and electronic symbols.
- Determine color of wires by manufacturer's code abbreviations.

This unit will cover automotive wiring circuits, schematics, schematic interpretation, symbols, wire color coding, and wiring harnesses. *Schematics* are diagrams that show the technician where unseen wires go, connect, and emerge.

SCHEMATICS

A typical automotive wiring schematic, containing about 1600 feet (488 meters) of wire in about 50 harnesses with more than 500 terminals, covers a page measuring about 18 inches by 36 inches (457 mm by 914 mm). Even at this size, the drawn wires are spaced only 0.125 inch (3.2 mm) apart. To reduce a schematic of this size to fit a page of this text would place the wires 0.03 inch (0.76 mm) apart. Such a schematic would be illegible. The purpose of this unit, then, is to display not a total electrical schematic, but the parts of a schematic, and to discuss schematic symbols and their interpretation related to typical actual schematics.

Schematics are used to follow (trace) electrical circuits. For example, when a wire enters a harness, where does it exit? If it is yellow where it enters, what is its color where it exits? Does the wire go directly from point A to point B, or does it enter a branch circuit and also go to point C? Is it a single- or double-controlled circuit? How many connectors are there, and where are they located? Schematics tell us all these things and more.

To a beginner, schematics appear difficult to read, and for those who do not understand schematic symbols and codes, they may be. But for those who have mastered the symbols, a schematic is a simple, clear road map. For example, the section of a schematic shown in Figure 11–1 means nothing to a layperson, but the experienced electrical technician recognizes that it shows part of a Chrysler circuit and is read as follows: Wire A is an accessory bus bar feed—12 gauge, red. Wire B is an exterior lighting circuit—12 gauge, black. Wire C is a gauge/warning lamp circuit—20 gauge, black with white stripe.

Schematic Symbols

Consider the circuit schematic shown in Figure 11–2. As you study the electrical schematic symbols (and

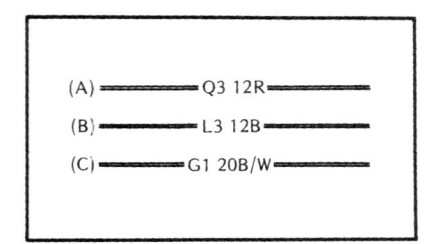

FIGURE 11–1 Wiring codes in a schematic

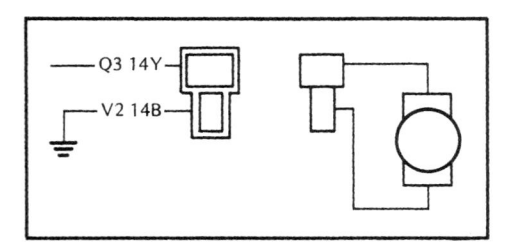

FIGURE 11–2 Portion of a typical schematic

FIGURE 11–3 Some of the schematic symbols used to depict a lamp

look ahead to the color codes in Figure 11–5), see if you can identify the circuit. If you noted the symbols for ground, male/female disconnect, and motor, you are correct. Assuming a Chrysler circuit, the coded symbol "V" denotes a windshield wiper and washer circuit. But since this is a single-speed motor, the circuit is probably for a windshield washer motor.

There is no standard verbal description for many electrical devices. For example, a bulb may also be called a lamp or a light. Symbolic representations also vary, as shown in Figure 11–3. There are single- and double-filament bulbs, as shown in Figure 11–4. Some common symbols used in automotive electrical diagrams and schematics follow. You may wish to index these pages with a paper clip or tab for easy reference throughout the remainder of this text.

TABLE 11–1 Abbreviations

A	Ampere	POS	Positive
ac	Alternating current	PRES	Pressure
ACC	Accessory	SOL	Solenoid
BAT	Battery	SPDT	Single-pole double-throw
C/B	Circuit breaker	SPST	Single-pole single-throw
dc	Direct current	TEMP	Temperature
DPDT	Double-pole double-throw	TOG	Toggle (switch)
MOM	Momentary	V	Volt
MOT	Motor	W	Watt
(n)	None	−	Negative
NC	(nc) Normally closed	Ω	Ohm
NEG	Negative	+	Positive
NO	(no) Normally open	±	Plus or minus
PB	Push button	%	Percent

TOGGLE SWITCHES	SPST (nc)	SPST (no)	DPDT (on-on)	DPDT (on-off-on)
	SPDT (on-on)	SPDT (on-off-on)		

PUSH BUTTON SWITCHES (MOMENTARY)	Normally Closed	Normally Open	SPDT	DPDT

PRESSURE ACTUATED SWITCHES	Open On Fall	Open On Rise	SPDT	DPDT

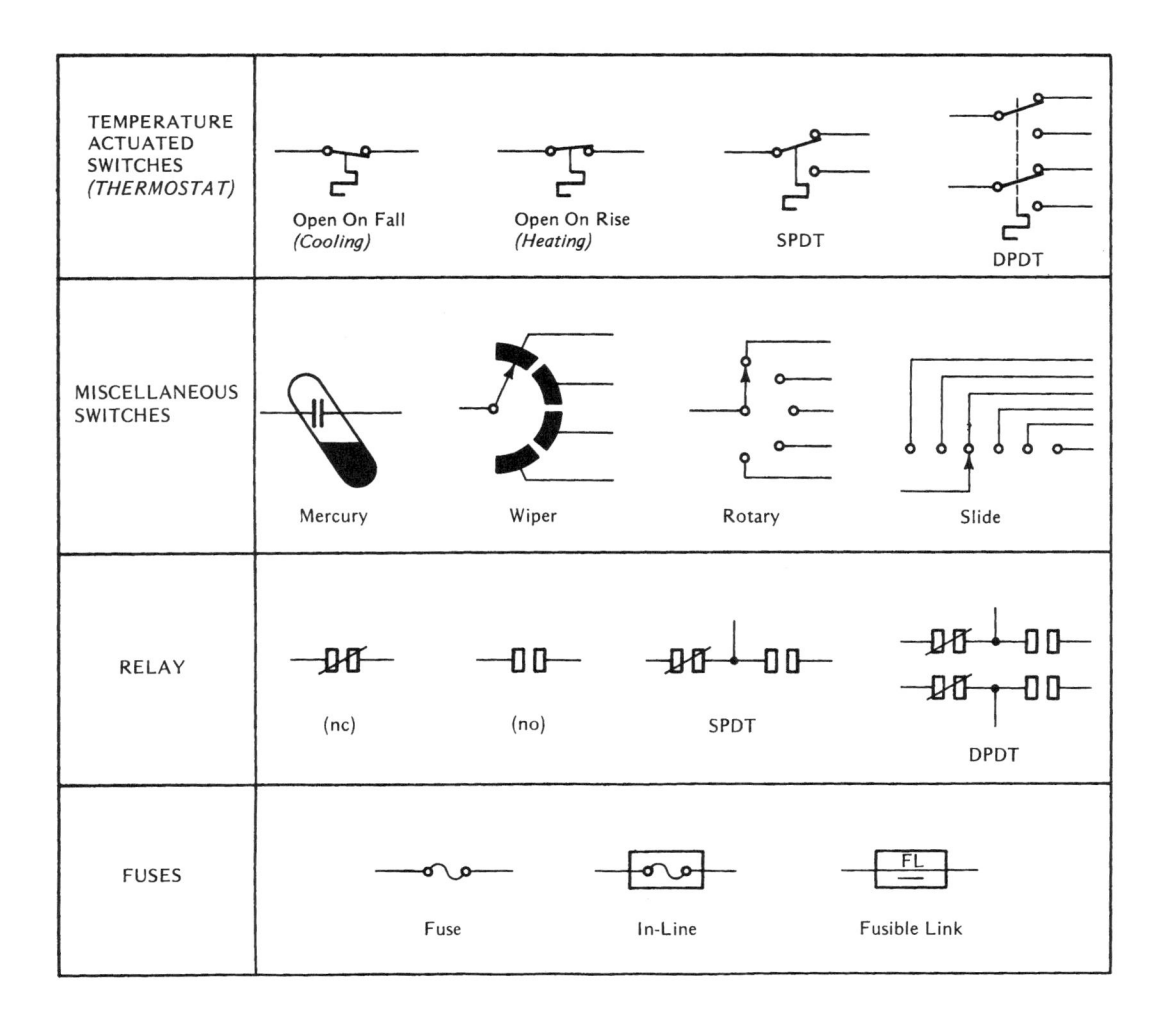

TEMPERATURE ACTUATED SWITCHES (THERMOSTAT)	Open On Fall (Cooling)	Open On Rise (Heating)	SPDT	DPDT

MISCELLANEOUS SWITCHES	Mercury	Wiper	Rotary	Slide

RELAY	(nc)	(no)	SPDT	DPDT

FUSES	Fuse	In-Line	Fusible Link

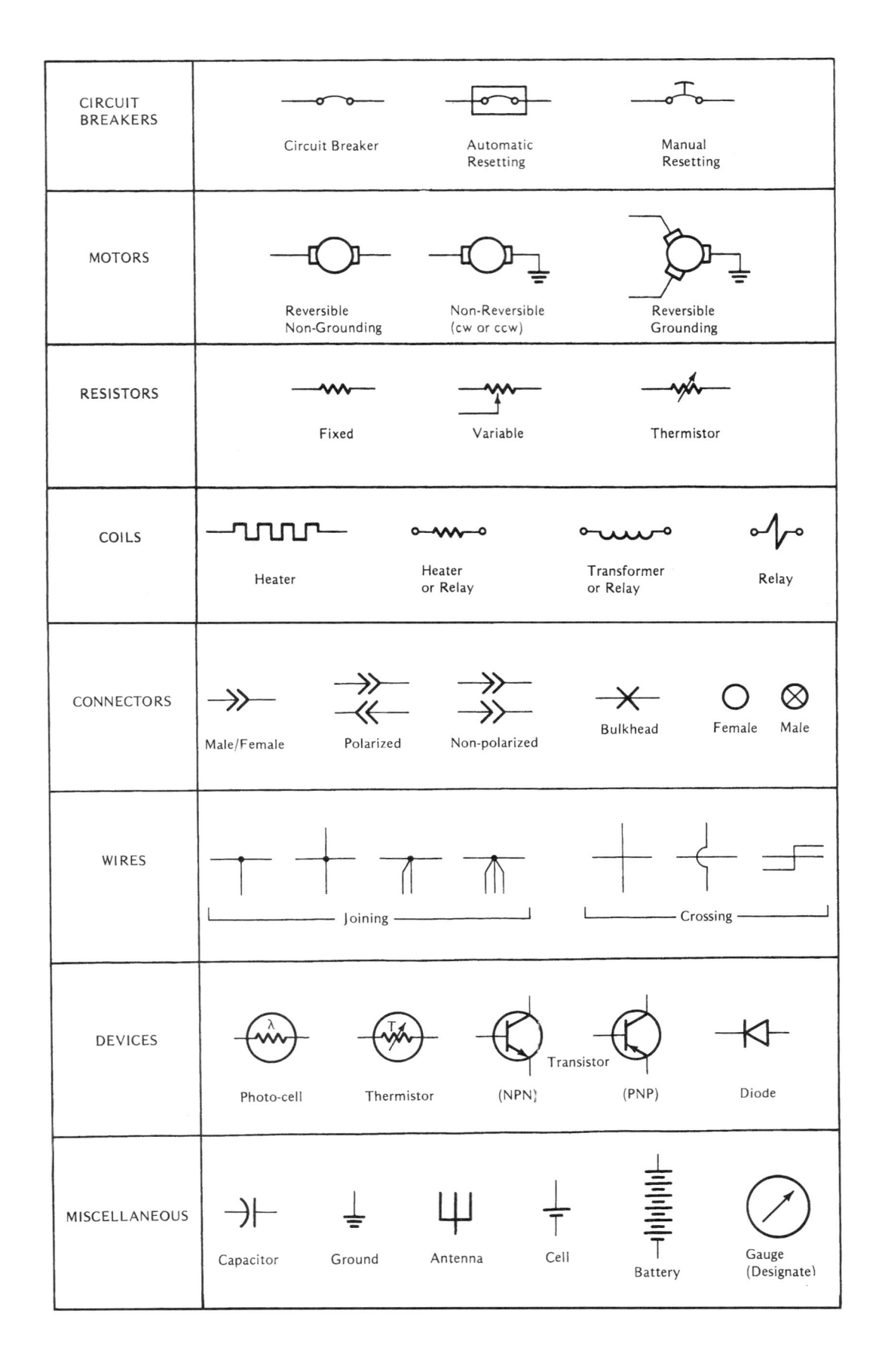

CIRCUIT BREAKERS	Circuit Breaker	Automatic Resetting	Manual Resetting	
MOTORS	Reversible Non-Grounding	Non-Reversible (cw or ccw)	Reversible Grounding	
RESISTORS	Fixed	Variable	Thermistor	
COILS	Heater	Heater or Relay	Transformer or Relay	Relay
CONNECTORS	Male/Female	Polarized	Non-polarized	Bulkhead — Female — Male
WIRES	Joining		Crossing	
DEVICES	Photo-cell	Thermistor	Transistor (NPN) — (PNP)	Diode
MISCELLANEOUS	Capacitor	Ground	Antenna — Cell	Battery — Gauge (Designate)

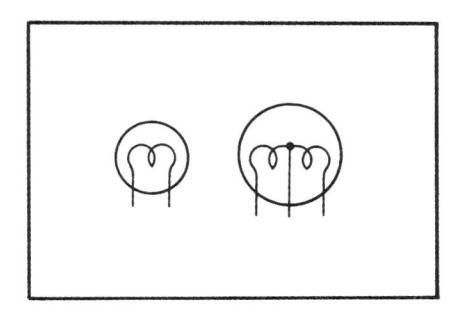

FIGURE 11–4 Typical symbols for single- and double-filament lamps

Color Coding

The color coding of wires provides quick and easy identification of a wire being traced in a harness. In schematic diagrams, the color codes are abbreviated in a number of ways, as shown in Figure 11–5. Wires are also coded in two colors by the use of striping, hash marks, or dots as shown in Figure 11–6, producing

Color	Abbreviation		
Aluminum	AL		
Black	BLK	BK	B
Blue (Dark)	BLU DK	DB	DK BLU
Blue (Light)	BLU LT	LB	LT BLU
Brown	BRN	BR	BN
Glazed	GLZ	GL	
Gray	GRA	GR	G
Green (Dark)	GRN DK	DG	DK GRN
Green (Light)	GRN LT	LG	LT GRN
Maroon	MAR	M	
Natural	NAT	N	
Orange	ORN	ORG	O
Pink	PNK	PK	P
Purple	PPL	PR	
Red	RED	RD	R
Tan	TAN	TN	T
Violet	VLT	V	
White	WHT	WH	W
Yellow	YEL	YL	Y

FIGURE 11–5 Typical color codes used to identify automotive wires

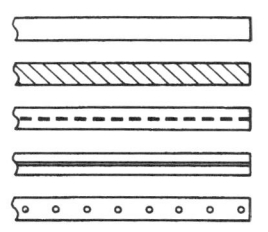

FIGURE 11–6 Some of the methods used to color code wires

more than 10,000 color combinations. The predominating color (base) is always given first; the stripe or dot color is given second. For example, a red wire with a white stripe may be designated RED/WHT, RD/WH, R/W, RED/W, or RD/W (that is, it may be any combination of the red and white abbreviations given in Figure 11–5).

Schematic Interpretation

Complete automotive electrical schematics are usually shown on more than one page. Therefore, to illustrate schematic interpretation, we will use one page of a multipage electrical schematic—page 16 of a 74-page Chrysler wiring diagram. This page shows the circuit for a headlamp switch. (The "See pg—" notes on the schematic refer to the pages of the wiring diagrams, not the pages of this text.)

Automotive schematics are usually shown with all switches in the open (OFF) position, and all components in their normal (deenergized) positions. However, for illustration, the headlamp switch schematic will be shown three times, each time with the headlamp switch in a different position: OFF (Figure 11–7), PARK (Figure 11–8), and ON (Figure 11–9).

Headlamp Switch OFF. Figure 11–7 shows the schematic with the headlamp switch in the OFF position, which is the standard electrical schematic representation for this circuit. Study the schematic briefly and trace the battery feed circuit. Note in the upper-left corner the wire labeled FM Charging System (Fm Chg Sys). This R6 wire comes through several junctions from the battery. It is a 12-gauge black wire that becomes L1 after the fuse block junction to feed the headlamp switch. This particular circuit does not pass through the fuse block. Headlamp circuit protection is provided by an internal circuit breaker in the headlamp switch, as noted in the schematic symbol.

Next, locate wire Q3 at the R6–L1 junction. This 12-gauge red wire feeds a four-circuit bus bar in the

122 Unit 11 WIRING CIRCUITS

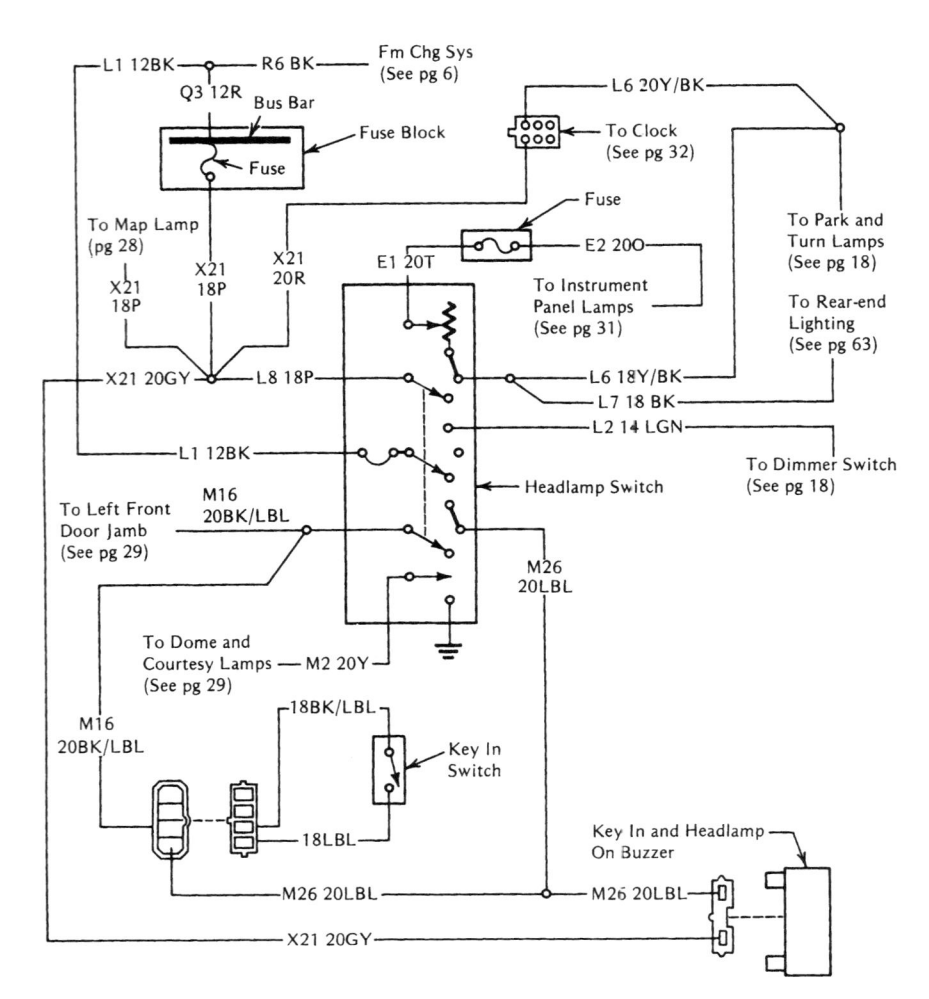

FIGURE 11–7 Headlamp circuit with switch in the OFF position

fuse block. Coming out of the fuse block, this circuit becomes X21 (18-gauge pink) and then becomes L8 (18-gauge pink) at the junction that feeds the parking lamp and instrument panel lamp circuits. Also at the X21–L8 junction, follow X21 (20-gauge gray) through the key in and headlamp on buzzer and key in switch to the headlamp switch. This is the reminder circuit to warn of exiting the car with the headlamps turned on or the key in the ignition switch. Note that X21 (20-gauge gray) becomes M26 (20-gauge light blue) after the buzzer, 18 LBL (18-gauge light blue) through the key in switch, and M 16 (20-gauge black with light blue tracer) after the connector. It is important to read color codes, wire sizes, and notes when tracing an electrical system on a schematic. In spite of the high cost of color printing, a few manufacturers provide full-color schematics.

Headlamp Switch in PARK. Figure 11–8 shows the headlamp switch in the PARK position. As in the OFF position, L1 feeds only the headlamp circuit. Since this switch section is still in a "vacant" position, the headlamp circuit is still not complete. However, the park lamp feed, L8, now completes several circuits: L6, L7, and E1.

Circuit L6 (18-gauge yellow with black tracer) feeds park and turn (side marker) lamps. (This circuit continues on other schematic pages, as applicable.) Circuit L7 (18-gauge black) feeds the rear-end lighting circuits: tail and rear side marker lamps. (This circuit also continues on other pages of the schematic.) Circuit E1 (20-gauge tan) is fed through the rheostat of the head-lamp switch. After the fuse, this circuit becomes E2 (20-gauge orange) to feed the instrument panel lamps.

Circuit M16 is complete through the headlamp switch to the buzzer on M26 (20-gauge light blue). The circuit of M16 also continues on another page to the left front door jamb switch.

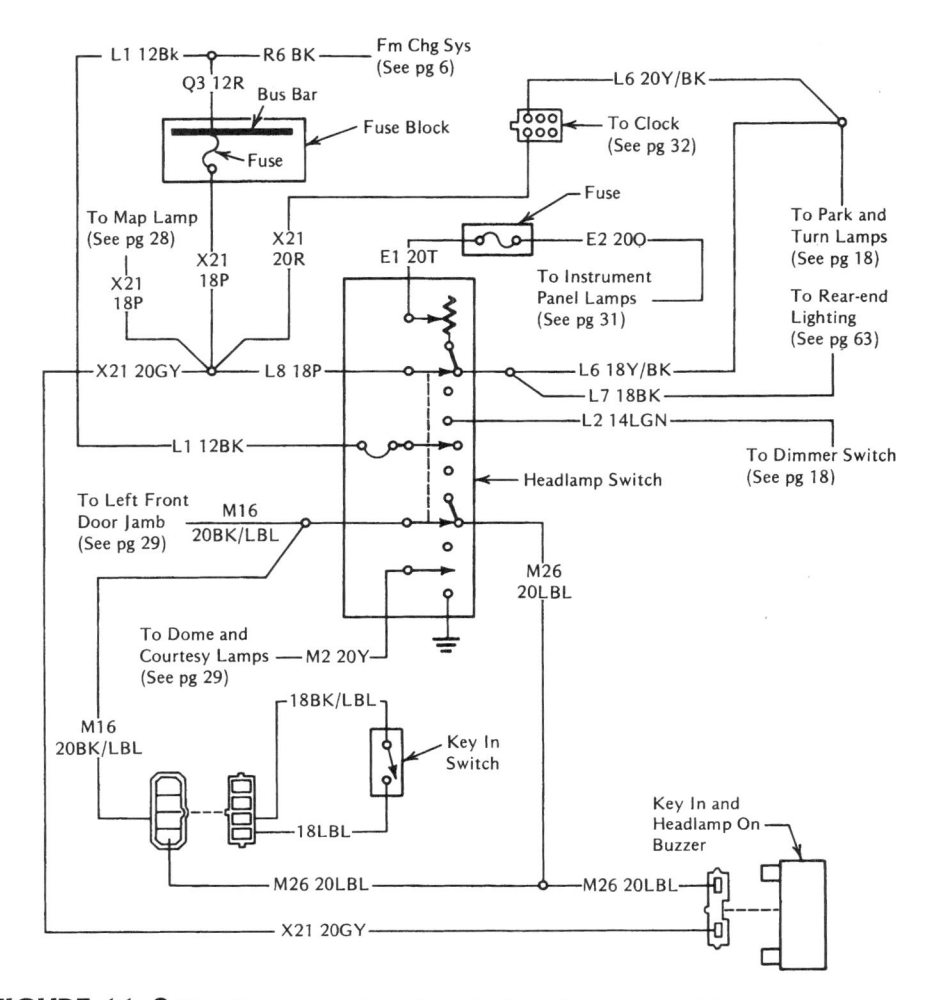

FIGURE 11-8 Headlamp circuit with switch in the PARK position

Headlamp Switch ON. The only circuit that is changed when the headlamp switch is placed in the ON position is the headlamp circuit. This may be noted by comparing Figure 11-8 with Figure 11-9.

Circuit L1 is now fed through the headlamp switch to circuit L2 (14-gauge light green). Circuit L2, traced on other pages of the schematic, feeds only the dimmer switch. After the dimmer switch (Figure 11-10), the circuit splits to high/low beam headlamp circuits. At this point, wire sizes and colors also change. Note that the dome and courtesy lamp circuit is not covered in the headlamp switch example.

The only function of the headlamp switch in this circuit is to provide a ground for turning these lamps ON or OFF. Not enough of this circuit is shown in the schematic example for tracing.

Schematic Sequence and Arrangement

The sequence and arrangement of wires in an actual harness or connector is not necessarily in the same order as shown in a schematic. For example, in Figure 11-11, the headlamp switch schematic and the actual connector represented by the graphic representation differ in sequence. Note, however, that eight wires are shown on both the schematic and the connector. Note also that wire identification, size, and color are the same on each. The graphic illustration is often given to show the actual location of wires in a connector.

Trying to trace an electrical circuit without a schematic is like digging up an entire yard to find a wa-

124 Unit 11 WIRING CIRCUITS

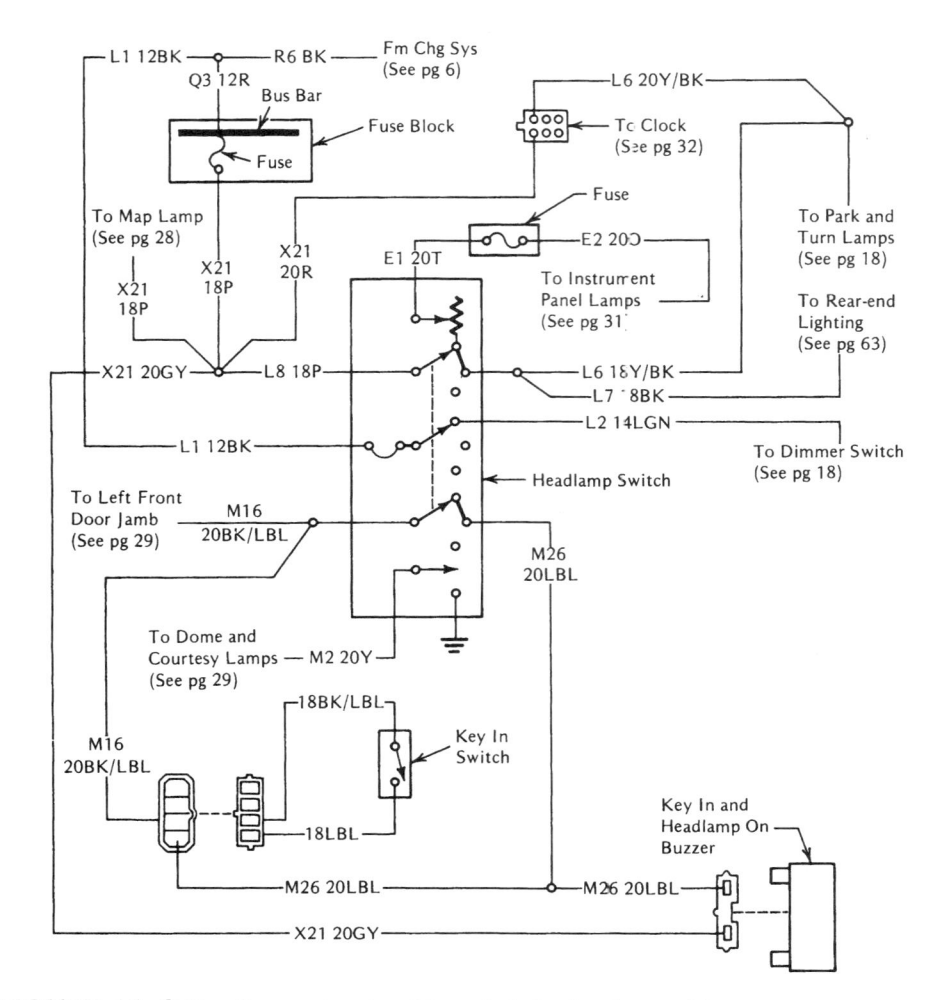

FIGURE 11-9 Headlamp circuit with switch in the ON position

ter pipe. The folks at the water company have no problem—they know where the pipes are because they have access to appropriate drawings or schematics. Automotive electrical technicians use schematics so that they do not have to rip into a wiring harness to find or trace a wire; they also have access to proper drawings.

Tracing a wire in a schematic can often be time consuming. In the long run, however, it saves a considerable amount of time over the "hunt and find" method. For example, suppose a technician traces a wire to a junction where it ties in with four or five other wires. What now? Tear out each of the wires to locate the continuing circuit? Not if the technician uses a schematic—three or four wires lead to an unwanted circuit; the correct wire is quickly located and traced, first in the schematic, then in the harness.

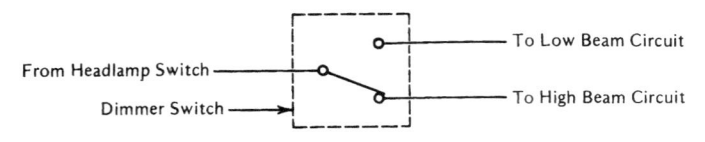

FIGURE 11-10 Simple headlamp dimmer switch circuit

WIRING HARNESSES

There are many wiring harnesses in the average car. Each may contain from 2 to 20 or more wires, but there is no standard number of wires for a harness. The following characteristics apply to wiring harnesses:

1. More than one circuit may be in one harness.
2. More than one harness may be in one circuit.
3. Wires in a harness may be of the same or of a different gauge and/or color.
4. Color and/or gauge of circuit wires may change within a harness.

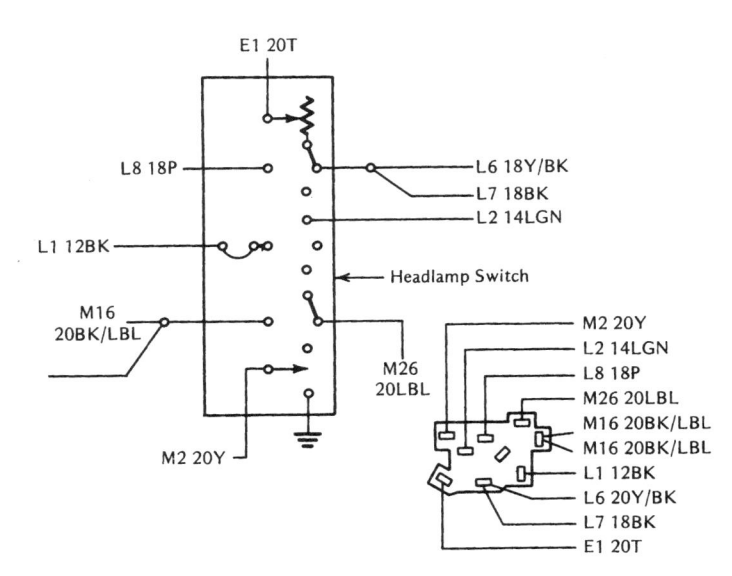

FIGURE 11-11 Headlamp switch schematic and connector graphic

5. Branch circuits may be joined (spliced) within a harness.

Most of the major harnesses originate behind the dash panel. Major harnesses are usually connected to branch harnesses by multiterminal connectors.

A major wiring harness may include any or all of the following:

1. Dash to fire wall and fire wall to front-end lighting.
2. Dash to fire wall and fire wall to engine.
3. Dash to rear-end lighting.
4. Dash to steering column.
5. Dash to car body and interior lighting.
6. Dash to car body and accessory circuits.
7. Dash to fuse block.

Individual wires may terminate at each end of a harness or anywhere along the "run" of the harness. The point of termination is determined by the placement of components and the routing of the harness.

Wiring harness routing (Figure 11-12) is designed as part of the total automotive electrical system. The routing must pass through and around channels, braces, components, and other parts in a manner that avoids damage to the wires. When replacing a harness, follow the original routing as closely as possible.

Avoid mechanical cables and linkage that could wear into the insulation of the harness and wiring. Another word of caution: When drilling a hole for wiring, first make sure that the other side of the panel is clear. It is often necessary to drill mounting holes for accessories such as after-market air conditioners, tape decks, CB radios, and accessory gauges. Take care not to drill into a wiring harness or, equally important, into a gas or brake line. Don't be surprised—be sure. Anything can happen. One technician, when drilling a hole in the right rear panel of a station wagon for an antenna, drilled into the spare tire.

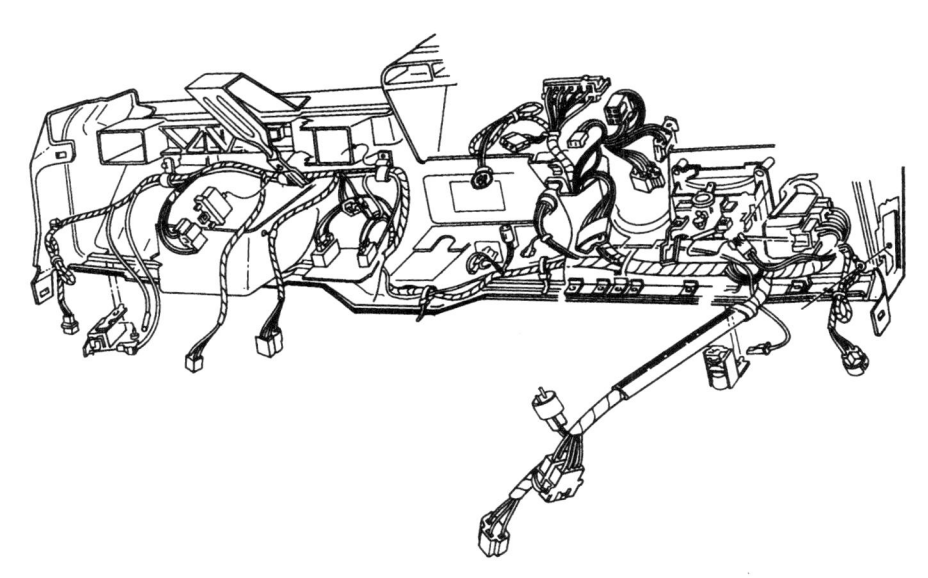

FIGURE 11-12 Careful attention must be given to the routing of the wiring harness to avoid damage to the wires

SUMMARY

- Modern motor vehicles are being increasingly controlled by electrical circuits.
- Electrical circuits and components are interconnected using wiring (cables).
- The effective automotive technician must be able to follow wiring from source to termination.
- Proper repair of electrical systems is as important as mechanical systems repair.

PRACTICAL EXERCISE

A manufacturer's electrical schematic manual for an available late-model vehicle is required for this exercise.

1. Locate the electrical wiring diagram for the headlamp circuits.
2. Follow the circuits on the diagram to become familiar with the color-coded wires and control components.
3. Determine what component directly precedes the headlamps (perhaps the HIGH-LOW-relay).
4. Compare the color codes given in the manual with the actual color of the wires found in the vehicle.
5. Compare the wiring disconnects with the schematic representations in the manual.

QUIZ

For open classroom discussion.

1. What electrical component precedes the headlamp switch?
2. What is the source of power for the headlamps?
3. What do you think would happen if the HIGH and LOW beam wires to the right headlamp were reversed?
4. What do you think would happen if the body ground wire to the left headlamp were disconnected?
5. According to the electrical schematic, what protection is provided for the headlamp circuit?

REVIEW

1. How many wires are there in a standard wiring harness? _____
2. Where do the wires in a wiring harness originate and terminate? _____
3. What are two common methods for joining wires to connectors? _____
4. What are two common methods for joining connectors to components? _____
5. What is the purpose of a schematic? _____
6. What is used to insulate noninsulated terminals? _____
7. In schematics, switches are usually shown in their (open) (closed) position.
8. In schematics, components are usually shown in their (energized) (deenergized) position.
9. A wire coded 20 BK/LBL means that it is
   a. 20 gauge, black with light blue tracer
   b. 20 gauge, light blue with black tracer
10. What type of illustration is used to show the actual location of wires in a component or connector?
    a. schematic   c. pictorial
    b. graphic   d. all of the above

UNIT 12

SEMICONDUCTOR DEVICES

OBJECTIVES

On completion of this unit you will be able to:

- Discuss semiconductor materials.
- Understand the effect of added impurities in useable semiconductors.
- Explain the action inside a semiconductor diode.
- Understand the electrical characteristics of a diode.
- Define zener voltage regulator diodes.
- Explain how transistors are constructed and how they operate.
- Discuss special semiconductor devices such as light emitting diodes, photo cells, and photo transistors.
- Discuss field effect transistors and their construction and operation.

SEMICONDUCTORS

Semiconductors are often used in modern automobile electrical systems. Semiconductor circuits perform operations such as switching, amplifying, and computing.

Electronic control of devices is increasing at a rapid rate. Timing circuits that used to include clock motor devices now use electronic timers. Power-level control devices that used to include variacs or rheostats now operate using semiconductors. The new automotive technician must be familiar with the electronic devices presently being used. These devices include semiconductors used as discrete components as well as in integrated circuits.

A *semiconductor* is a material that falls somewhere between the good conductors—most metals—and the poor conductors, or insulators—such as glass or wax. Semiconductors function in electronic circuits because of their crystalline structure and because they are "doped" with materials called impurities. The impurities provide characteristics to the semiconductor material that make the material useful in rectifiers and amplifiers.

The basic semiconductor materials are germanium (Ge) and silicon (Si). Silicon, as the most popular, will be covered here. When the pure silicon (Si) material is doped with an impurity, it becomes either P-type silicon or N-type silicon, depending on the material added as the impurity. The P stands for positive (+), and the N for negative (−). The impurity combines with the silicon crystals and modifies the crystalline structure. If P-type impurity is added, the crystals of silicon form improperly, leaving a space for an electron in the structure; this space is called a *hold,* and it is movable in the silicon. If N-type impurity is added, the crystals of silicon form with an extra electron available; this extra electron is movable.

Silicon Diodes

A silicon diode is formed when P-type silicon and N-type silicon are joined, forming a junction (Figure 12–1). The diode will pass current in one direction only, as shown in Figure 12–2. If the battery is reversed, no current will flow, as depicted in Figure 12–3.

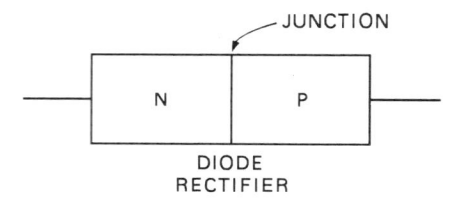

FIGURE 12–1 Diode rectifier

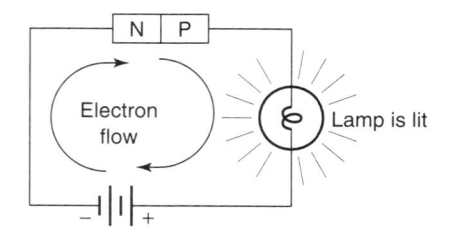

FIGURE 12–2 Current flow through a diode

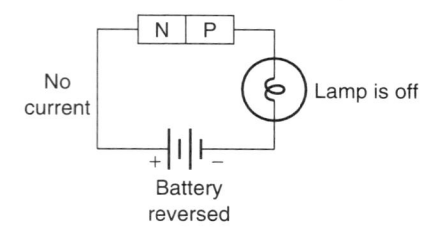

FIGURE 12–3 Reverse biased diode—no current

Testing a Diode

A diode may be tested using an ohmmeter. The small electronic-type diode, the stud-type diode, or the press fit-type diode (Figure 12–4) used in alternators may be tested using the same procedure.

The small diodes used in electronic circuits are usually marked N to indicate the negative (−) and P to indicate the positive (+) terminal. This is usually accomplished by a painted band around the negative end of the diode, as shown in Figure 12–5. The schematic symbol for a diode is sometimes painted on the stud-type diode or press fit diode. The base termi-

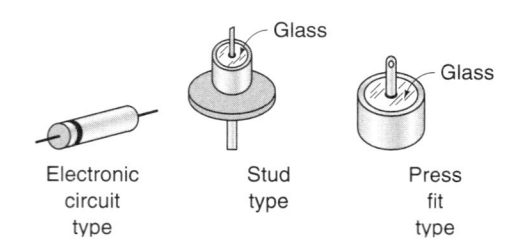

FIGURE 12–4 Diode types

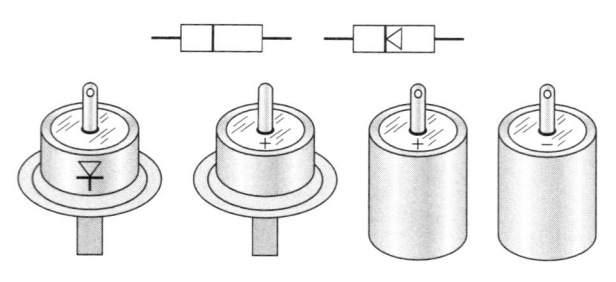

FIGURE 12–5 Diode markings

nal may be marked with a + or − to indicate terminal polarity.

Procedure
1. Select an ohmmeter scale R × 100 or higher.
2. Place the red positive (+) lead of the meter on one terminal of the diode, and connect the black negative (−) lead of the diode to the other lead (Figure 12–6).

   NOTE: With the meter connected as shown in Figure 12–6, the diode will not conduct. Current will not flow through the diode in this direction. The meter will indicate a high resistance (megohms).

3. Reverse the meter connections to the diode as shown in Figure 12–7.

   NOTE: With the meter connected as shown in Figure 12–7, the pointer should indicate low resistance.

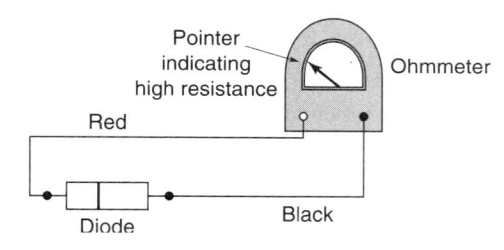

FIGURE 12–6 Testing a diode—reverse bias (high resistance)

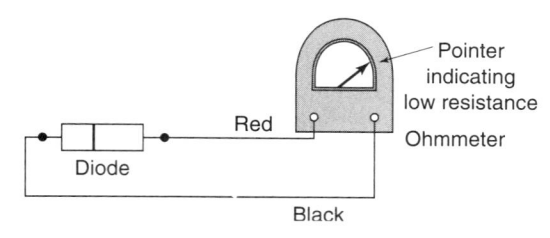

FIGURE 12–7 Testing a diode—forward bias (low resistance)

Resistance measurements as given in this procedure indicate a good diode. If the resistance measured is high (megohms) in both directions, the diode is open. If the resistance measures low in both directions, the diode is shorted.

Diode Uses

One of the main uses of diodes is in rectifier circuits. *Rectifiers* convert alternating current (ac) systems to direct current (dc) systems.

A simple rectifier circuit (half wave) is shown in Figure 12–8. When line X is positive (+), during the first alternation of the input, line Y is negative (−). Current will flow through this diode and the load. When line X is negative (−), during the second alternation of the input, line Y is positive (+). The diode will not conduct. There is no voltage across the load, and there is a gap in the output. Only half the input appears at the output. This is half-wave rectification.

By using four diodes in what is called a *bridge circuit*, full-wave rectification becomes available (Figure 12–9). In the full-wave system, all the input is available at the output; There are no gaps in the output. As illustrated in Figure 12–9, when line X is positive (+), line Y is negative (−) (first alternation of the input). Diodes A and C will conduct. The first alternation appears across the load. Follow the path of current.

When line X is negative (−), line Y is positive (+) (second alternation). Diodes B and D will conduct. The second alternation appears across the load. There are no gaps in the output. The current flow through the load is always in the same direction.

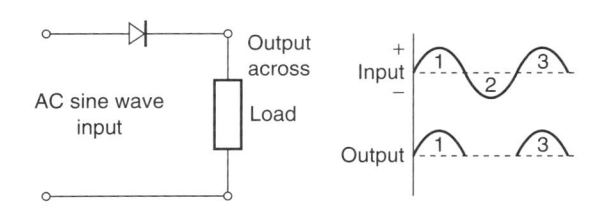

FIGURE 12–8 Half-wave rectifier

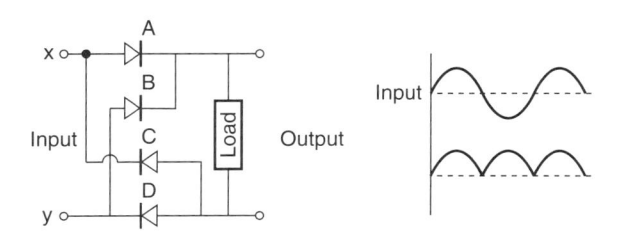

FIGURE 12–9 Full-wave bridge rectifier

Zener Diodes

Zener diodes are specially designed diodes that operate with reverse voltage applied. They are used to provide a constant output voltage where the input voltage varies.

Regular diodes are designed to operate with the maximum reverse voltage under a fixed limit. For example, the diode designated IN4001 has a maximum reverse voltage of 50 volts. The manufacturer warrants that the IN4001 will not break down if the applied reverse voltage is limited to below 50 V. *Breakdown* of the diode is the point where the resistance of the diode changes rapidly. In the case of the IN4007, the manufacturer warrants that the diode will withstand 1000 volts in the reverse direction before breakdown. The manufacturer does not indicate, however, the exact voltage at which either of these diodes will break down. All diodes break down at some reverse voltage. As long as current flow is limited, the breakdown of a diode does not harm the diode.

Zener diodes are manufactured to break down at specified voltages. Their purpose is to operate in the breakdown mode. They are available with voltage levels below 5 volts and to levels above 300 volts. Zener diodes are available with power levels from less than one watt to hundreds of watts.

EXAMPLE

A 5-volt, 1-watt zener diode can handle current up to 0.2 amperes. When connected to a dc source voltage of 13.5 volts through a 50-ohm, 2-watt resistor, the zener will regulate at 5 volts, as shown in Figure 12–10.

Solution

If the value of R_2 is above 32 ohms, the voltage across R_2 will be +5 volts. The circuit operates because the zener diode changes resistance automatically.

Without the load resistance R_L, the zener has a resistance of 29.4 Ω. If a 50-Ω resistor were placed in the

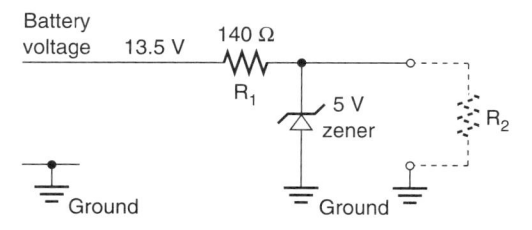

FIGURE 12–10 Zener diode voltage regulator

circuit at R_2, the resistance of the zener would increase to 71.4 Ω. The voltage across R_2, 50 Ω, would be +5 volts.

If R_2 were changed to a 100-Ω resistor, the resistance of the zener would change to 466.67 Ω. The voltage across R_2, 100 Ω, would be +5 volts. Actually, the output voltage and therefore the voltage across the zener diode would change only slightly (decrease) as resistance (R_2) is placed across the output. It is this slight change in voltage that causes the diode resistance to change.

Zener diodes are used in the voltage regulator circuit controlling the alternator in an automobile. A zener diode in combination with an amplifier can provide voltage regulation of high amounts of power.

TRANSISTORS

Transistors are three-terminal devices made from semiconductor materials. There are two main types of transistors—bipolar and field-effect transistors. *Bipolar transistors* have low input resistance. They are current-controlled devices. A small current flow through the input circuit controls a large current flow in the output circuit.

Field-effect transistors have a high input resistance. They are voltage-controlled devices. A small input voltage controls a large current flow in the output. Both types are used extensively in modern automotive circuits.

Bipolar Transistors

Bipolar transistors are shown in Figure 12–11. The three terminals of a bipolar transistor are called the emitter, the base, and the collector. Bipolar transistors are effective because a small current flow between the emitter and base can control a much larger current flow between the emitter and collector. This relationship between emitter–base current and emitter–collector current makes the transistor useful as a switch or amplifier.

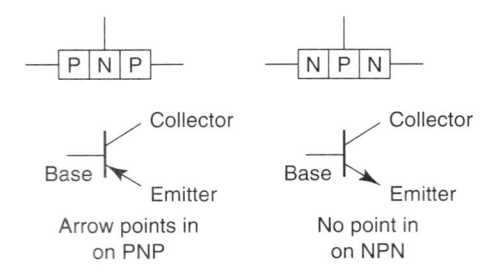

FIGURE 12–11 Bipolar transistor

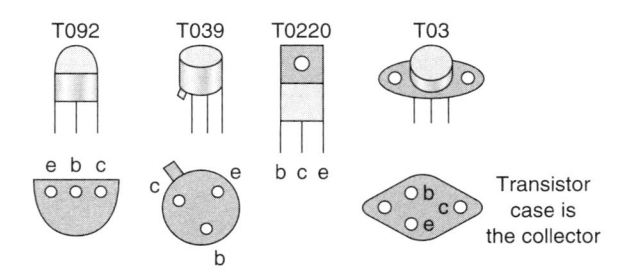

FIGURE 12–12 Common transistor types

Transistors are available in many shapes and sizes. Some of the more common transistors are shown in Figure 12–12.

Note the layout of the terminals on each transistor. These layouts are standard. A transistor may be tested using a standard ohmmeter (Figure 12–13). The following procedures are appropriate for testing transistors.

Test Procedure I: Checking the Emitter–Base Circuit

1. Select R × 100 or a higher scale.
2. Connect the negative, black (−) lead of the ohmmeter to the transistor emitter.
3. Connect the positive, red (+) lead of the ohmmeter to the transistor base.
4. Record the resistance reading.
5. Interchange (swap) the meter leads.
6. Record the resistance readings.

Notes

- If the resistance reading is low in Step 4 and high in Step 6, the transistor is an NPN transistor.
- If the reading is high in Step 4 and low in Step 6, the transistor is a PNP transistor.
- If the readings were low in both directions, the transistor is shorted.
- If the resistances are high in both directions, the transistor is open.

Test Procedure II: Checking the Base–Collector Circuit

1. Connect the negative, black (−) meter lead to the collector.
2. Connect the positive, red (+) meter lead to the base.

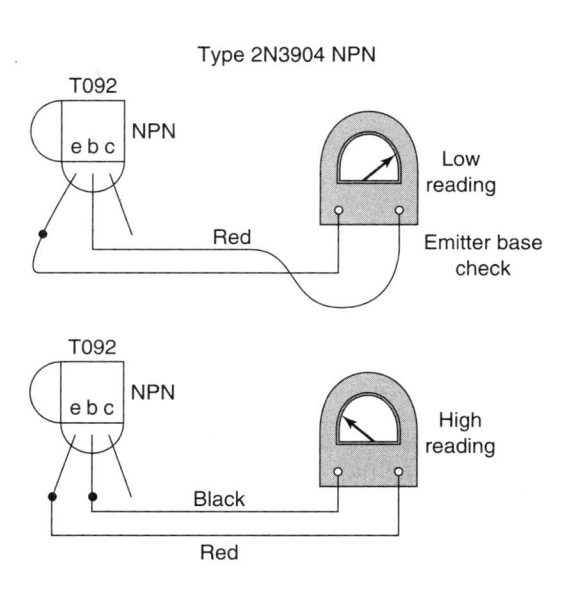

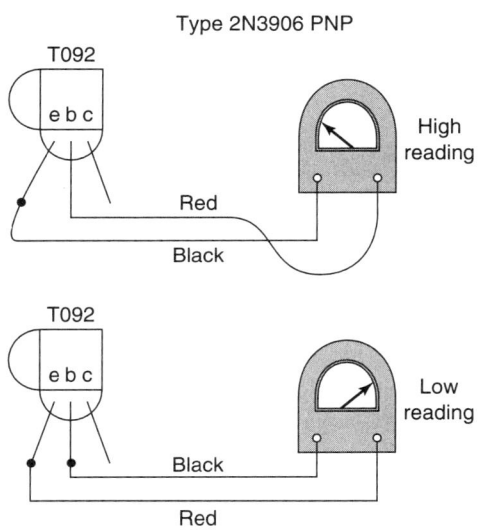

FIGURE 12–13 Using an ohmmeter to check a transistor

3. Record the resistance reading.
4. Interchange (swap) the meter leads.
5. Record the meter reading.

Notes

- If the reading is low in Step 3 and high in Step 5, the transistor is an NPN type.
- If the resistance is high in Step 3 and low in Step 5, the transistor is a PNP type.
- The test for short and open is as given previously.

Test Procedure III: Checking the Emitter–Collector Circuit. Connect an ohmmeter between the emitter and collector of the transistor.

Notes

- An ohmmeter connected between the emitter and collector of a transistor should indicate high resistance in either direction.
- A bipolar transistor is an OFF device. If there is no current flow between the emitter–base circuit, there is no current flow between the emitter–collector. The emitter–collector is open. In order to turn the transistor ON, (emitter–collector current), a circuit must be connected providing current flow between the emitter and base.

The Transistor as a Switch

It is sometimes required that a switch at one location must turn on a device at another location. Relays have been used for this purpose. Relays are expensive and subject to many failures. A transistor may be used as a remote switch; see Figure 12–14.

In the modern automobile, a signal coming from the computer turns on a transistor acting as a switch that allows for current flow through the ignition coil. The transistor takes the place of the points (Figure 12–15).

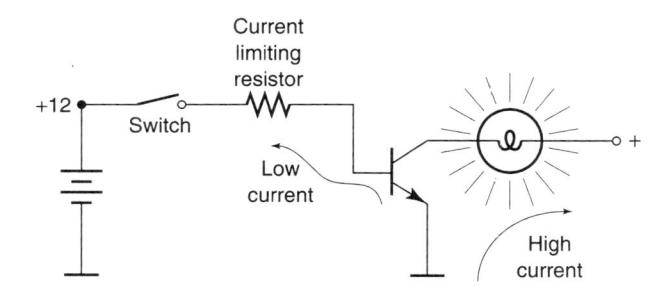

FIGURE 12–14 Transistor switch

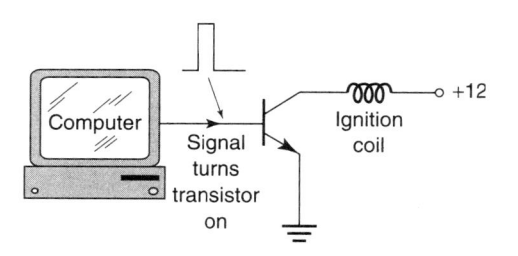

FIGURE 12–15 Practical transistor switch

Unit 12 SEMICONDUCTOR DEVICES

Transistor Amplifier

Basically, when a transistor is being used as a switch, it is amplifying. A small emitter–base current is causing a larger emitter–collector current to flow. When used as a switch, the transistor has two conditions—OFF and ON. These are called *cutoff* and *saturation*.

When a transistor is operating as a signal amplifier, it operates between the OFF and ON condition, cutoff and saturation, at all times.

Current Gain

Before discussing transistor amplifiers, the current gain of a transistor must be considered. Transistors are available with current gain between a low value of 10, for example, to a high value of 300 or more.

A simple circuit may be used to check the current gain of a transistor. A 12-V power source, a resistor, a potentiometer, two meters, and the transistor are needed. The circuit is connected as shown in Figure 12–16.

A 500-kΩ resistor is connected in the circuit for protection of the transistor. If R_2 were adjusted for zero resistance and R_1 were not in the circuit, the emitter–base circuit of the transistor would burn out.

Test Procedure
1. Connect the circuit as shown in Figure 12–16.
2. Adjust R_2 for a current indication of 10 microamps emitter–base current (M_1).
3. Read and record the collector current (M_2).
4. Divide the collector current by the base current to obtain current gain.

For example, if the collector current was 1 milliamp when the base current was adjusted to 10 microamps, the current gain of the transistor is 100 (1 milliamp divided by 10 microamps).

$$\text{Current Gain} = \frac{\text{Collector current}}{\text{Base current}}$$

$$h_{FE} = I_c/I_b$$

$$= \frac{0.001 \text{ A}}{0.00001 \text{ A}} = 100$$

where h_{FE} = dc current gain

I_c = collector current

I_b = base current

A transistor with a current gain of 100 is used in the following example. The term used for current gain is h_{FE}. Consider the circuit of Figure 12–17. How were the component values selected?

1. The supply (battery) voltage is measured at 12 V.
2. The emitter–base voltage is measured and found to be 0.6 Vdc (voltage across a silicon diode).
3. The voltage across R_1 is measured at 11.4 volts. (The emitter–base circuit and the base resistor R_1 are in series. 11.4 Vdc + 0.6 Vdc = 12 volts)

NOTE: It is desired that a collector current of 1 ma be used (a common amount of collector current in small-signal amplifiers). Since the transistor has an hfe of 100, a base current of 10 µA is needed.

4. The voltage across R_1 will be 11.4 volts. This voltage, divided by 10 µA, equals the value of R_1.

$$R = \frac{E}{I} = \frac{11.4 \text{ V}}{10 \text{ µA}} = 1.14 \text{ MEG}\Omega$$

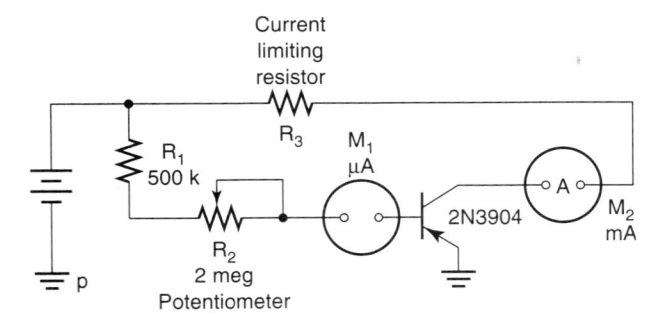

FIGURE 12–16 Checking current gain

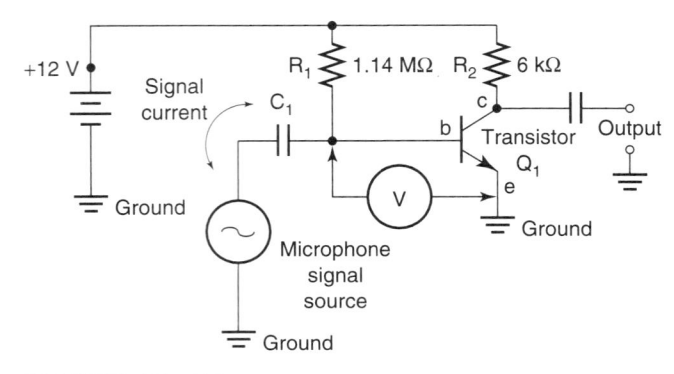

FIGURE 12–17 A small-signal amplifier

NOTE: The collector resistor has 1 mA flowing through it. If a maximum ac output is to be available with a 12 VDC supply, the signal can be 12 V peak-to-peak or 6 volts positive (+) and 6 volts negative (−) from the reference. In order to accomplish this, the steady-state dc voltage at the collector must be +6 Vdc.

5. Select a 6-kΩ resistor as R_2. With 1-ma collector current flowing through R_2 (6 kΩ), the voltage across R_2 will be 6 volts. The voltage from the collector to ground is 6 VDC.

Application of a Signal to the Emitter–Base Circuit

The input resistance (emitter–base) of a small-signal transistor drawing an emitter–collector current of 1 mA is less than 3000 Ω. Considering 3000 Ω, a 9-millivolt signal will cause a change in emitter base current of 3 microamps.

A 3-μA change in emitter–base current causes a 300-μA change in collector current (h_{FE} = 100). The collector voltage will change thus:

$$I_c = I_b \times h_{FE} = 3\ \mu A \times 100 = 300\ \mu A$$
$$E = I \times R = 300\ \mu A \times 6\ k\Omega = 1.8\ V$$

A signal input of 0.009 V provides an output signal of 1.8 V. This is a voltage gain of 200. The output signal will be an inverted replica of the input × 200 (Figure 12–18).

The input signal may be increased in amplitude. As the input level approaches 30 millivolts peak, the output will include distortion (that is, output will not be a replica of the input). As the output approaches 0 volts at the collector (saturation) and 12 volts at the collector (cutoff), distortion occurs (Figure 12–19).

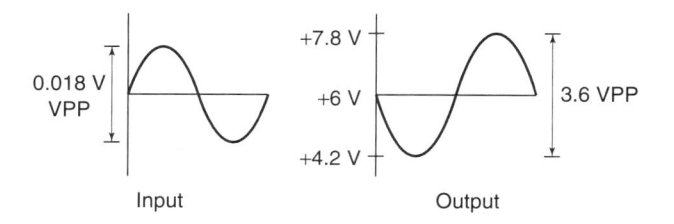

FIGURE 12–18 Input and output signals

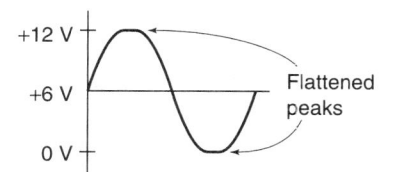

FIGURE 12–19 Distortion

Maximum Ratings and Electrical Characteristics

The manufacturers' data on transistors provide important information on transistor operation. A partial data sheet is shown in Figure 12–20. It provides the type of information that is considered when a transistor is to be used in a circuit. Make a special note of the range of current gain on the single transistor type. It is common for transistors of the same type to have a range of current gain.

Light-Emitting Diodes (LEDs)

All semiconductor diodes produce light (photons) when conducting. The light is not seen because the materials silicon (Si) and germanium (Ge) is opaque. The photons are produced but do not escape the device.

Light-emitting diodes (LEDs) are constructed of materials that are translucent to light. The light can escape and therefore can be seen.

One of the earliest LEDs available was colored red. Shortly after the red LED came green LEDs. LEDs are now available in white, red, orange, yellow, green, and the latest addition, blue.

LEDs are connected in a circuit such that the proper polarity of voltage will cause the diode to conduct and the LED to produce light. A series resistor is usually included to limit current flow. The diode's forward voltage is, as with almost all diodes, approximately 0.6 volts. The circuit is shown in Figure 12–21.

Photoconductive Cells

Photoconductive cells are usually made using cadmium sulfide (CaS) or cadmium selenium (CdSe) as the cells' active material. The light-sensitive material is deposited on a glass surface. Two electrodes are fixed over the glass substrate and provide a controlled path for light to act on the active surface.

TYPE 2N3904 SILICON
Maximum Ratings

Rating	Symbol	Value	Unit
Collector-emitter voltage	V_{ceo}	40	Vdc
Collector-base voltage	V_{cb}	60	Vdc
Emitter-base voltage	V_{eb}	6.0	Vdc
Collector current	I_c	200	ma

Electrical Characteristics on Condition

	Symbol	Min	Max
Current gain dc	h_{FE}	100	300
I_c = 10 ma V_{ce} = 1.0 Vdc			
Current gain ac	h_{fe}	100	400

FIGURE 12–20 Type 2N3904 transistor data sheet

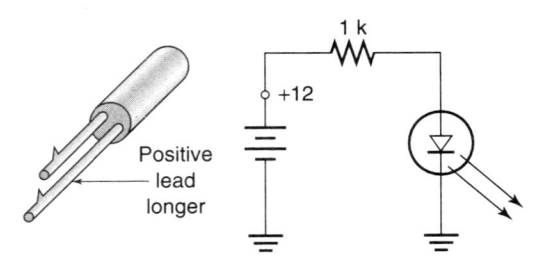

FIGURE 12–21 Light-emitting diode

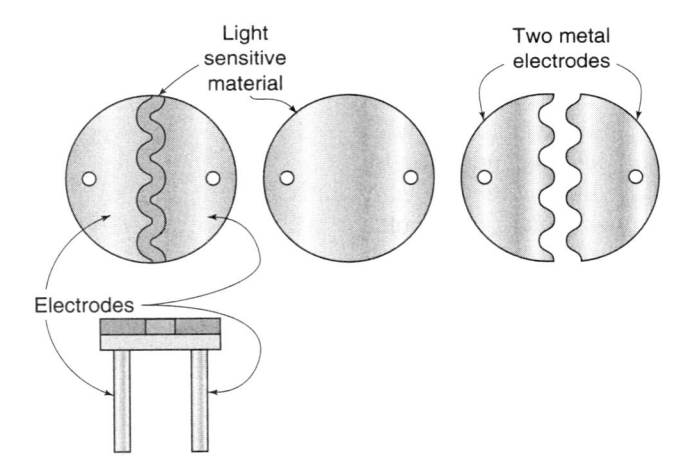

FIGURE 12–22 Photoconductive cell

The resistance of a photo cell is high in darkness and low when light strikes its surface. The symbols used for photo-conductive cells are shown in Figure 12–22. The arrows in Figure 12–23 indicate the light to which the device is sensitive. The second symbol used

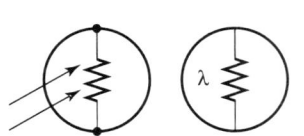

FIGURE 12–23 Photoconductive cell symbols

is the Greek letter lambda (Λ or) to represent the wavelength of light.

Photovoltaic Cells

The photovoltaic cell (Figure 12–24) converts light energy directly into electrical energy. It is a junction device made from a P-type layer and an N-type layer. When light strikes one of the junction surfaces while the other is isolated from the light, a voltage is produced across the junction.

The amount of current available from a photovoltaic cell depends on the surface area. The larger the area, the larger the current.

Large arrays of photovoltaic cells called solar panels have been used to power experimental electric vehicles. If the photovoltaic cell can be made highly efficient, solar (sun) powered cars could become a reality.

Phototransistors

Incorporating a photovoltaic cell in the same body as a transistor produces a light-sensitive transistor. The equivalent circuit is shown in Figure 12–25.

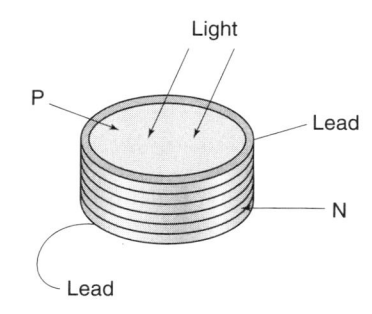

FIGURE 12–24 Photovoltaic cell

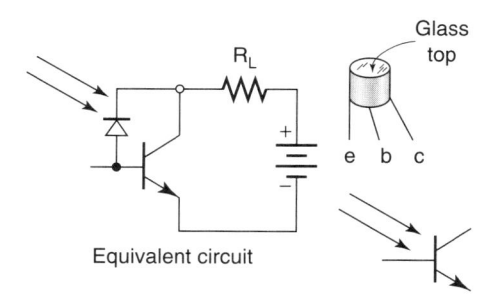

FIGURE 12–25 Phototransistor

When light strikes the surface of the photodiode within the transistor, it produces a voltage that turns the transistor ON. The greater the amount of light striking the phototransistor, the greater the collector current.

Field-Effect Transistors

In the 1950s, the effect of an electric field on current flow in a semiconductor was discovered. Field-effect transistors (FETs) thus came into being.

The transistor works because an electric field can cause the cross-sectional area of a conduction path to change. A change in the cross-sectional area of conduction in a transistor changes the resistance of the device. A field-effect transistor is a resistive device where the resistance changes according to the application of voltage to the terminal, called a *gate*. A FET is, in other words, a voltage-variable resistor. In application with supply voltages and other resis-tors, the FET may be used as an amplifier or a switch.

MOSFETs

Metal oxide semiconductor field-effect transistors (MOSFETs) are presently the leading FETs in use. They have great application in digital computers. Their use is also increasing as power amplifiers and as switches. Recent developments have provided MOSFETs with an ON resistance of less than 0.009 Ω. It is obvious that such a device would be useful as a switch. Voltage levels of these devices range from low-voltage devices of 50 volts up to values above 1000 volts.

MOSFET Construction. The most popular MOSFET transistors manufactured today are the enhancement type. This simply means a transistor that is normally OFF (NO). It will turn ON with the application of proper voltage. Depletion-mode transistors are available but are not as popular. The two enhancement-mode transistors are the N-channel and the P-channel. The letter indicates the conduction path main material, either P- or N-type silicon. In general, the construction of an N-channel MOSFET is as shown in Figure 12–26. The P-channel MOSFET is shown in Figure 12–27.

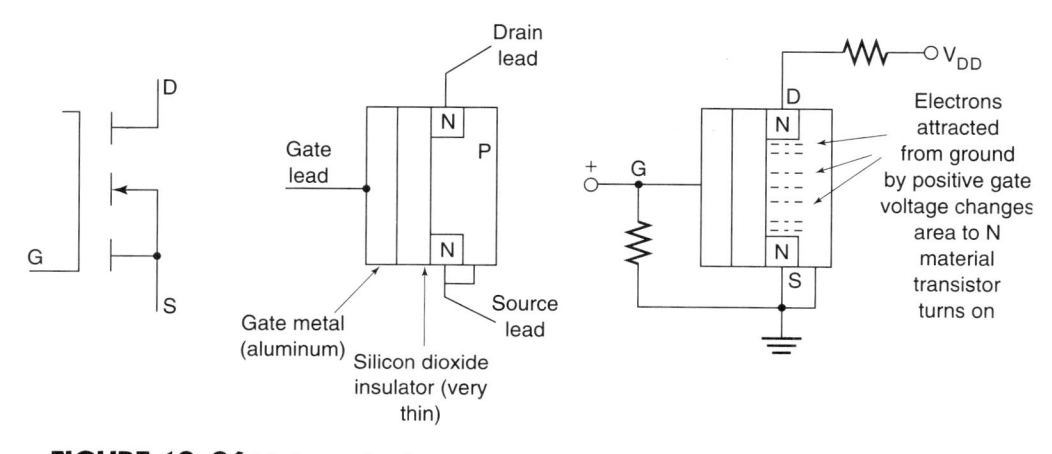

FIGURE 12–26 N-channel enhancement

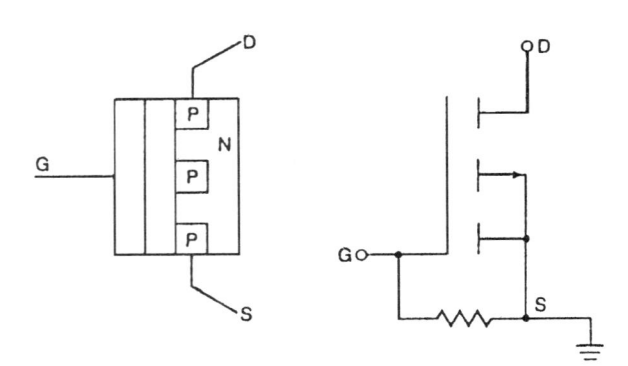

FIGURE 12-27 P-channel enhancement

Enhancement mode means the transistor must be turned ON. The construction shows open areas between the source N material and the drain N material. To turn the device ON, the gate must be made positive (+) with respect to the source. A positive (+) gate attracts electrons into the area between the N blocks, effectively turning the whole area into N material. This completes the path from source to drain through N material. The magnitude of the positive (+) voltage on the gate determines the number of electrons attracted, and therefore the amount of change in resistance between source and drain. In power FETs, this resistance can approach zero ohms (0 Ω).

SUMMARY

- Semiconductors have resistance that falls between good conductors and good insulators.
- "Doped" semiconductor materials are useful in the manufacture of diodes and transistors.
- Diodes have low resistance in one direction and high resistance in the other direction.
- In a transistor a small base current controls a larger collector current.

PRACTICAL EXERCISE

A late-model vehicle with appropriate electrical service manual, a small-signal transistor, such as a 2N3904, and an analog multimeter are required for this exercise. It is intended to stimulate classroom participation and discussion.

EXPERIMENT 1

1. Select resistance on the analog multimeter and select the highest R scale, probably R × 10,000.
2. Connect the black lead (negative) to the transistor emitter.
3. Connect the red lead (positive) to the transistor collector.
4. Record the meter reading. (See Note 1.)

5. Wet one of your fingertips.
6. Place your fingertip between the transistor collector and the base lead (Figure 12-28).
7. Record the meter reading. (See Note 2.)

8. In Step 3 the transistor is turned (OFF/ON).

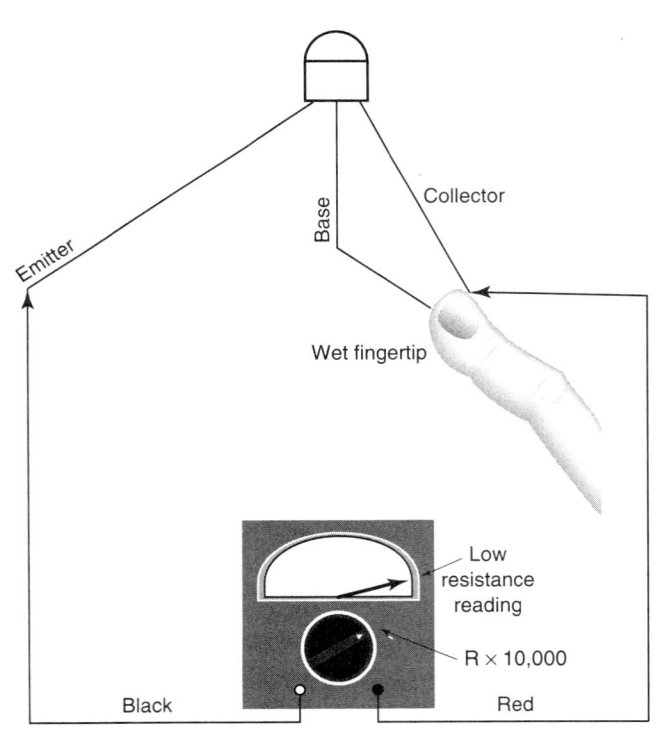

FIGURE 12-28 Current flow through wet fingertip and emitter base current turns transistor ON

9. In Step 6 the transistor is turned (OFF/ON).

10. Why, in your opinion, is there a difference in Steps 4 and 7? (See Note 3.) _____

NOTES

1. The meter should indicate full scale, or close to full scale, in Step 3, since there is no emitter–base current to turn the transistor ON.

2. The meter should jump upscale in Step 6, indicating a lower resistance. The transistor is turned ON.

3. When the wet fingertip is placed between the collector and base, current flows through the skin, from the positive (+) meter lead through the transistor base to the emitter is reduced. When emitter–base current flows the transistor turns ON.

4. The battery in the analog meter provides the electrical power necessary to conduct this experiment.

EXPERIMENT 2

1. Raise the hood of the vehicle. Do not start the engine.

2. Consult the electrical schematic of the manual and locate the following components in the vehicle. Make a note of where they are found.

 a. Approaching vehicle headlamp sensor (automatic dimmer) _____

 b. Daylight sensor (headlamp delay) _____

 c. Instrument panel lamps _____

QUIZ

1. What type sensor is used to signal an approaching vehicle with headlamps on? _____

2. What type sensor is used to provide the daylight signal-controlling headlamps? _____

3. Are there any light-emitting diodes (LEDs) used in the instrument panel? _____ Explain. _____

REVIEW

Answer the following questions true (T) or false (F).

(T) (F) 1. The resistance of pure silicon (Si) is a little over 1.34 times that of copper (Cu).

(T) (F) 2. Doping of pure silicon (Si) with a material having three electrons in its outer orbit provides P-type silicon.

(T) (F) 3. Doping of pure silicon (Si) with a material having five electrons in its outer orbit provides N-type silicon.

(T) (F) 4. The two types of material used in semiconductors are cobalt (Co) and silicon (Si).

(T) (F) 5. Diodes are used to change dc to ac.

(T) (F) 6. Diodes normally conduct in both directions.

(T) (F) 7. The output of a full-wave rectifier is easier to filter than the output of a half-wave rectifier.

(T) (F) 8. Transistors are two-terminal semiconductors.

(T) (F) 9. In a transistor, a small signal current in the collector controls a large signal current in the base.

(T) (F) 10. Transistors may be used to make amplifiers.

(T) (F) 11. The breakdown voltage of a diode is dependent on the size of the diode.

(T) (F) **12.** If one of the breakdown voltages across a diode is exceeded the diode will be destroyed.

(T) (F) **13.** One of the main uses of a zener diode is as a voltage regulator.

(T) (F) **14.** Making the base of an NPN transistor more positive (+) will increase the collector current.

(T) (F) **15.** When the base current of a transistor changes from 12 µA to 24 µA and the collector current changes from 1.4 mA to 2.9 mA, the transistor has hfe of 125.

◆ UNIT 13 ◆

SEMICONDUCTOR INTEGRATED CIRCUITS

OBJECTIVES

On completion of this unit you will be able to:

- Explain the development of transistor circuits from individual components to integrated circuits (ICs).
- Give the names and functions of some of the more common integrated circuits in operational amplifiers.
- Describe basic digital circuits.
- Explain some individual digital integrated circuits.
- Perform binary counting.
- Define large-scale integration (LSI) and tell why it is popular.

SEMICONDUCTOR DEVICES

Transistors and transistor circuits were developed and improved rapidly between the early '50s and the present. The circuits contained the transistor, resistors, and capacitors connected to provide necessary output signals. Manufacturers soon recognized that the individual components could be produced in a single process, providing a functioning device in one package.

The diagram in Figure 13–1 is that of an operational amplifier. An operational amplifier provides extremely high gain and is a relatively stable device. Originally, the device was made using individual components. In Figure 13–1(a) there are 22 transistors, 11 resistors, 1 capacitor, and a diode. Constructing an operational amplifier from individual components was a tedious and expensive operation.

Manufacturing processes improved from the '50s to the present. It became practical to produce integrated circuits (ICs). The transistor resistors and capacitors in integrated circuits are produced using semiconductor materials. In integrated circuits, the transistors are produced using N- and P-type silicon. The resistors are produced using silicon strips, with the doping levels providing for the desired amount of resistance. Capacitors are provided by making a PN junction diode and having it connected in the IC as a reversed bias diode. A reverse biased diode provides capacitance. Manufacturing processes provide for the inclusion of two complete operational amplifiers in an eight-pin package, as shown in Figure 13–1(b). The package is called *dual in-line*, since there are lines of pins on both sides.

The symbol for an operational amplifier is given in Figure 13–2. Note the two inputs marked + (positive) and − (negative). These markings have to do with the input's relation to the output. The terminals are properly called the *inverting* and *noninverting* inputs.

The relation between input and output is described by four statements:

1. If a positive (+) voltage is applied to the inverting (−) input, the output will be negative (−).

140 Unit 13 SEMICONDUCTOR INTEGRATED CIRCUITS

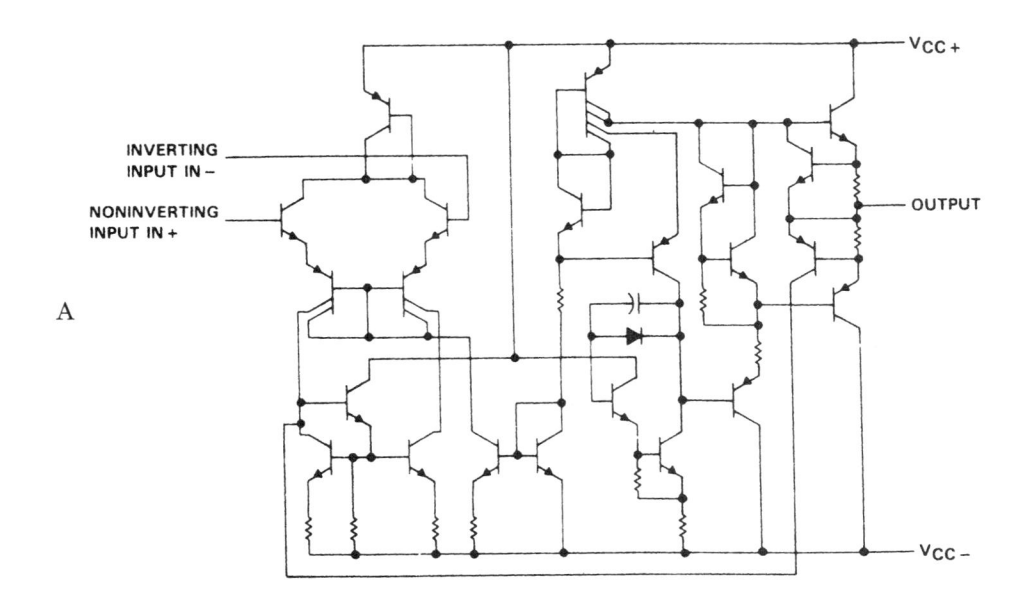

FIGURE 13-1 (a) Operational amplifier schematic with (b) dual in-line package

2. If a positive (+) voltage is applied to the noninverting (+) input, the output will be positive (+).
3. If a negative (−) voltage is applied to the inverting (−) input, the output will be positive (+).
4. If a negative (−) voltage is applied to the noninverting (+) input, the output will be negative (−).

As indicated earlier, an operational amplifier has high gain capabilities. Open loop (without feedback) gains of well over 200,000 are possible.

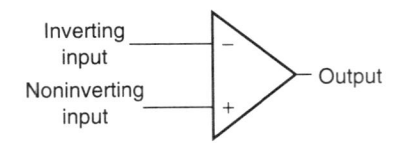

FIGURE 13-2 Operational amplifier symbol

The operational amplifier is normally used with a few external components in order to provide the required output. Examples of some of these circuits are shown in Figure 13-3.

GATE CIRCUITS

In modern computers, including those used in automobiles, it is often necessary to determine whether things are happening either together or individually. For example, it may be decided that the air conditioning fan is to come on when the temperature is above 78° F and the compressor is operating. This may be done with series switches or in the computer with an AND circuit. An example of an AND circuit is shown in Figure 13-4. Here two transistors are connected in series.

In either the bipolar circuit or the MOSFET circuit, there must be +5 volts at both A and B in order to obtain +5 volts at the output (both transistors must be turned on in order to obtain an output).

GATE CIRCUITS

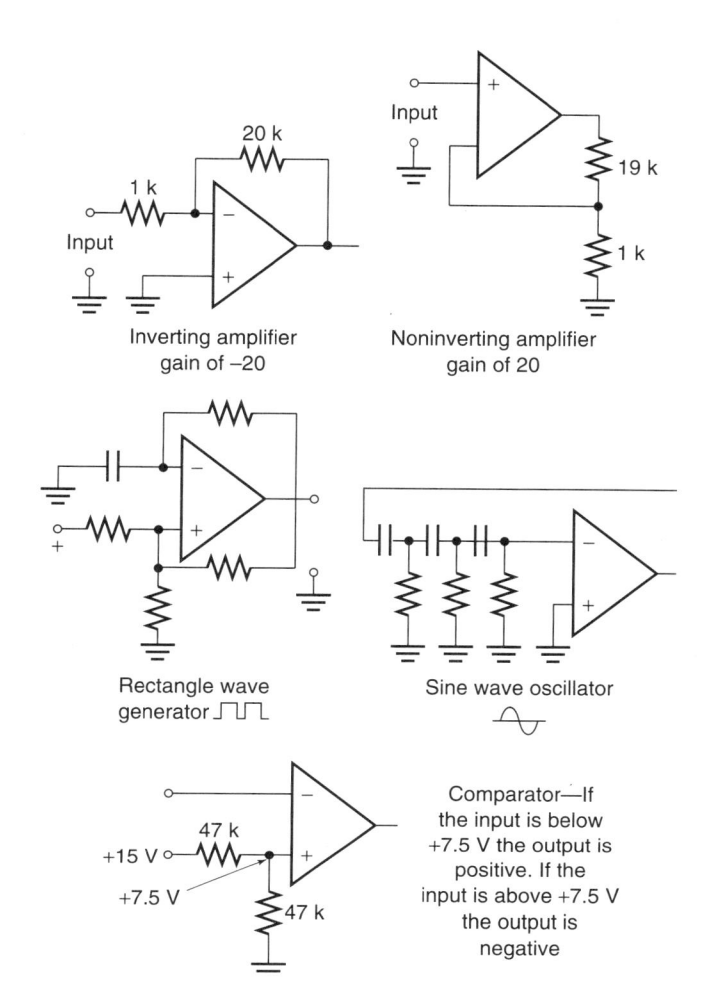

FIGURE 13-3 Operational amplifier circuits

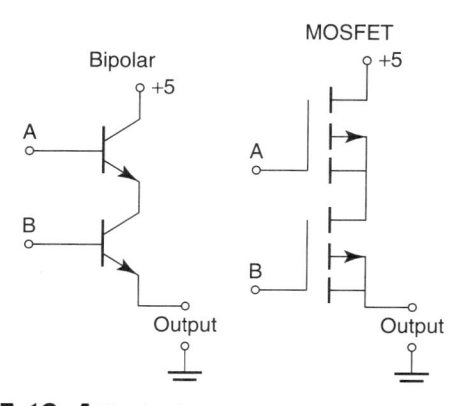

FIGURE 13-4 Typical AND gates

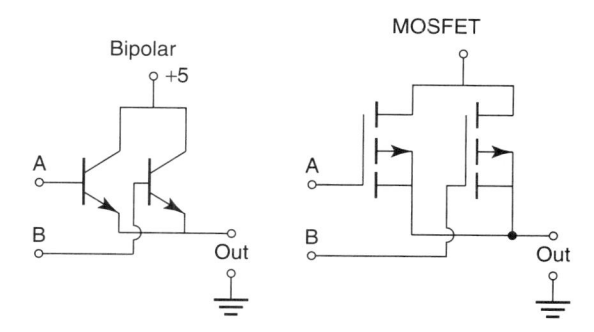

FIGURE 13-5 Typical OR gates

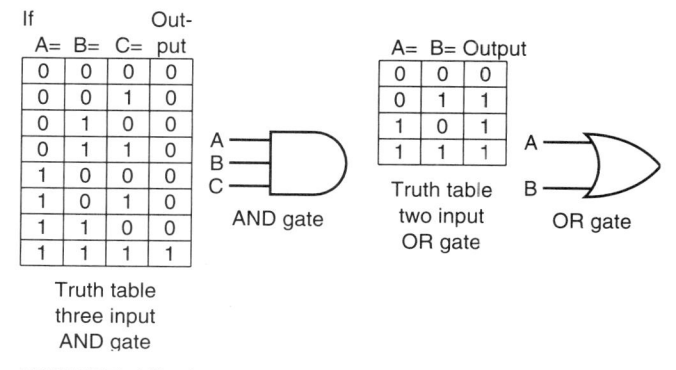

FIGURE 13-6 AND and OR gates with truth tables

In the case of the OR gate, if either A or B is positive (+5), there will be +5 volts at the output. In the OR gate, if either A or B (or both) are +5 V, there will be +5 V at the output. Examples of OR gates are shown in Figure 13-5. The symbols used in diagram for AND and OR gates are shown in Figure 13-6.

Another important gate circuit used in computers is the exclusive OR(XOR) gate. In this gate circuit, the output is positive (+) if *either* A or B input is positive (+), but not both. A simple circuit for an exclusive OR gate is shown in Figure 13-7.

The exclusive OR gate consists of a standard OR gate, transistors Q_1 and Q_2, a pass transistor (Q_3), and

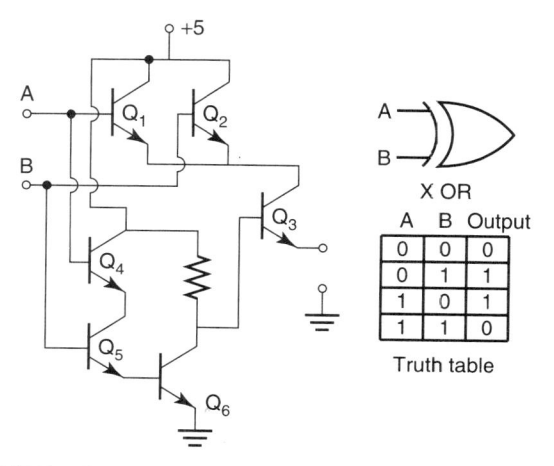

FIGURE 13-7 Exclusive OR (XOR) gate

an inverted AND gate transistor (Q_4 and Q_5), with Q_6 as the inverter.

In order to have a positive output from this circuit, transistor Q_3 must be turned on. Transistor Q_3 is controlled by transistor Q_6, which in turn is controlled by transistors Q_4 and Q_5. If both A and B are present, Q_6 will turn on and Q_3 will turn off. If both A and B are not present, Q_6 is off and Q_3 is on.

If A is present, Q_1 is on. If B is present, Q_2 is on. If A is present or B is present but not both, then Q_3 is on and there will be an output.

ONE-SHOT MULTIVIBRATOR

A one-shot multivibrator (MV) is an electronic circuit that produces a rectangular output pulse for every trigger input. A dual one-shot multivibrator is shown in Figure 13–8.

The output rectangular pulses from the one-shot are of constant amplitude and time duration and are independent of the input wave shape. This fact makes the one-shot particularly useful in tachometer circuits.

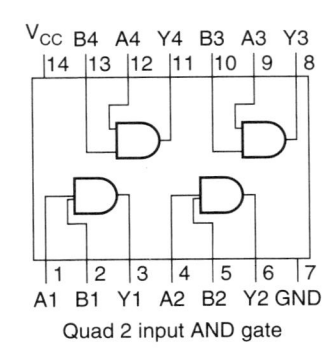

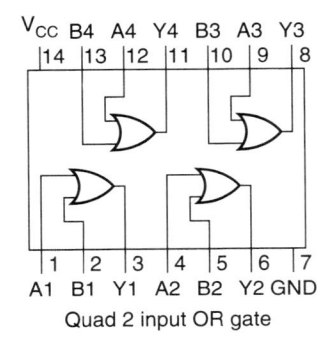

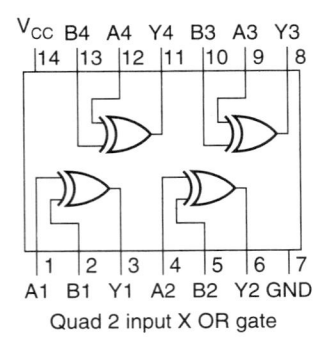

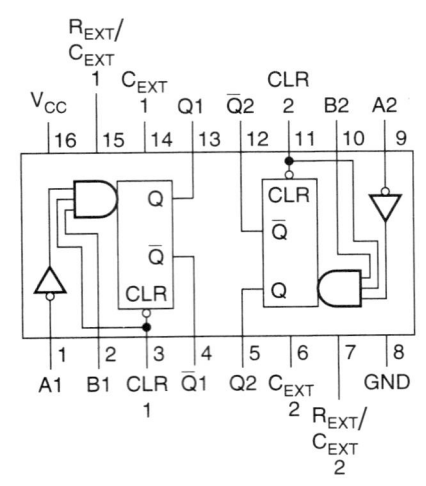

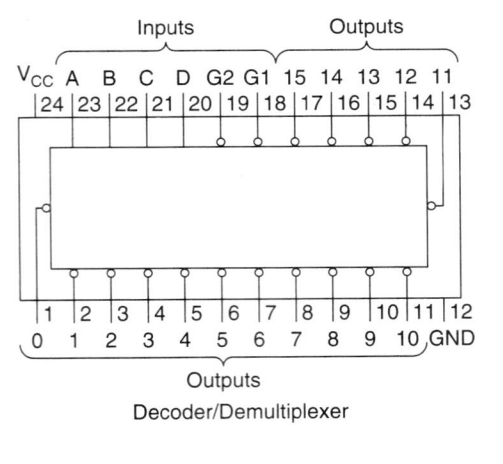

FIGURE 13–8 Digital IC Layout

If the input is taken from the same source as the ignition coil timing, there will be one pulse input for every plug firing. Since the output of the one-shot is rectangular and of fixed time duration, the average dc output is an indication of the number of plug firings. A circuit such as that of Figure 13–9 will provide a reasonable engine rpm indication.

FLIP-FLOP CIRCUIT (+5 V SYSTEM)

The flip-flop circuit is the basic counting circuit used in digital computers. In the flip-flop circuit, the output changes from 0 to +5 or from +5 to 0 with the application of a trigger pulse. Figure 13–10 shows typical flip-flop wave patterns.

In Figure 13–11 the 0 at the trigger input indicates that the flip-flop is triggered by a negative-going wavefront (voltage changing from +5 to 0). The output is said to be set when Q is high (+5 V) and $\overline{Q}$ is low (0 V).

One type of flip-flop circuit that contains most of the controls needed in computer application is called the JK flip-flop. The JK flip-flop is shown in Figure 13–12.

The truth table associated with the JK flip-flop provides an easy-to-understand relationship between the input and output signals. With a flip-flop of this type, almost any counting task or information storage required can be handled. Figure 13–12 shows examples of standard JKs.

DIGITAL COUNTING

The digital circuits covered in previous examples have shown output voltages of either 0 volts or +5 volts. It is by use of 0 volts for 0 and +5 volts for 1 that almost all computations are accomplished in the on-board digital computer. The following simple examples of digital systems provide some understanding of what is going on within the computer.

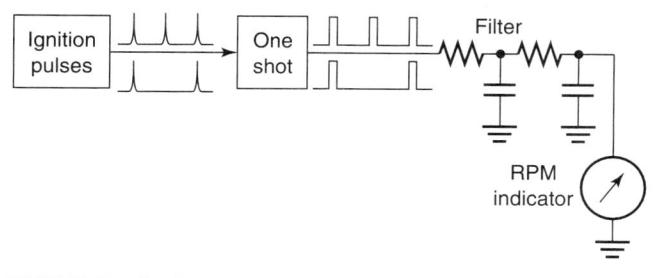

FIGURE 13–9 Tachometer circuit

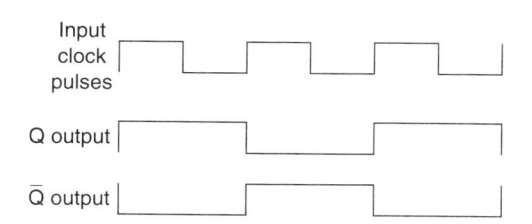

FIGURE 13–10 Flip-flop wave forms

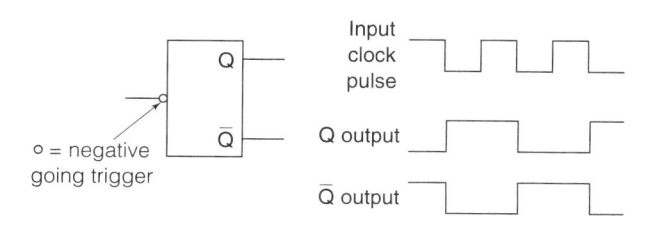

FIGURE 13–11 Flip-flop negative trigger

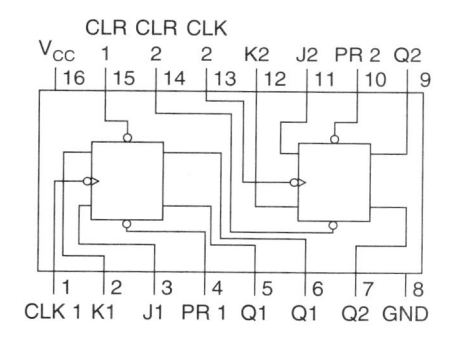

Function table

Inputs					Outputs	
PR	CLR	CLK	J	K	Q	$\overline{Q}$
L	H	X	X	X	H	L
H	L	X	X	X	L	H
L	L	X	X	X	H*	H*
H	H	↓	L	L	Q_0	$\overline{Q}_0$
H	H	↓	H	L	H	L
H	H	↓	L	H	L	H
H	H	↓	H	H	Toggle	
H	H	H	X	X	Q_0	$\overline{Q}_0$

* Unstable condition, the condition will not persist when preset and/or clear is returned to high state.

Q_0 The output logic level of Q before input conditions were established.

FIGURE 13–12 JK flip-flop with truth table

Digital systems operate using a counting system called the binary system. There are only two units in the binary system: zeros and ones, as shown in chart 1. A given 0 or 1 is weighted by its position in a line representing the total number. See chart 1.

	11	10	9	8	7	6	5	4	3	2	1	Position
← Etc.	1024	512	256	128	64	32	16	8	4	2	1	Weight

Chart 1

A 1 in a column indicates the weight is present. A 0 indicates the weight is not present. Consider the binary number 101001, shown in chart 2.

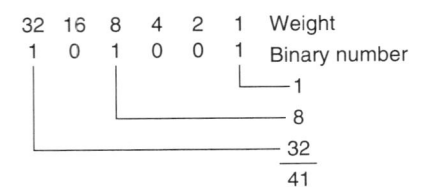

```
32  16  8   4   2   1   Weight
 1   0  1   0   0   1   Binary number
                    └── 1
                ┌────── 8
        ┌────────────── 32
                       ───
                        41
```

The equivalent number in the decimal system is 41. Within the computer, binary 0 is represented by 0 volts. Binary 1 is represented by +5 volts. The operations that take place in a computer are accomplished by manipulating the zeros and ones, that is 0 volts and +5 volts, in order to obtain the desired results. This is accomplished on individual ICs and through the interconnection of these ICs, and also in larger components called large-scale integration (LSI).

LARGE-SCALE INTEGRATION (LSI)

Integrated circuits are popular because they are compact and relatively inexpensive. Still, with most integrated circuits, most of the space is needed to get the connections for input and output signals to the chip (Figure 13–13).

The connecting pins are the largest electrical parts of the IC. It became obvious that the next step was to incorporate a good part of a given computer system on a single chip and eliminate all the large interconnecting parts. When this was accomplished, the devices became large-scale integration (LSI) circuits. Most modern computers contain a number of LSI circuits. The large chips in on-board computers are examples of LSI. An example of large-scale integration is shown in Figure 13–14.

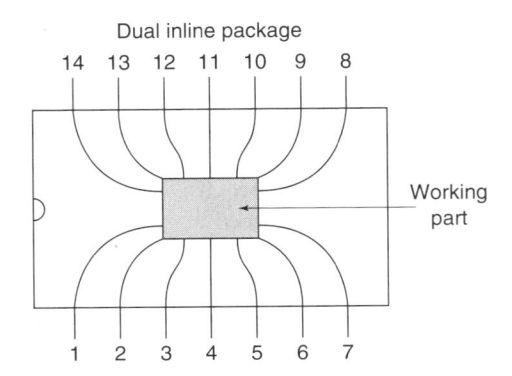

FIGURE 13–13 Integrated circuit connection to pins

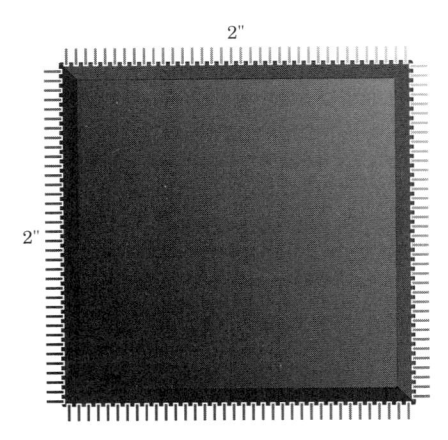

FIGURE 13–14 Computer integrated circuit with 144 input/output pins, large-scale integration (LSI)

SUMMARY

- Integrated circuits are the popular choice in modern systems.
- It is possible to provide high gain in a small package.
- Common gate circuits are the AND gate, the OR gate, and the XOR gate.
- Binary counting using 0s and 1s is used in digital computers.
- As production systems developed, it became practical to provide large-scale integrated circuits (LSI).

PRACTICAL EXERCISE

This exercise is intended to stimulate classroom and shop participation and discussion. A late-model vehicle and appropriate electrical service manual are required.

1. Raise the vehicle hood. DO NOT start the engine.
2. With the aid of the electrical schematics, locate the on-board computer components. Note their location. Notes: _____
3. Locate the alternator and alternator voltage regulator. Note their location. _____

QUIZ

1. Is the alternator regulator internal or external to the alternator? _____
2. Is the regulator made up of individual parts, such as relays, or is it of integrated circuits? _____

EXERCISE SUMMARY

1. The on-board computer(s) contain many integrated circuits.
2. The symbol shown in Figure 13–15 cautions of the danger of damage, by static electricity to delicate electronic components.
3. The symbol in Figure 13–16 cautions of the danger of personal injury. Exercise caution when servicing a component or system.

Danger of static electricity

FIGURE 13–15 Static electricity hazard warning symbol

Danger of personal injury

FIGURE 13–16 Personal injury hazard warning symbol

REVIEW

1. The gain of an operational amplifier is:
   a. high
   b. low
   c. slow
   d. fast

2. If both inputs to a two-input AND gate are +5 volts, the output will be:
   a. +5 V
   b. −5 V
   c. 0 V
   d. 2.5 V (+ or −)

3. If either input to a two-input OR gate is +5 volts, the output will be:
   a. +5 V
   b. −5 V
   c. 0 V
   d. 2.5 V (+ or −)

4. If both inputs to a two-input OR gate are binary 0, the output of the OR gate will be:
   a. binary 1
   b. binary 0
   c. positive (+)
   d. negative (−)

5. If one input to an XOR gate is binary 1 and the other is binary 0, the output will be:
   a. binary 1
   b. binary 0
   c. positive (+)
   d. negative (−)

(T) (F) 6. There are usually two or three transistors in an operational amplifier.

(T) (F) 7. If both inputs to a two-input AND gate are binary codes, the output will be binary.

(T) (F) 8. A one-shot multivibrator produces an output pulse with a time duration that is independent of the input pulse time.

(T) (F) 9. The Flip-flop circuit is the basic counting circuit used in computers.

(T) (F) 10. The binary number 101110 is equivalent to decimal 48.

Sketch the electrical symbol that depicts the following:

11. diode
12. two-input AND gate
13. exclusive OR gate (XOR)
14. operational amplifier
15. two-input OR gate

◆ UNIT 14 ◆

ALTERNATING CURRENT AND THE ALTERNATOR

OBJECTIVES

On completion of this unit you will be able to:

- Understand how electricity is produced by rotating coils past magnets and/or rotating magnets past coils
- Explain how generators and alternators produce electricity in the form of a sine wave
- Describe how modern alternators produce electricity in three phases
- Discuss the effect of capacitors as filters
- Explain how oscillations take place in a tank circuit
- Describe the use of diodes to rectify alternating current (ac) to produce direct current (dc)
- Demonstrate an understanding of the use of voltage regulators

Alternating current (ac) is the most common form of electricity. It is the electricity generated at power plants and distributed around the world. It is also the form of electric power produced in the automotive alternator. This ac is immediately converted into direct current (dc) within the alternator.

GENERATING AC VOLTAGE

Unit 5 discussed the fact that an electric current flowing in a wire produces a magnetic field about the wire. Another relationship between magnetic fields and electricity is that whenever a wire passes through a magnetic field, a voltage is induced in the wire. The amount of voltage that will be induced in the wire depends two factors: the strength of the magnetic field and the speed at which the wire cuts through the field.

Two basic principles apply in generating electric voltages:

1. Stronger magnetic field = higher voltage
2. Higher speed = higher voltage

GENERATED ELECTROMOTIVE FORCE (VOLTAGE)

A *generator* is a device that converts mechanical energy into electrical energy by mechanically rotating coils of wire in a magnetic field. An *alternator* is a device that converts mechanical energy into electrical energy by rotating magnetic fields past coils of wire.

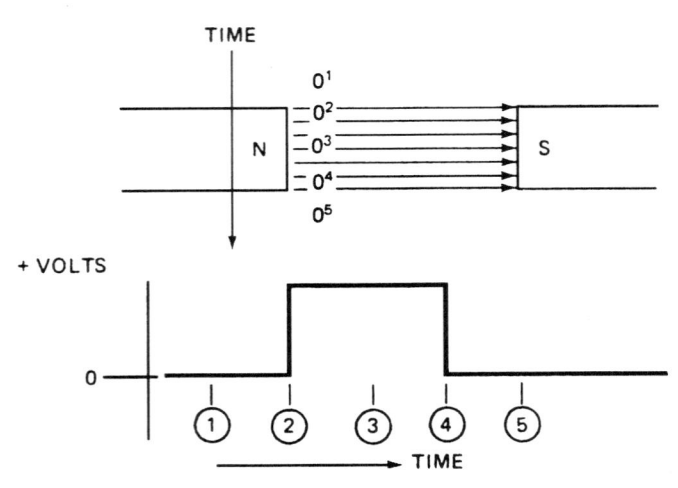

FIGURE 14–1 Passing a wire through a field

In Figure 14–1 a piece of wire is shown moving down through a uniform magnetic field. In position 1, the wire is not cutting through the field, and no voltage will be generated. As the wire enters the magnetic field (position 2), a level of voltage is established. As the wire moves through the uniform field (position 3) at a constant speed, the level of voltage remains constant. At position 4, the wire leaves the magnetic field, and the voltage drops to zero. In position 5, the wire does not cut through the magnetic field, and the voltage remains at zero.

The voltage generated, as shown in Figure 14–1, is a positive (+) voltage. The positive (+) voltage is generated as the wire is moved down through the field. If the wire were moved *up* through the magnetic field the opposite voltage polarity—negative (−)—would be induced.

ROTATING COILS

For practical mechanical reasons, generators are made up of coils rotating in a magnetic field. When a coil rotates in a magnetic field, an alternating voltage is produced. In Figure 14–2, an example of a coil section, one wire is shown. As the wire rotates through a complete revolution, a complete cycle of alternating voltage is produced. The complete revolution, divided into twelve equal parts, is shown.

In position 0, the wire is moving parallel to the magnetic lines, so the wire does not cut through them. No voltage is produced. As the wire moves from position 0 to position 1, it cuts through a few magnetic lines. The (+) indicates the direction of electron movement and produced voltage. The graph in Figure 14–3 indicates the level of voltage. As the wire moves from 1 to 2, it cuts through an increasing number of lines. A corresponding increase in voltage is indicated on the graph. At position 3, the wire is cutting directly down through the magnetic lines. The maximum voltage is generated, as indicated in the graph. As the wire moves to positions 4 and 5, it cuts through fewer lines until, at position 6, it is again moving parallel to the magnetic lines. The graph shows voltage decreasing to a zero value at position 6.

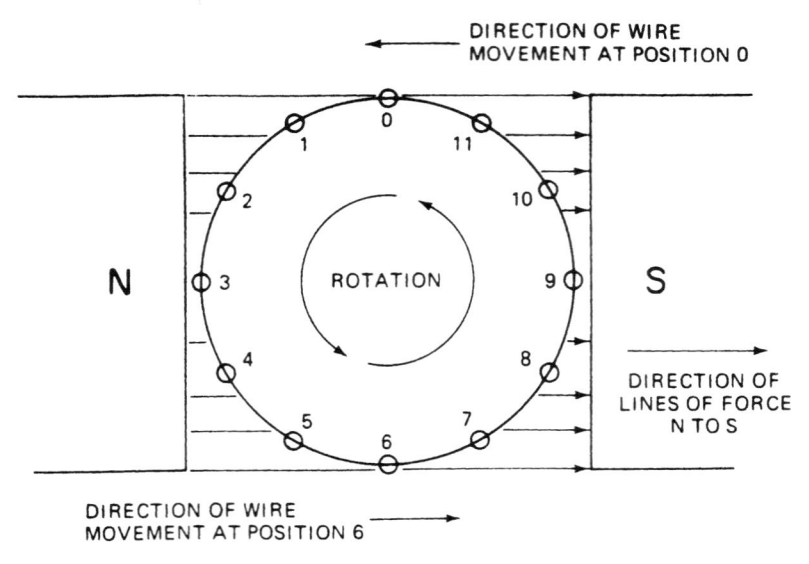

FIGURE 14–2 Rotating wire

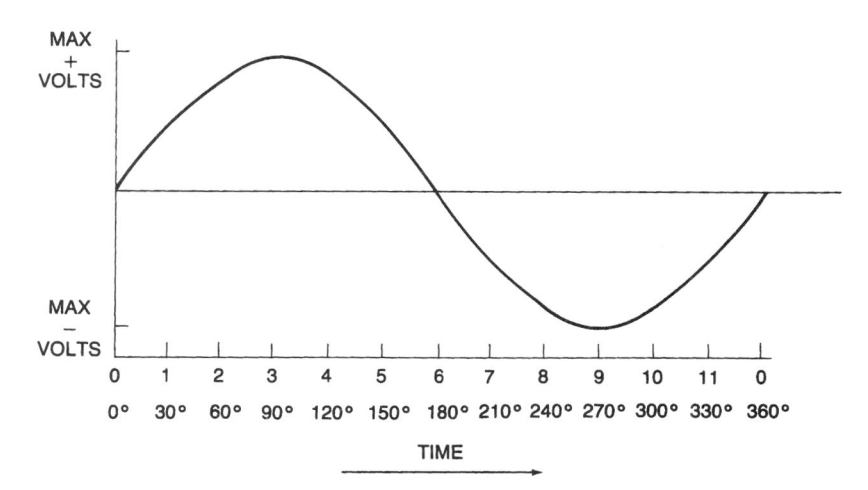

FIGURE 14–3 Sine wave

As the wire moves from position 6 to position 7, the direction of movement through the magnetic lines changes. The level of voltage starting at 7 is increasingly negative, reaching a maximum at position 9. The voltage decreases at 10 and 11, falling to 0 volts at position 0. This completes one revolution and one complete cycle of voltage. As the wire continues to rotate, another cycle starts.

The voltage produced in this manner is called *alternating* voltage because the output of the generator will first be positive (+) and then negative (−) at each of the output terminals.

SINE WAVE

The output voltage varies in the form of a sine wave. The generator winding cuts through the magnetic field at various rates for each position of rotation. As a measure of rotation, the circle is divided into 360 degrees. This is shown in Figure 14–3.

Most power companies in the United States generate electricity with a frequency of 60 hertz (Hz) also referred to as 60 cycles per second (cps). There are 60 complete cycles of positive (+) and negative (−) alternations during each time period of one second. Equipment designed to operate with voltage at this frequency is marked on the nameplate as 60 Hz. Many foreign countries produce voltages and operate equipment at 50 cycles per second (50 cps or 50 Hz).

In the automotive alternator the speed of rotation of the alternator rotor varies with engine speed. The frequency of the voltage produced also varies with engine speed.

THREE-PHASE POWER

Automotive alternators produce electricity in the form of three-phase (3-ϕ) alternating current (ac). There are two main systems of electrical connection of three-phase power. They are the delta (Δ) and wye (Y) connections, shown in Figure 14–4. When considering the operation of a generator or alternator, remember that it is the connection that determines output voltage and current.

In the delta (Δ) connection, the line voltage is equal to the coil voltage. The line current is equal to the coil current times the square root of three (1.732). The line current is the vector sum of the currents of the two coils joined at the line connection. A vector addition is necessary since the currents are 120° out of phase with each other.

In the wye (Y) connection, the line current is equal to the coil current. The line voltage is equal to the coil voltage times the square root of three (1.732). The line voltage is the vector sum of the coil voltages between two lines.

Vector summary is the method of adding forces (voltage or current) that are not acting at the same strength at the same time.

RESISTIVE AC CIRCUITS

The effect of ac voltage and current in resistive circuits is the same as the effect of dc voltage and current. Ohm's law holds true, as does the power formula:

$$(V = I \times R) - (I = V/R) - (R = V/I)$$
$$W = I \times V$$

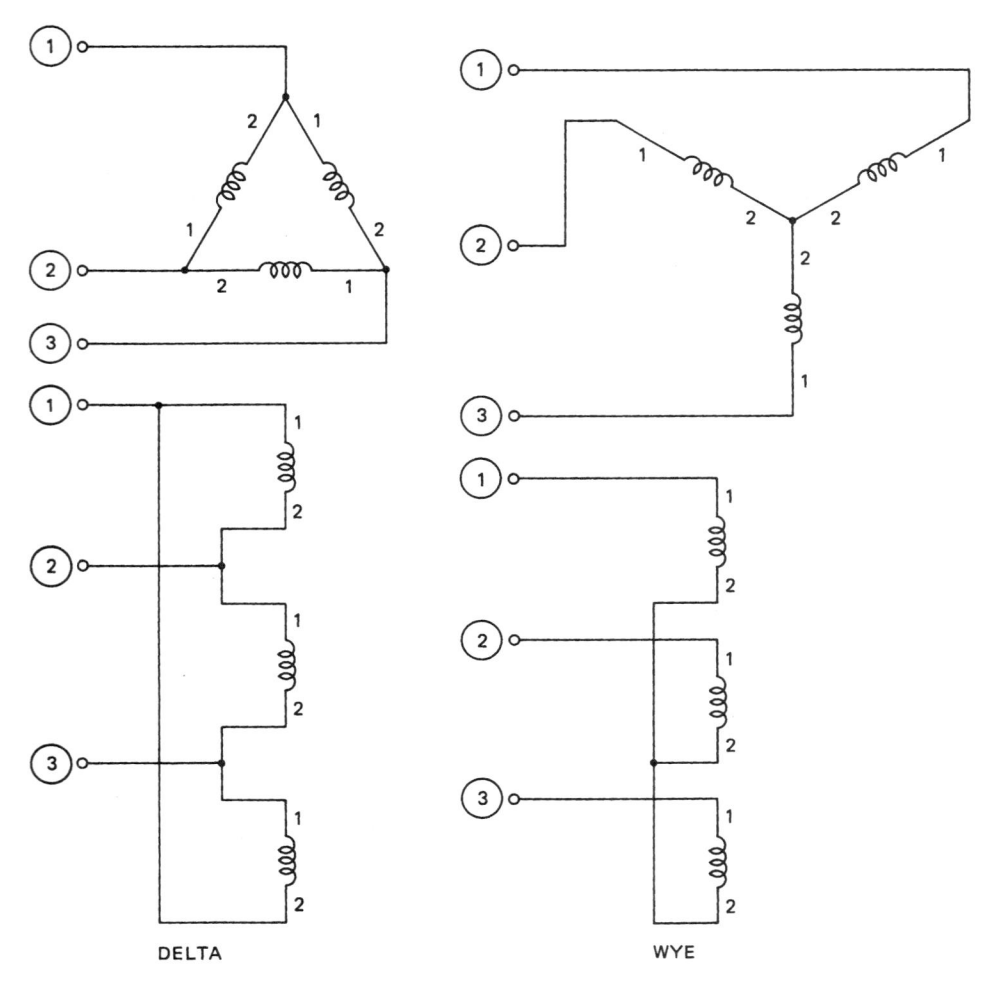

FIGURE 14–4 Three-phase delta (Δ) and wye (Y) connections.

CAPACITANCE IN AC CIRCUITS

In Unit 8 the charging of a capacitor with dc voltage was shown (see Figure 8–4). When an ac voltage is applied to a capacitor current will also flow.

Since ac voltage is continuously changing, current is always flowing, trying to charge the capacitor to a new voltage level. In Figure 8–4, charging a capacitor with dc voltage at time period 1, when the voltage is 0, the current is maximum. As the voltage across the capacitor rises, the current decreases. Finally, when the capacitor is fully charged, the current is 0. A similar situation exists with ac voltage.

In Figure 14–5, when the voltage is at 0 the current is maximum. When the voltage is maximum the current is 0. This condition is called a *leading current*. The current leads by 90° of one cycle.

INDUCTANCE IN AC CIRCUITS

The ac current and voltage in inductors (coils of wire) are as shown in Figure 14–6. In an inductor the current lags the voltage by 90°. This is similar to the situation found in dc circuits (see Figure 8–10).

TANK CIRCUITS

A special condition exists when a capacitor and coil are connected together. This circuit is called a *tank circuit* (Figure 14–7), and it oscillates.

When the switch is pressed (closed), power is applied to the circuit for a short period of time. The capacitor charges, and current rises through the coil.

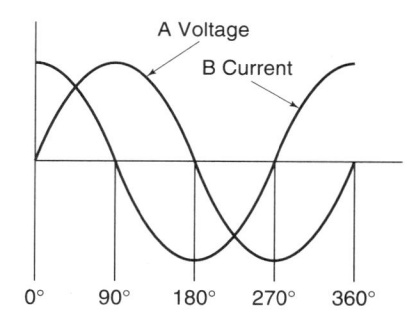

FIGURE 14–5 Current leading voltage in a capacitor

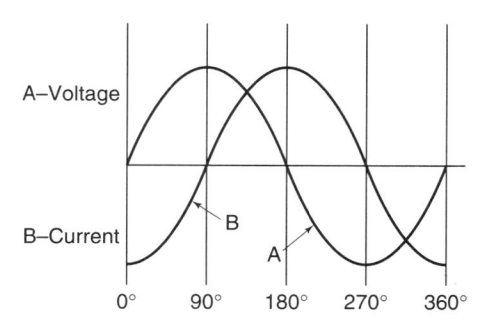

FIGURE 14–6 Current lagging voltage in an inductor

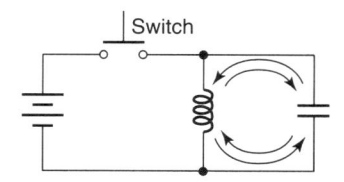

FIGURE 14–7 Tank circuit (parallel resonance)

When the switch is released (opened), power is removed from the circuit. The capacitor discharges through the coil. A magnetic field sets up about the coil. The magnetic field then collapses, charging the capacitor. Oscillation takes place as shown in Figure 14–8. Since a fixed amount of energy is provided to the circuit the amplitude of the oscillation decreases as electrical energy is converted into heat in the circuit resistance. The oscillations are "damped," that is, of decreasing amplitude.

These oscillations are similar to what takes place at the engine ignition coil. The frequency of the oscillation is controlled by the value of the capacitor and inductor, as given by the formula:

$$f_o = \frac{1}{2\pi\sqrt{LC}}$$

FIGURE 14–8 Oscillation in a tank circuit

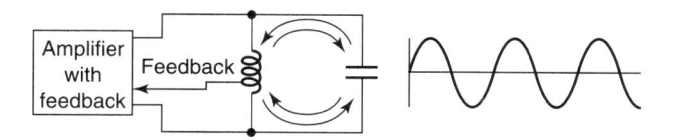

FIGURE 14–9 AC constant amplitude oscillator

When an electronic power amplifier device is connected with a tank circuit, constant amplitude oscillations can be maintained. The frequency of the oscillations, which is an ac voltage, is controlled by the value of the coil (L) and the capacitor (C). This is shown in Figure 14–9.

A pressure-sensing device, a variable capacitor, is used to sense manifold pressure in modern vehicles. Its variable capacitance with pressure is used to cause frequency variations of an oscillator. These frequency variations indicate pressure changes to the on-board computer.

ALTERNATORS

The alternator, shown in Figure 14–10, is an alternating current (ac)-generating device used on the modern automobile to provide the direct current (dc) needed to recharge and supplement the battery. The ac is converted to dc by rectifiers called *diodes*.

There are many types of alternators in use, but all have the same basic components and operating principles. Generally, there are two end frames, one stator, one rotor, one rectifier (circuit), and a brush set. End frames are usually made of die-cast aluminum (Al), with air vents or slots to provide for internal ventilation. The back end frame, often called the slip ring end housing, contains the brush assembly, the heat sink, and the diodes. The voltage regulator is also often found in this housing. A doughnut-shaped laminated steel frame with three sets of wire windings (coils) makes up the alternator stator, shown in Figure 14–11. These coil windings are arranged and connected to provide a three-phase wye (Y) or delta (Δ) connection, covered later. The rotor consists of two pole pieces and a slip ring on a rotating shaft member. Nested between and inside the pole pieces is a coil of wire, the field coil. Power transmission to the field coil

FIGURE 14–10 A typical alternator

FIGURE 14–11 A typical alternator stator assembly

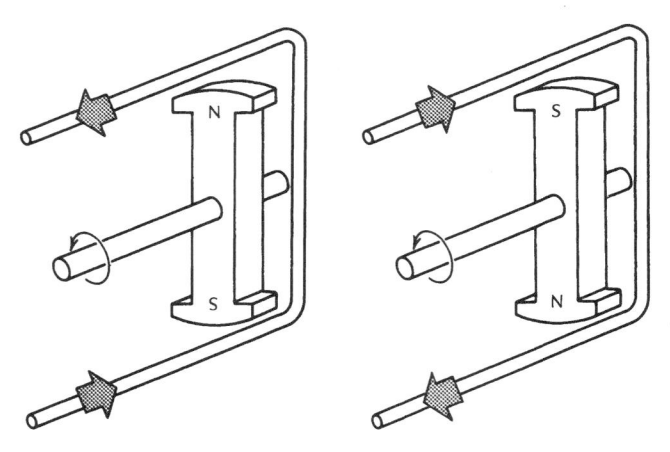

FIGURE 14–12 Induced voltage polarity reverses each half-revolution

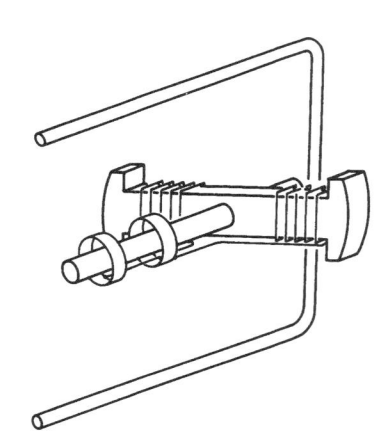

FIGURE 14–13 Coil added to rotating member.

is through the slip ring and brushes, located in the end housing.

Six diodes, three positive (+) and three negative (−), are usually used as rectifiers. A seventh diode is often used to isolate the alternator circuit.

INDUCED POLARITY

Earlier we showed that whenever a permanent magnet is rotated inside a coil of wire, a voltage is induced. This voltage is first in one direction and then in the other, depending on pole piece position, as illustrated in Figure 14–12. The voltage reverses twice each revolution (360°), or once every half-revolution (180°).

THE ROTOR

The strength of a permanent magnet is more or less fixed, and offers little control of the level of generated voltage. Picture an electromagnet with many turns of fine copper wire, as shown in Figure 14–13. The ends of

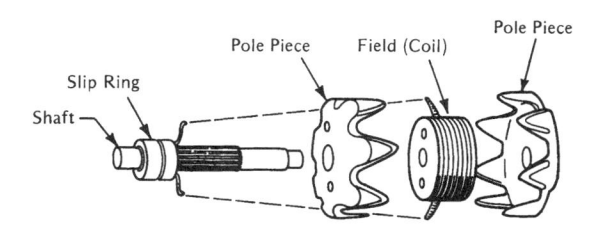

FIGURE 14–14 Rotor coil between pole pieces.

the wires (coil) are attached to flat copper rings (slip rings). A carbon composition brush against each ring transfers a controlled voltage to the coil, thereby controlling its electromagnetic strength. To make it more efficient, this coil (the field) is located between two pole pieces, Figure 14–14. Each pole piece can have 6 or more poles, all north (N) on one piece and all south (S) on the other piece. Regardless of the total number of poles, there is always an equal number of N-S poles. The coil of wire, the pole pieces, the slip rings, and the shaft make up the rotor assembly, which is held in place by bearings in the end housings. The end of the shaft opposite the slip ring has a pulley that is used to belt drive the alternator off the engine.

THE STATOR

The coil of wire through which the field (rotor) rotates is called the *stator*. There are actually three sets of wires interwoven in the stator, and each set has an output current independent of the others.

Since each winding produces a single-phase ac output, the alternator is actually a three-phase ac producing device. The three windings are connected in one of two ways: wye (Y) or delta (Δ), as shown in Figure 14–15. The stator is located between the two end housings and is held in place by through bolts. It is in a fixed position so that it is properly positioned around the rotor.

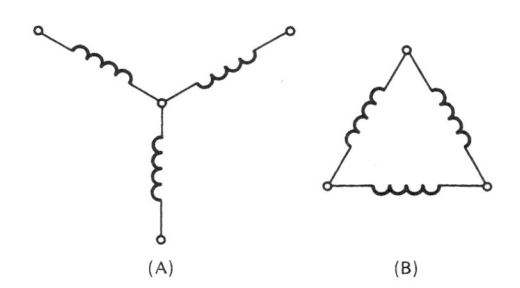

FIGURE 14–15 (a) Wye (Y) and (b) delta (Δ) connections

ALTERNATOR OUTPUT

The voltage produced by an alternator changes from positive (+) to negative (−) potential, illustrated by a sine wave. Note the sine wave produced by a simple two-pole rotor and single coil stator (Figure 14–16). The curve shows the amount of change in voltage through one revolution (360°) of the rotor. The voltage generated in the first half-revolution is positive (+) at 180°. It increases to its peak and then decreases. In the second half-revolution (180°), the voltage is negative (−). This is single-phase alternating current (ac). If rectified to direct current (dc), the voltage would vary as shown in Figure 14–17.

This pulsating voltage would not be suitable for automotive use. To flatten the pulses to steady dc, large filter capacitors would be required. The more practical solution, then, is to use a three-phase alternator. When operating, the ac voltage produced in the alternator is as shown in Figure 14–18. When rectified, the three-phase produces a dc output as shown in Figure 14–19. This voltage is relatively level. Compare the dc output in Figure 14–17 to the dc output in Figure 14–19.

DIODES

Diodes are devices that allow current to flow in one direction only. A kind of an electrical check valve, they allow current to pass through in one direction but block current in the opposite direction. Diodes are available in many sizes and shapes.

Diodes used as rectifiers in alternators are not always interchangeable. As mentioned earlier, diodes are two-terminal devices. The two terminals are the anode and the cathode (Figure 14–20). Electron flow is always from the cathode to the anode. Some circumstances require that the anode of the diode be connected to the diode case. In other situations the cathode must be connected to the case (see Figure 14–21). This requirement is demonstrated in Figure 14–22, where the three diodes on the left have their anodes connected together, whereas the three diodes on the right have three cathodes connected together. The three diodes on the left could be press-fit directly into the alternator end housing or frame, providing a ground. The three diodes on the right would be press-fit into a metal heat sink insulated from the alternator case. This heat sink is the positive (+) terminal of the alternator.

Some alternators have an isolation diode installed inside the alternator case. This diode protects the alternator should anyone reverse the battery connections. In alternators without the isolation diode, a

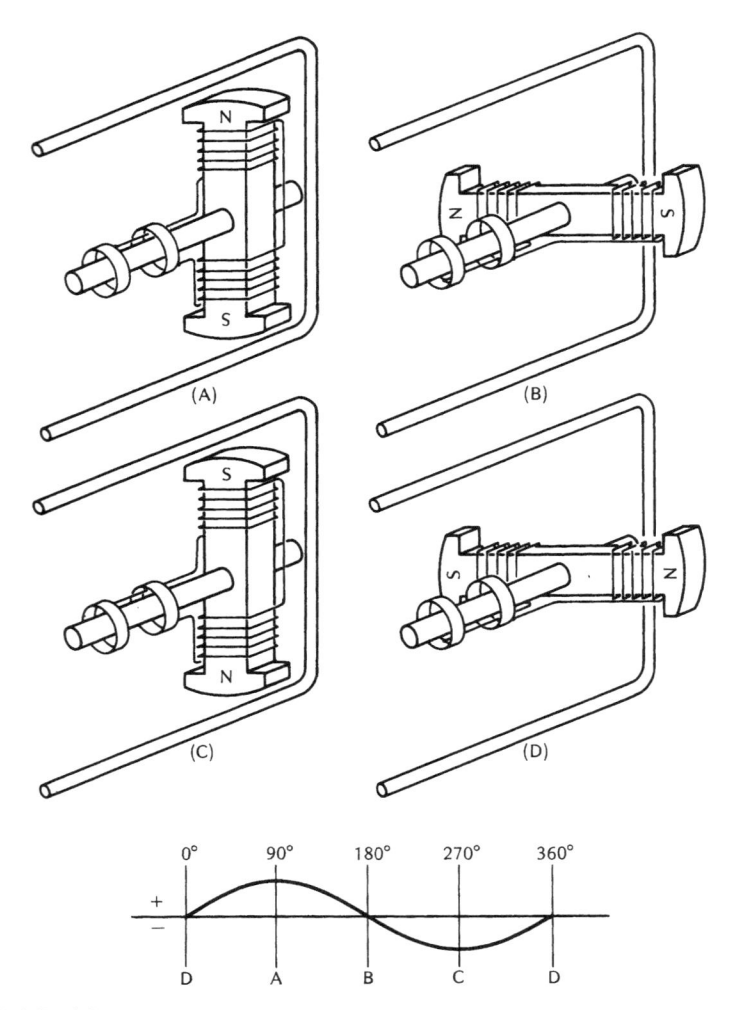

154   Unit 14 ALTERNATING CURRENT AND THE ALTERNATOR

FIGURE 14–16 One revolution (360°) of operation

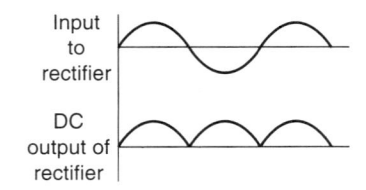

FIGURE 14–17 Rectifier input and output

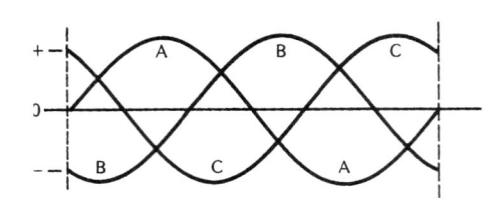

FIGURE 14–18 Two added windings produce phases B and C

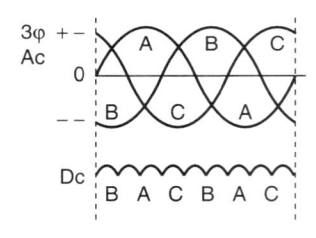

FIGURE 14–19 Rectified three-phase alternator

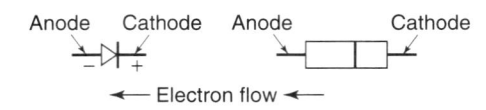

FIGURE 14–20 Simple diode showing cathode and anode

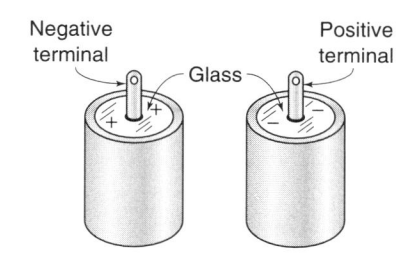

FIGURE 14–21 Positive (+) and negative (−) diodes for an alternator

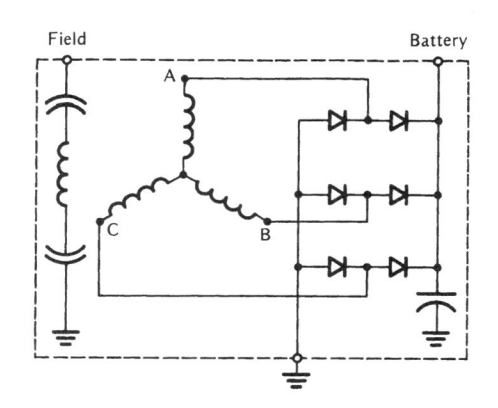

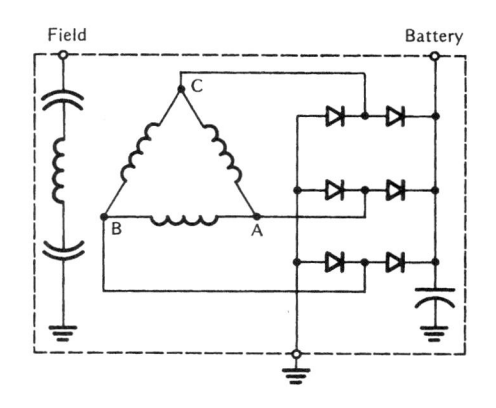

FIGURE 14–22 Wye (Y) and delta (Δ) alternators with six diode rectifiers

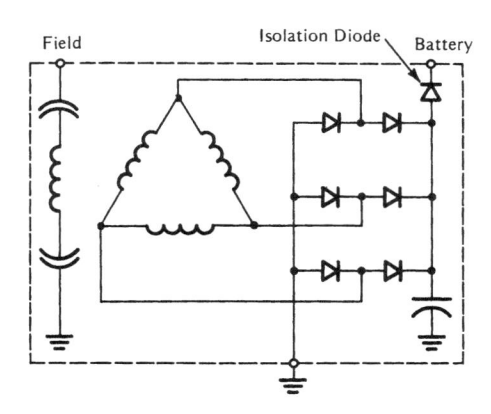

FIGURE 14–23 Isolation diode

reverse battery connection would cause most of, if not all, the diodes to burn out. (See Figure 14–23.)

VOLTAGE REGULATORS

Voltage regulators have been miniaturized in recent years so that they are now often matchbook-sized components inside the alternator slip ring end housing. Although their design and size have changed, the basic principle remains about the same.

The purpose of the voltage regulator is to sense the output from the alternator and to regulate the current flow to the alternator rotating field (rotor) so that the output of the alternator remains at 13.5 volts.

SUMMARY

- Alternators are used in modern vehicles to produce electricity.
- Alternators produce three-phase power in the form of sine waves. The three-phase power is rectified to provide direct current for general automotive use.
- Capacitors are used to provide filtering, that is, to keep the voltage constant.
- Voltage regulators provide current flow to the alternator field coils that maintain the voltage output of the alternator within the required limits.

PRACTICAL EXERCISE

A motor vehicle and a voltmeter are required for this exercise.

1. Open the hood. DO NOT start the engine.
2. Connect a voltmeter across the battery.
3. Note and record the voltage. It is _____ volts.
4. While observing the voltmeter, have a classmate start the engine.
5. Did the voltage rise or fall during starting? _____
6. What voltage was observed during Step 4? _____volts
7. While observing the voltmeter, ask your assistant to increase the engine speed.
8. Did the voltage rise or fall during Step 7? _____
9. What was the voltage during Step 7? _____ volts
10. While observing the voltmeter, turn the headlamps to ON and turn the heater blower motor to high speed.
11. The voltage across the battery is now _____ volts.
12. Increase the engine speed to between 1400 and 1500 rpm.
13. The voltage across the battery is now _____ volts.
14. Is the voltage in Step 13 higher or lower than that noted in Step 7? _____ In your opinion, why? _____
15. Reduce the engine speed to idle.
16. Turn off the heater blower motor and headlamps.
17. Disconnect the voltmeter from the battery.
18. Turn off the engine.

EXERCISE SUMMARY

1. The voltage should have dropped slightly during the starting procedure of Step 4. The amount of voltage drop depends on battery condition.
2. A voltage regulator controls the output from the alternator.
   a. If the voltage output is too high (over 14.5 volts) suspect the regulator or a wiring problem.
   b. If the voltage output is too low, inspect the belt(s), battery cable terminals, voltage regulator, alternator, and battery.
3. The voltage measurement of Step 13 should be slightly higher than the measurement taken in Step 7.

REVIEW

1. A device used to provide a specific voltage drop in an electrical system is called a
   a. resistor
   b. diode
   c. transistor
   d. capacitor

2. A device that will not permit the flow of dc current but will, under certain conditions, permit the flow of ac current is called a
   a. resistor
   b. diode
   c. transistor
   d. capacitor

3. The primary purpose of the regulator is to supply voltage to the alternator
   a. field
   b. stator
   c. armature
   d. frame

4. The solid-state device that has replaced the vacuum tube is the
   a. thermistor
   b. diode
   c. transistor
   d. zener diode

5. Based on information given in this section for one pair of rotor pole pieces, how many sine wave cycles (per revolution) would there be in a wye circuit with eight pairs of pole pieces?
   a. 6
   b. 12
   c. 24
   d. 48

6. Which of the following is NOT a term associated with alternators?
   a. commutator
   b. stator
   c. rotor
   d. field

7. The rotating member of an alternator is called the
   a. armature
   b. stator
   c. rotor
   d. commutator

8. The stator is made up of _____ winding(s).
   a. one
   b. two
   c. three
   d. four

9. There are usually _____ diodes in an alternator rectifier circuit.
   a. two
   b. four
   c. six
   d. eight

10. There are _____ pole pieces in an alternator.
    a. two
    b. four
    c. six
    d. eight

◆ UNIT 15 ◆

LIGHTING

OBJECTIVES

On completion of this unit you will be able to:

- Identify the types of headlamps used in modern vehicles.
- Understand how headlamps circuits are protected.
- Aim headlamps.
- Discuss automotive headlamp control.
- Do general automotive lighting service.
- Understand the operation of turn signals and hazard flashers.

The automotive lighting system consists of two general groups: lights required by law and lights provided for convenience. The law requires headlamps, tail lamps, stop lamps, turn signal lamps, license plate lamps, and marker lamps. Convenience lights include under-hood, trunk, map compartment, interior, courtesy, and instrument lamps. The total lighting system includes hundreds of feet of wiring with many bulbs, fuses, circuit breakers, switches, and controls. We cover typical standard and optional lighting systems, but lighting circuits, like other automotive circuits, vary from car to car. Therefore, manufacturer's schematics and diagrams must be consulted for a specific system.

HEADLAMPS

The sealed-beam headlamp, pioneered in 1928 by Delco-Remy, was introduced in 1939. Federal safety legislation of 1940 made sealed-beam headlamps mandatory in all states. The four-lamp headlamp system was introduced in 1958. Until a few years ago, all headlamps were round (Figure 15–1); now, rectangular headlamps (Figure 15–2) are common. But both round and rectangular headlamps are used, depending on design. Other than the shape of the lamp, few changes have been made since its inception.

Perhaps the most significant development in recent years is Tung-Sol's Guard-Glo ™. Tung-Sol, a division of Wagner Electric Corporation, has developed a headlamp with a third filament. Known as the reserve filament, it is of a higher resistance than, and is shunted by, the main filament. If the main filament burns out, the reserve filament immediately burns to provide a beam that can be seen by oncoming drivers at least 500 feet (152.5 meters) away. Although the reserve filament does not provide normal lighting, it does serve as emergency lighting until a replacement can be made.

There are currently two types of sealed-beam lamps: the single-filament lamp, which serves as the high beam of a four-lamp system, and the two-filament lamp, which serves as both high and low beam of a two- or four-lamp system. Regardless of the type, the filaments are hermetically sealed in an all-glass enclosure. The reflective surface, usually of parabolic shape, is aluminized. The filaments are placed forward of the center of the reflector in the correct position for optimum focusing. The lens is designed to direct the light to form the required beam pattern. Current two- and four-lamp systems are shown in Figure 15–3.

The brightness of a lamp is expressed in candlepower (cp). By definition candlepower is "Luminous intensity expressed in terms of standard candles," as determined by the National Bureau of Standards.

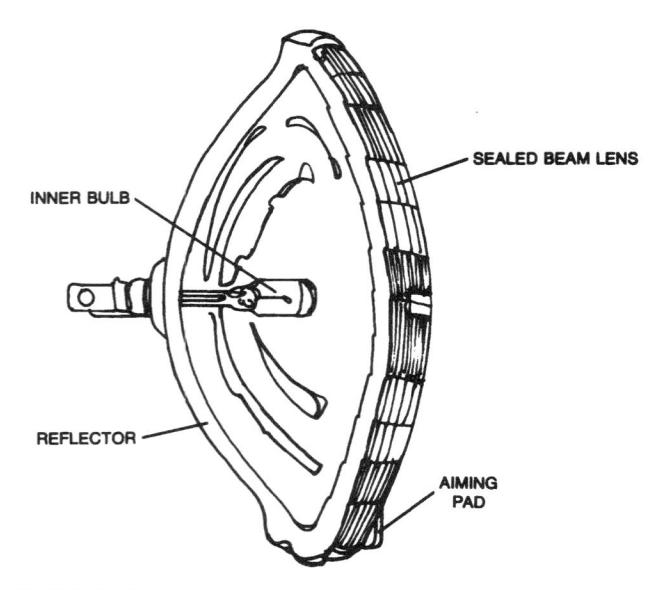

FIGURE 15–1 Round headlamp

Halogen sealed-beam headlamps (Figures 15–1 and 15–2) provide about 150,000 cp. Except for the inner bulb, they are similar in construction to the tungsten sealed-beam headlamp, which provides about 75,000 cp. The filaments of the halogen lamp are in a bulb within a bulb; the filaments of the tungsten lamp are in a single glass enclosure (bulb).

Headlamp Circuit Protection

Headlamps draw surprisingly high current. Consequently, the headlamp switch is rated for high current service and can withstand brief inrush current and momentary overloads. The headlamp circuit is protected by a quick-reset circuit breaker built into the switch. Fusing the circuit would leave a car blacked out, should a momentary short circuit or overload occur. The circuit breaker has a set of normally closed (NC) contacts, one of which is affixed to a bimetallic strip. High current, beyond its critical limit, causes the bimetallic strip to bend, which opens the contacts to interrupt the circuit. By design, the bimetallic strip then cools rapidly to return to its original position, closing the contacts.

The bimetallic strip is composed of two dissimilar metals, such as brass and invar, fused together. The two metals expand and contract at different rates when heated or cooled. Since they are bonded together, this action causes the strip to bend (Figure 15–4).

Headlamp Aiming

Proper aiming of headlamps is essential to maintain driver visibility and to prevent "blinding" drivers of approaching cars. States that have mandatory vehicle inspections check for properly aimed headlamps. Poorly aimed headlamps top the list of defects found during those inspections.

Although headlamp aiming is a mechanical procedure, it is important that the automotive electrical technician understand it. If headlamps are not properly aimed after replacing a burned-out lamp, the job is only half-done. The method of aiming depends on the equipment used, and the technician should follow the instructions included with the equipment. Aiming headlamps is a simple service, easily accomplished in a short period of time. Figure 15–5 shows the location of adjusting screws for both horizontal and vertical adjustments.

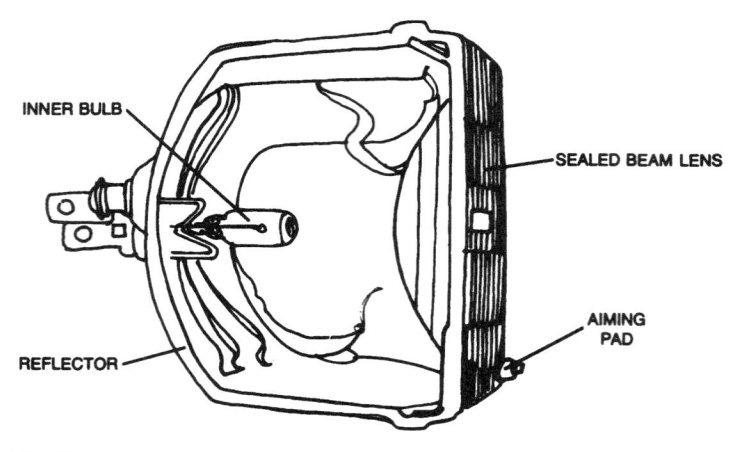

FIGURE 15–2 Rectangular headlamp

HEADLAMPS

HORIZONTAL 4-HEADLIGHT SYSTEM

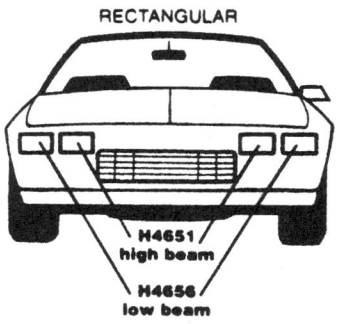

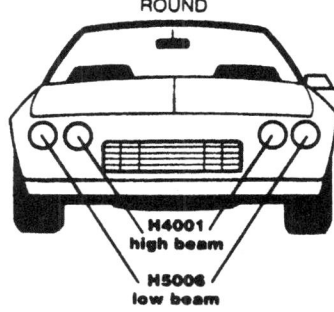

RECTANGULAR
- H4651 high beam
- H4656 low beam

ROUND
- H4001 high beam
- H5006 low beam

VERTICAL 4-HEADLIGHT SYSTEM

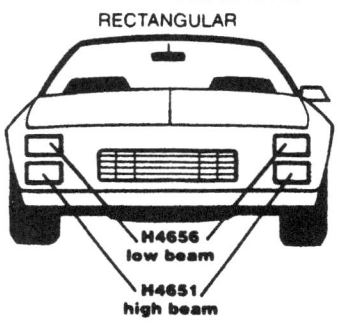

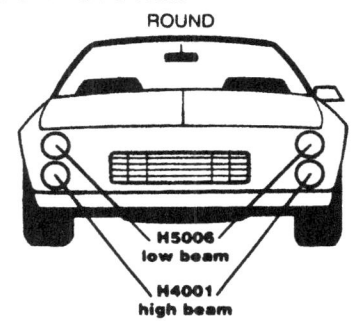

RECTANGULAR
- H4656 low beam
- H4651 high beam

ROUND
- H5006 low beam
- H4001 high beam

2-HEADLIGHT SYSTEM

RECTANGULAR
- H6054 high low beam

ROUND
- H6024 or H6014 high low beam

THIS GE POWER PLUS™ HALOGEN HEADLAMP	REPLACES THIS STANDARD HEADLIGHT
H4001	4001 5001
H4651	4651
H4656	4652
H5006	4000
H6024 H6014	6014
H6054	6052

FIGURE 15-3 Headlamp systems

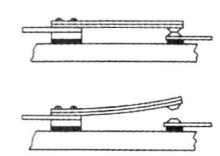

FIGURE 15–4 Bimetallic circuit breaker

Headlamp Dimmer Switches

The dimmer switch, when a foot-operated device, is found at the extreme left of the floorboard. When hand operated, it is located on the steering column, usually as a part of the turn signal lever. The dimmer switch is an ON-ON switch used to shift the headlamps from the low-beam circuit to the high-beam circuit, or vice versa. A simple schematic of the dimmer switch circuit is shown in Figure 15–6. The switch can be electrically checked using a test lamp, as shown in Figure 15–7. If current is available at point A when the headlamp switch is turned on, but not at point B (on low beam) or at point C (on high beam), the switch is defective and must be replaced. If there is no current available at point A, the problem is before the dimmer switch.

Automatic Headlamp Control

Solid-state technology, responsible for many of today's creature comforts, provides for the automatic dimming of headlamps when meeting an approaching car. The approaching headlamps are sensed by a photoelectric cell. When the light intensity reaches a predetermined level, an amplifier section triggers a power relay, causing the headlamps to switch automatically from high beam to low beam. The system holds on low beam until the photocell is darkened after the oncoming car has passed. This mechanism also prevents shifting to high beam if the approaching car also shifts to low beam. A typical wiring schematic for the automatic dimmer is shown in Figure 15–8.

There is a relay in this system that is the main power control to the high- or low-beam headlamps. If the relay is energized, the low-beam headlamps are on. If the relay is not energized, the high-beam headlamps are on. The power relay is located in the lower left of Figure 15–8.

Follow a circuit from the fused headlamp power source through the headlamp switch to the dimmer switch. If the dimmer switch is in the LO position, power is fed through the switch to the power relay, which energizes. The common movable contact mates with the LO contact. Power is fed from the headlamp switch through the relay to the LO beam headlamps.

When the dimmer switch is actuated to the HI AUTO position, the power relay deenergizes. The connection from the headlamp switch is now through the deenergized contact of the power relay (HI) to the high-beam headlamps.

With the dimmer in HI AUTO switching of the headlamp is through the auto headlamp control. This device is activated by a photocell that senses the headlamps of an approaching vehicle. Light striking the photocell causes the cell resistance to lower. This change is sensed and amplified, which causes +12 volts to appear at the DIM output terminal of the auto headlamp control. The 12 volts is fed through the dimmer switch, now in the HI AUTO position, to the power relay. The power relay energizes, switching power to the low-beam headlamps.

When the approaching vehicle's light no longer strikes the photocell, the +12 volt output from the auto headlamp control is reduced to zero and the power relay deenergizes. The headlamps switch back to high beam.

Twilight Sensing

Twilight sensing for automatic turn-on of headlamps is also controlled by a photocell. The "twilight" photo cell is placed so as to "look up" to determine the brightness of daylight. When twilight is encountered, the sensor bypasses the regular headlamp switch and turns the headlamps on. Power to the sensor is through the ignition switch. A suitable delay is incorporated into the twilight sensing to eliminate the turning off of headlamps when stopped for a short time under a street lamp, for example (Figure 15–9).

Headlamps Delay Off

Headlamps delay off is operated with control from twilight sensing. If twilight sensing calls for headlamp power with the ignition switch on, the delay timer will keep the headlamps on even after the headlamp switch is turned off. The time the headlamps remain on is selected through a delay time potentiometer, as shown in Figure 15–9.

Headlamp Switches

Although commonly called a headlight switch, this device controls parking lamps, tail lamps, dash lamps, marker lamps, and interior lamps. It is a driver-operated switch, usually a two-position push-pull type that can also be rotated. There are many variations of

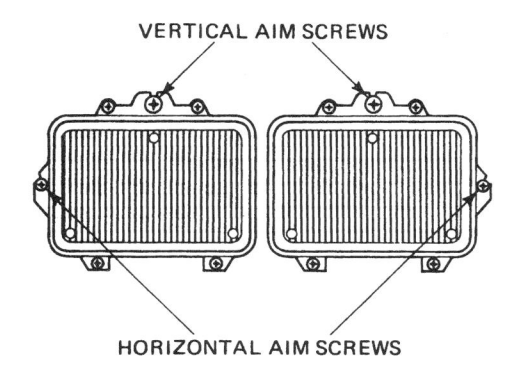

Location of headlight aiming screws. (Courtesy of Pontiac Motor Division, GMC.)

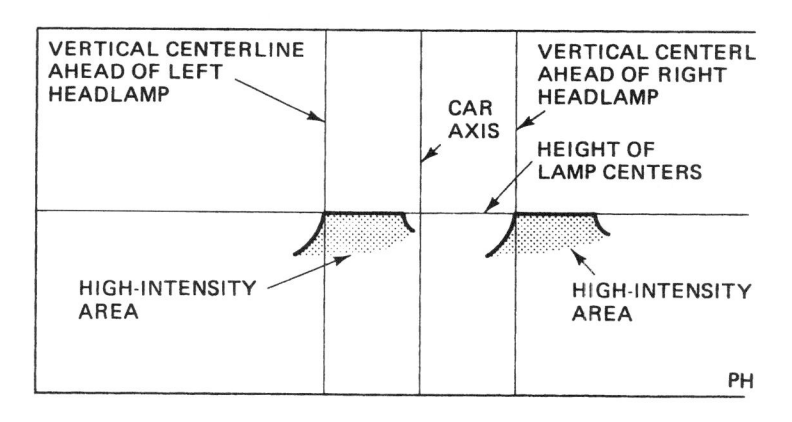

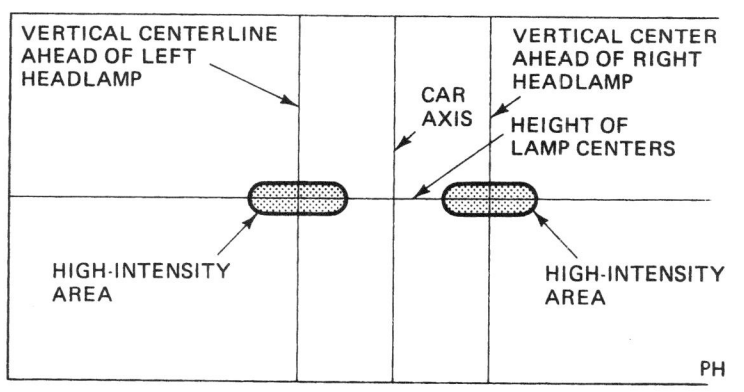

The upper pattern represents what the headlights should look like 25 ft (7.6 m) from a wall on low beam. The lower pattern represents a properly aimed high-beam pattern. (Courtesy of Chrysler Corporation.)

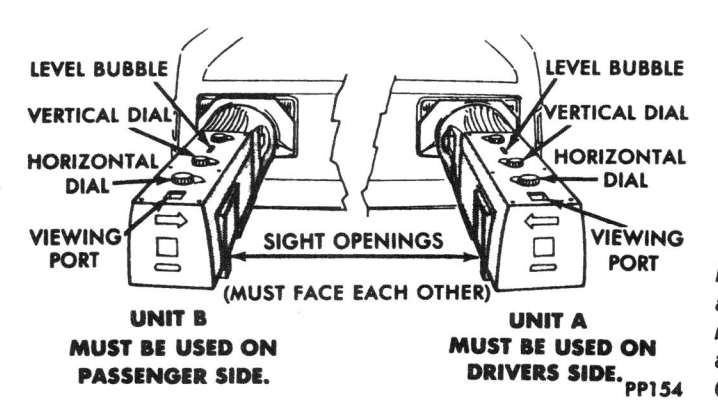

Typical mechanical headlight aimers. Always follow aiming equipment manufacturer's instructions for correct calibration and aiming procedures. (Courtesy of Chrysler Corporation.)

FIGURE 15–5 Headlight aiming

164 Unit 15 LIGHTING

FIGURE 15–6 Simple headlamp dimmer switch circuit

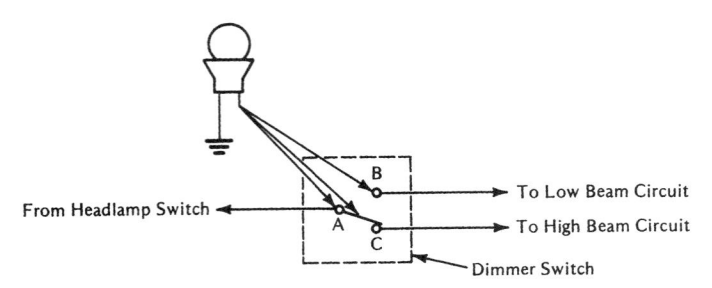

FIGURE 15–7 Testing a dimmer switch with a test lamp

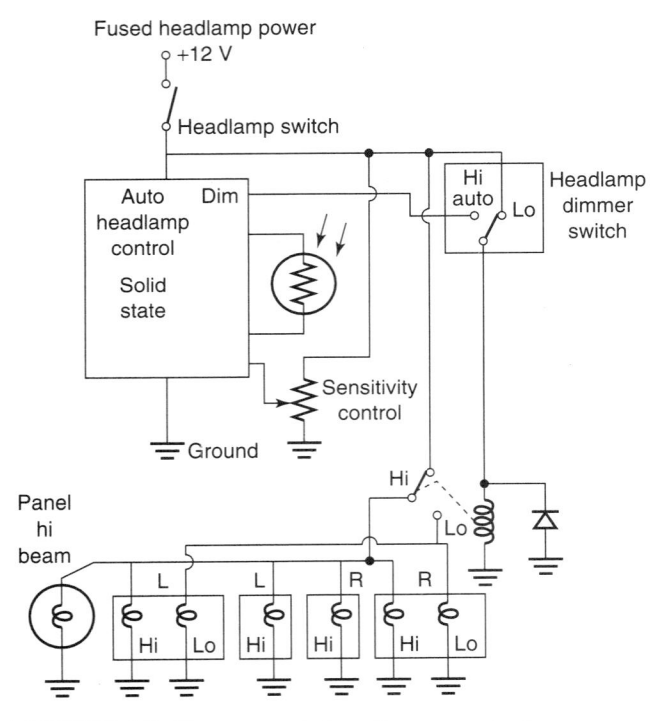

FIGURE 15–8 Automatic headlamp control

tion for the running lights. In this position the headlamps, tail lamps, dash lamps, and side marker lamps are on.

Rotating the switch knob in either ON position brightens or dims the dash lamps. Rotating the knob fully clockwise (cw) or counterclockwise (ccw), depending on design, with the switch in either ON or OFF position turns the interior lamps on and off.

The knob and shaft of the headlamp switch is usually a one-piece assembly—the knob cannot be removed from the shaft for switch removal. Therefore, to remove the switch from the dash panel for service or replacement, proceed as follows:

1. Disconnect the battery ground (−) cable.
2. Pull the headlamp switch to the full ON position.
3. Locate and press the spring-loaded lock.
4. While pressing the lock, pull the stem from the switch.
5. Remove the switch retainer. This is usually accomplished by rotating the retainer counterclockwise (ccw) from the front of the dash panel.
6. Remove the switch from the rear of the dash panel.

the headlamp switch used on late-model vehicles, however. We concentrate here on the push-pull type, which is by far the most popular.

When the switch is pushed fully in it is in the OFF position. Pulled out to the first ON position, the park lamps, tail lamps, dash lamps, and side marker lamps are activated. Pulled fully out is the ON posi-

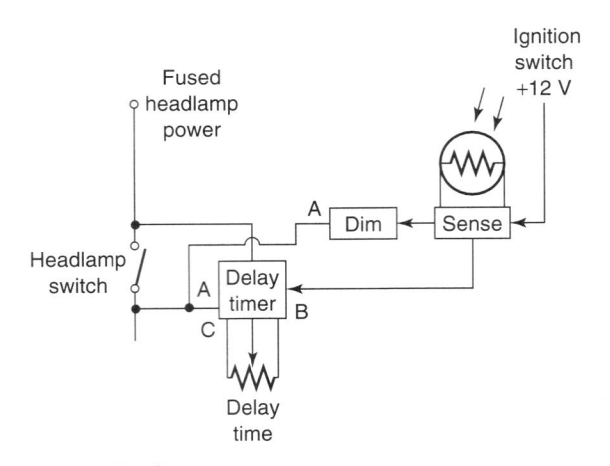

FIGURE 15–9 Twilight sensing

For reassembly it is not necessary to push the spring-loaded lock. After replacing the switch retainer removed in Step 5, insert the stem/knob assembly and push it into the switch. It should automatically lock into position. WARNING: Take care when working under the dash. Even with the battery disconnected, carelessness can deploy the air bag. Follow all precautions noted in appropriate manufacturer's service manuals.

PARKING LAMPS

Parking lamps are provided on the right and left front of the vehicle to provide illumination while parked. In many states it is illegal to use these lamps for primary illumination while driving. They do, however, offer secondary emergency lighting should the headlamps fail.

All parking lamps are illuminated when the headlamp switch is in the first ON position. Some may also be illuminated when the switch is in the full ON position. Parking lamps Figure 15–10, are usually of the double-contact bayonet dual-filament type.

The second filament of the parking lamp is used in the turn signal circuit. The "park" filament provides four candlepower (4 cp) of illumination and draws just over 0.5 amperes at 123 volts. Whenever the headlamp switch is in the first ON position, the park lamps and tail lamps share a common circuit. The parking circuit is protected either by a fuse in the main fuse block or by a circuit breaker in the headlamp switch.

TAIL LAMPS

Tail lamps are provided on the right and left rear of the vehicle for illumination while parked or driving. They are illuminated whenever the headlamp switch is in either ON position. There may be one, two, or more tail lamps on each side of the rear of the vehicle, depending on style and body design. At least one lamp on each side is of the double-contact, bayonet, dual-filament type. The second filament is used in the turn signal and brake warning circuit.

If there is more than one lamp on each side, the other lamps may be of the single-filament type with no brake warning or turn signal function. Figure 15–11 shows a single-filament, single-contact, bayonet-base bulb. On the other hand, two (or all) lamps may serve the dual function. The tail lamps generally provide 4 cp of illumination.

LICENSE PLATE LAMP

Most states require that the license plate be illuminated after dark. This is generally accomplished by one or more low-candlepower-rated bulbs. These bulbs, usually of the single-filament, single-contact, bayonet-base type, are part of the taillight circuit. As part of this circuit they are illuminated any time the headlamp switch is in either ON position.

SIDE MARKER LAMPS

Side marker lamps are located on each side of the vehicle, front and rear. The front lamps are either clear with amber lens or amber with clear lens, providing amber illumination in either case. The rear lamps are clear with red lens, providing red illumination. The lenses are designed to have a reflective quality should the bulb burn out. The side marker bulbs burn continuously whenever the headlamp switch is in either ON position. They are often connected with the turn signal circuit, and flash on and off when the turn signal is activated.

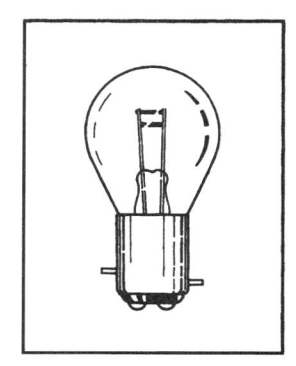

FIGURE 15–10 Typical parking lamp—double contact, indexing (grounding is accomplished through the case)

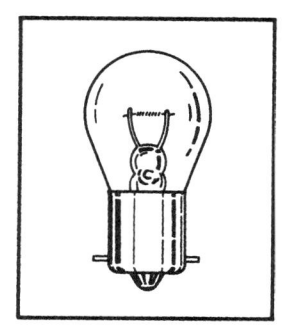

FIGURE 15–11 Typical single-contact, bayonet-base bulb (grounding is accomplished through the case)

CORNERING LAMPS

Cornering lamps, not to be confused with side marker lamps, are found on most luxury-class vehicles. They share a common circuit with the tail lamps, but are generally controlled through a separate set of contacts in the turn signal switch assembly. Their purpose is to provide additional "cornering" illumination when making a turn. When turning, either right or left, the respective cornering lamp is illuminated if the turn signal switch is activated. It does not flash with the turn signals, but remains illuminated until the turn signal switch turns it off after making the turn.

CLEARANCE LAMPS

Clearance lamps are usually associated with trucks. They are used to mark the upper front and rear extremes of a vehicle, and their number and location is a matter of law in most states. Specific information can generally be obtained from state law enforcement agencies or may be found in manufacturers' specification manuals.

Some luxury-class and imported vehicles may have clearance lamps for appearance. Their use and circuits vary. Therefore, for specific information the technician must consult the appropriate service manual.

BACKUP LAMPS

Backup lamps are provided on the rear of the vehicle to provide visibility when backing at night. These lamps are, however, illuminated day or night whenever the vehicle's transmission is placed in reverse. This, in effect, gives a pedestrian or person in a vehicle stopped behind you visual warning that your vehicle is primed to go into reverse. Some vehicles are also equipped with an audible, as well as a visual, warning.

The backup lamp (and/or alarm) switch is activated by the transmission shift linkage. On some car lines this switch is a part of the neutral safety switch; on others it is a separate switch. The location and type switch and linkage arrangement varies. It is therefore necessary to refer to applicable service manuals for specific information.

Lamps (Bulbs)

Lamps, or bulbs, as they are commonly called, used in automotive service (Figure 15–12) have an undetermined life expectancy. Some have been known to give

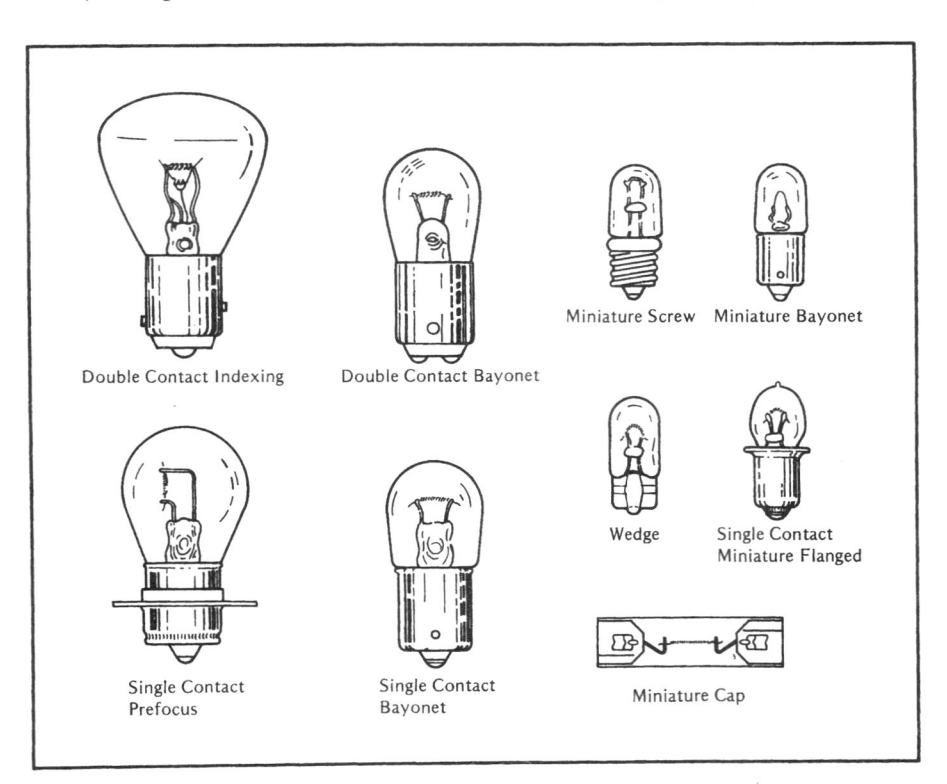

FIGURE 15–12 Some typical bulb types used in automotive service

many years of service; others have lasted only a few minutes or hours. Most bulbs have a predetermined service life, but, because of rough automotive service, this life expectancy may not be valid.

Automotive bulbs are identified by number. For example, a typical park/turn signal bulb is known as a number 1157. This is pronounced in the trade as "eleven fifty-seven." Also, it is an 1157 regardless of whether it is marketed by General Electric, Sylvania, Tung-Sol, Westinghouse, or any other manufacturer. Trade references to bulbs, used by parts counter representatives and automotive technicians, help to avoid confusion and set a standard. Some examples of these references are given in Figure 15–13.

TURN SIGNALS

Turn signals warn other drivers that a vehicle is to make a right or left turn. The turn signal circuit (Figure 15–14) consists of at least four double-filament lamps, two dash indicator lamps, one flasher, and one turn signal switch.

The turn signal circuit is usually integrated with the hazard and brake light circuits, covered later in this unit. For clarity, however, only the turn signal circuit will be covered at this time. The turn signal switch selects the right or left side front and rear lamp circuit, depending on driver action. When the turn signal switch lever, located at the left-center of the steering column just below the steering wheel, is pulled down, the left turn signal circuit is activated. When it is pushed up, the right turn signal circuit is activated.

OFF Position

Power is available from the ignition switch ACC terminal, through the fuse block, to the flasher. Since the contacts of the flasher are normally closed (NC), power is also available at the turn signal switch. And since neither circuit of the turn signal switch is closed, none of the lamps will light. This is the neutral or OFF position of the switch, as shown in Figure 15–15.

LEFT Turn

When the turn signal switch lever is pulled down to signal a left turn, the switch center contact "makes" with the left-front and left-rear contacts and completes a circuit (Figure 15–16) through the turn signal switch to the left-front and left-rear bulbs. The flasher then becomes sensitive to current flow and "breaks" the circuit. Consequently, the flasher repeatedly "breaks" and "makes" the circuit until the turn signal switch is canceled.

RIGHT Turn

When the turn signal switch lever is pushed up to signal a right turn, the switch center contact "makes" with the right-front and -rear contacts and completes a circuit (Figure 15–17) through the turn signal switch to the right-front and right-rear bulbs.

Number	Spoken As	Not
97	ninety-seven	nine seven
257	two fifty-seven	two five seven
1034	ten thirty-four	one zero three four
4001	four double-oh one	four zero zero one
4416	forty-four sixteen	four four one six

FIGURE 15–13 Trade reference terms for some common automotive lamps

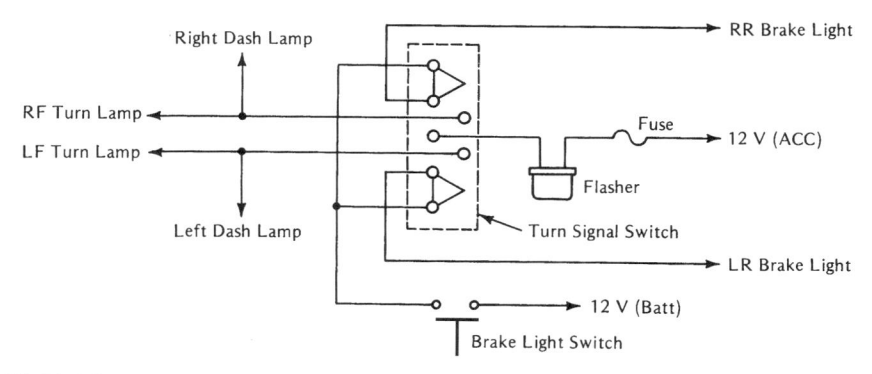

FIGURE 15–14 A typical turn signal circuit (switch is shown in the OFF position)

168 Unit 15 LIGHTING

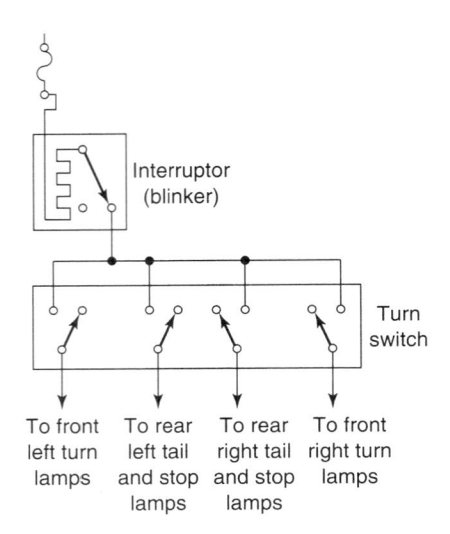

FIGURE 15-15 Turn signal switch in the OFF position

Front Side Marker Lamps

In some designs, the front side marker lamps are part of the parking lamp and turn signal circuit. The side marker lamps in the circuit shown in Figure 15-18 are grounded either through the parking lamp filament or the turn signal lamp filament, depending on the position of the headlamp and/or turn signal switch. These variations are illustrated in Figures 15-19 through 15-21. When the turn signal switch is activated with the parking lamps OFF, the front turn signal lamp and side marker lamp flash simultaneously.

Trace the circuit in Figure 15-19 from the headlamp switch through the park lamp filaments to ground and through the side marker lamp filaments to ground via the turn signal lamp filaments.

Trace the circuit of Figure 15-20 from the turn signal switch through the turn signal lamp filament to ground and through the side marker lamp filament to ground via the park lamp filament. The side marker lamp will illuminate simultaneously with the turn signal lamp.

Trace the circuit of Figure 15-21 from the turn signal switch through the turn signal lamp filament to ground and from the headlamp switch through the park lamp filament to ground. The side marker filament is grounded through the turn signal lamp filament only during the interval the turn signal lamp is not illuminated. The side marker lamp will illuminate alternately with the turn signal lamp. The left side marker lamp

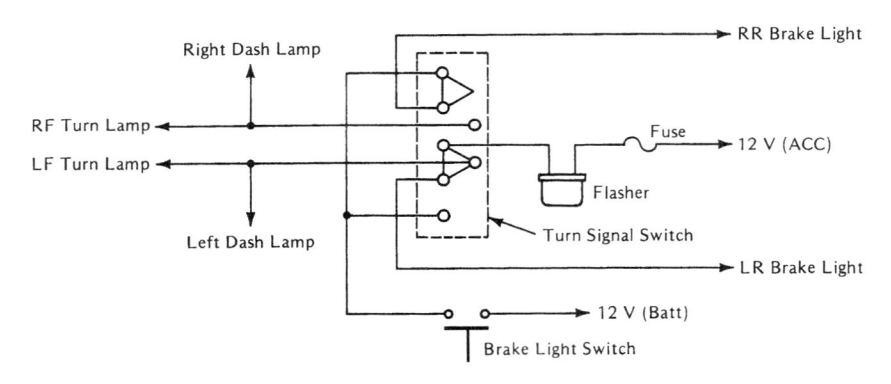

FIGURE 15-16 Turn signal switch position for a left turn

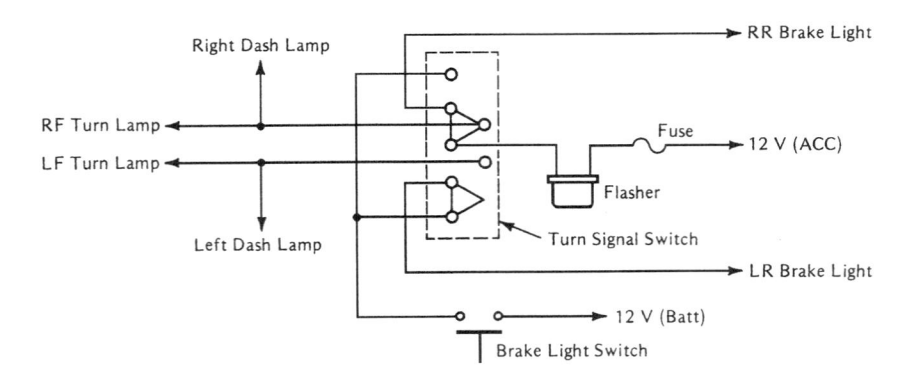

FIGURE 15-17 Turn signal switch in position for a right turn.

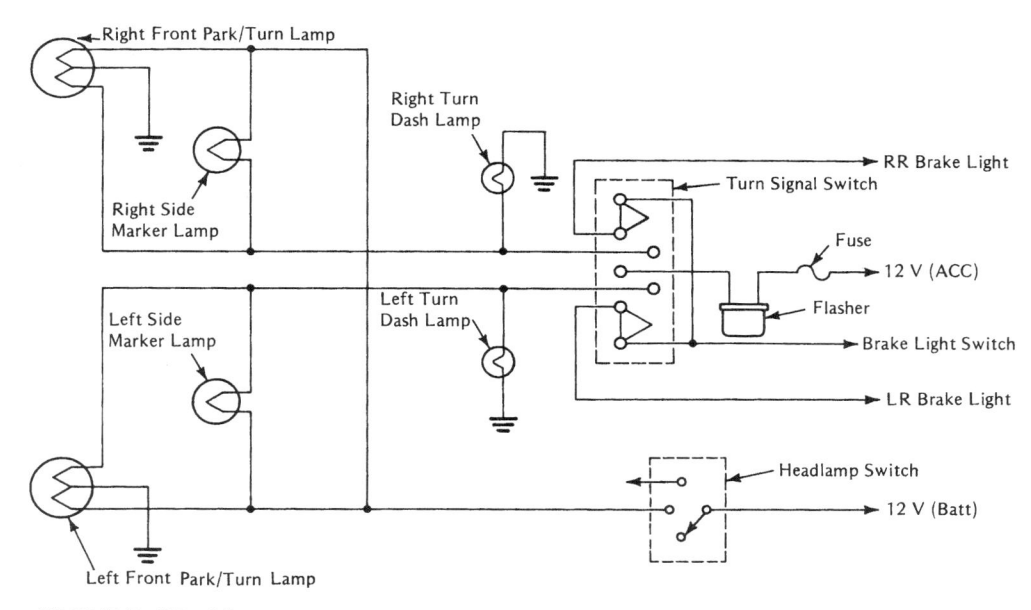

FIGURE 15–18 Typical side marker lamp schematic with all switches in the open or OFF position

will be illuminated because it is grounded via the right turn signal lamp filament.

Power is available from the turn signal switch, and the circuit of the front turn signal lamp is complete to ground through its own filament. At the same time, the circuit of the front side marker lamp is also complete to ground through the parking lamp filament.

When the turn signal switch is activated with the parking lamps on, the front turn signal lamp and side marker lamp flash alternately. Power is now available from the turn signal switch and the headlamp switch. The circuit of the front turn signal lamp is complete to ground through its own filament. Now, however, the same voltage is available to both sides of the side marker lamp and, since there is no ground path, it will

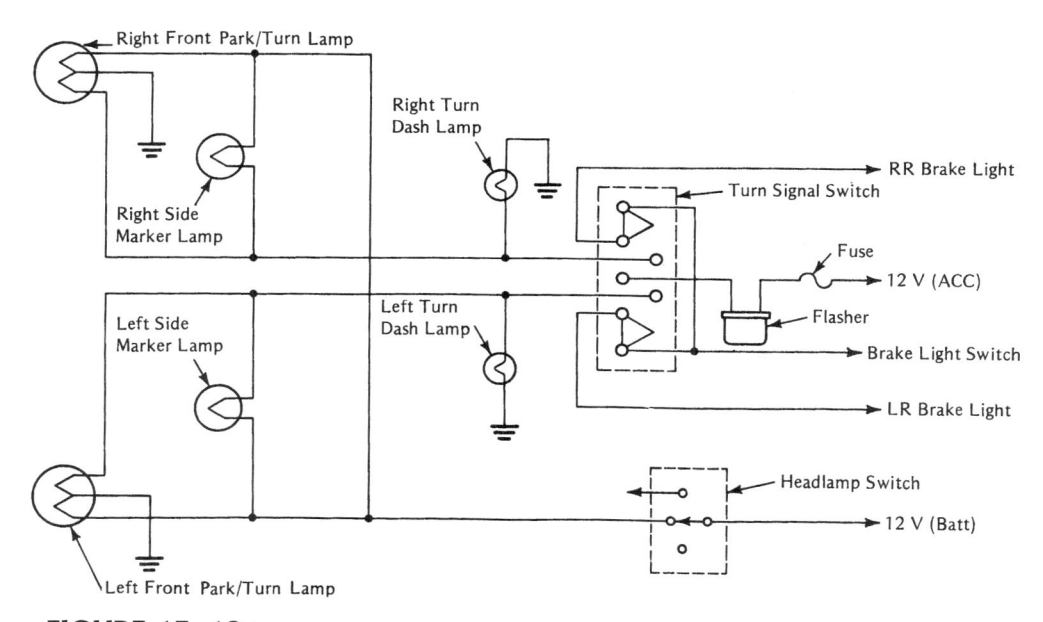

FIGURE 15–19 Side marker lamp schematic with turn signal switch in center position and headlamp switch ON.

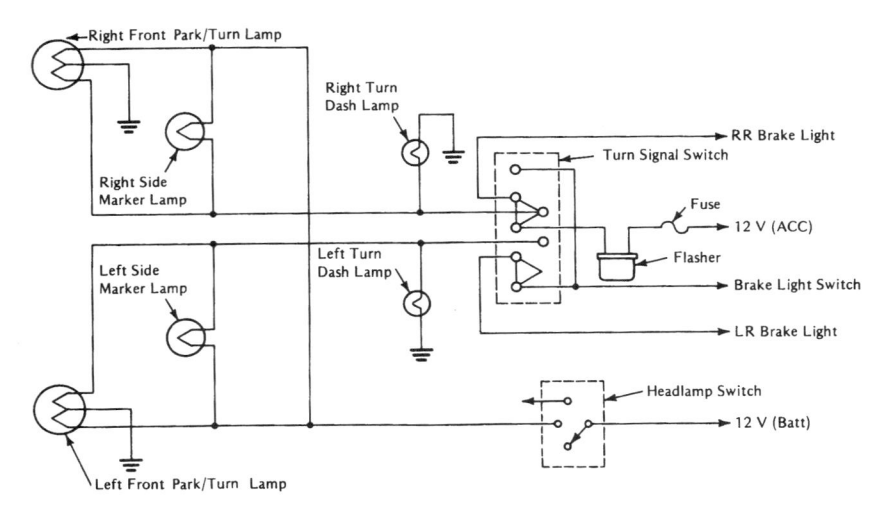

FIGURE 15-20 Side marker lamp schematic with turn signal switch in right turn position and headlamp switch off

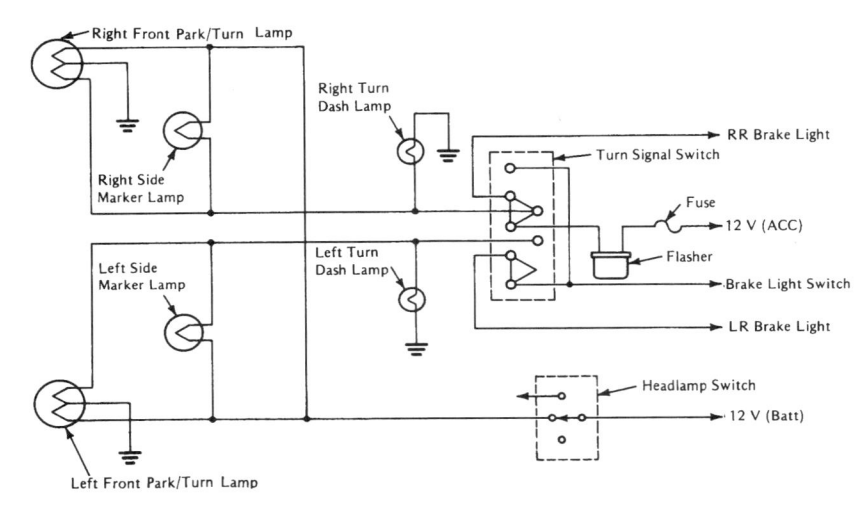

FIGURE 15-21 Side marker lamp schematic with turn signal switch in right turn position and headlamp switch on

not burn. When the flasher opens, however, the parking lamp circuit is completed through the side marker lamp and through the turn signal lamp filament to ground; consequently, the side marker lamp will burn.

Turn Signal Flashers

The turn signal flasher is actually a thermally actuated circuit breaker switch. It is balanced to the electrical load in the system for which it is designed. Its normally closed (NC) contacts are designed to open and close from 60 to 120 times per minute. This action "breaks" and "makes" the electrical feed circuit to the turn signal switch. The flasher shown in Figure 15-15 operates on the expansion-contraction bimetallic strip principle previously described.

The flasher is often difficult to locate. It may be found in the fuse block, clipped to the under-dash panel, above the glove compartment, or even taped to the wiring harness. Replacement guides usually give the location of a flasher as well as the proper replacement part number. Flashers are sensitive to loads. For example, a flasher designed to operate on an eight-lamp load will not function properly on a four-lamp load.

HAZARD LAMPS

Beginning in the mid-1960s all cars were equipped with hazard warning lamps; that is, all turn signal lamps flashing simultaneously, or alternately, to signal a hazardous condition. A good example of a hazardous condition is a disabled car in a traffic lane. Although the hazard circuit is integrated into the turn signal circuit and uses the same wiring harness and lamps, it has a separate power source, switch, and flasher. Power is supplied to the hazard circuit from the battery through the fuse block and flasher. The ON-OFF switch (usually a push-pull type) is usually located on the lower right section of the steering column, just under the steering wheel.

A typical hazard warning system schematic is shown in Figure 15–22. Figures 15–23 through 15–25 show this circuit with the hazard switch in the ON position and the turn signal switch in the OFF, RIGHT turn, and LEFT turn positions. When ON, this system overrides the function of the turn signal switch, but it is often affected by the brake light circuit. Trace the circuit of Figure 15–23 from the flasher to all four lamps.

Note, by tracing the schematic of Figure 15–24, that the hazard circuit overrides the turn signal circuit to illuminate all four lamps.

By tracing the schematic of Figure 15–25 you may note that the hazard flasher is now bypassed to illuminate all four lamps.

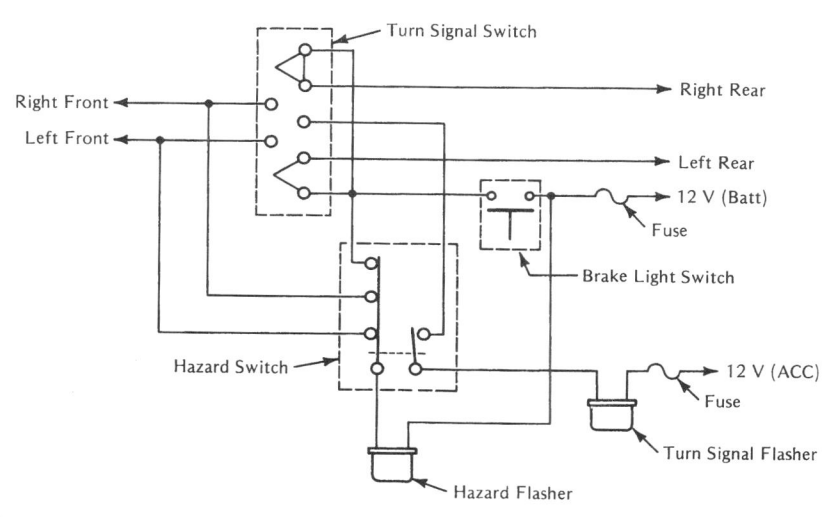

FIGURE 15–22 Typical schematic of hazard signal circuit

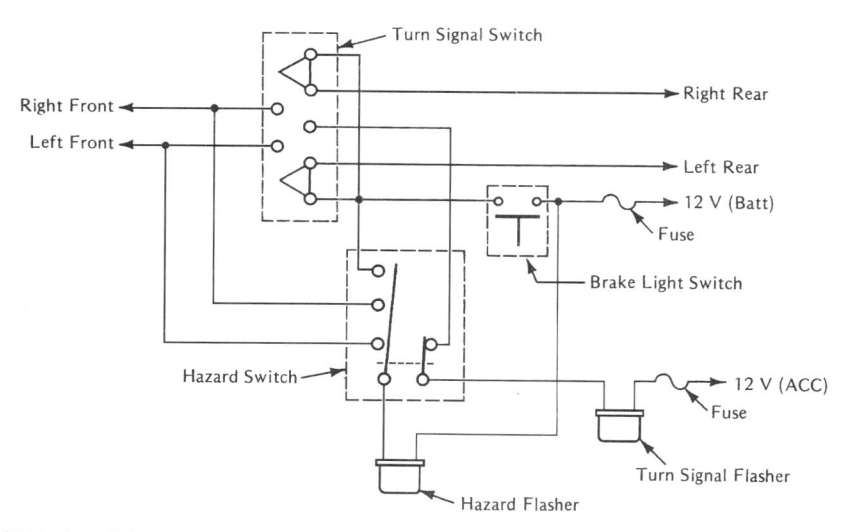

FIGURE 15–23 Hazard signal circuit with turn signal switch in the OFF position and hazard switch ON

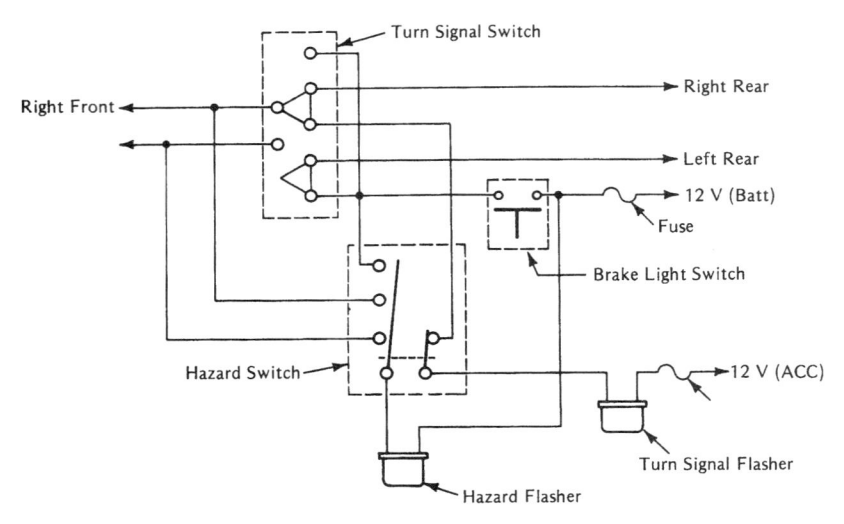

FIGURE 15-24 Hazard signal circuit with turn signal switch in RIGHT turn position and hazard switch ON

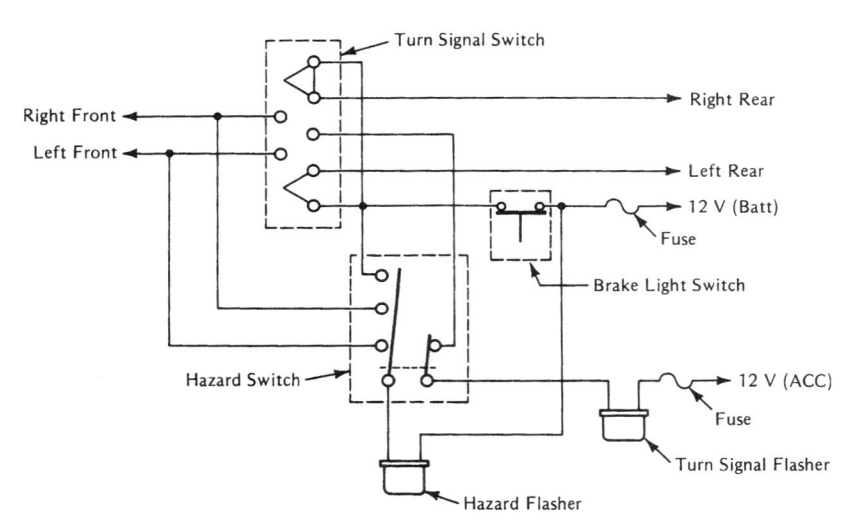

FIGURE 15-25 Hazard signal circuit ON with brake light switch energized

Hazard Flashers

Although the hazard flasher is similar in appearance to the turn signal flasher, it is not the same. The two are not interchangeable. The contacts of the hazard flasher are normally open (NO) and are paralleled by a coil of calibrated resistance wire. When a load is placed on the hazard flasher, the coil heats to close the bimetallic strip-controlled contacts. This turns all lamps on and takes the load off the coil, which now cools. The bimetallic strip also cools, which opens the points. This action (Figure 15-26) is repeated as long as the switch is closed.

BRAKE LIGHTS

Brake lights, or stop lights, as they are often called, are an important signaling device. They allow a driver following another car to know that the car ahead is slowing or stopping. Brake lights are credited with preventing many rear-end accidents. The brake light circuit (Figure 15-27) is relatively simple. It consists of the rear lamps, also used for turn signals, and a switch, and shares a common wiring harness with the turn signal circuit. When the brake pedal is pressed, the brake light's normally open (NO) switch is closed. This action completes a

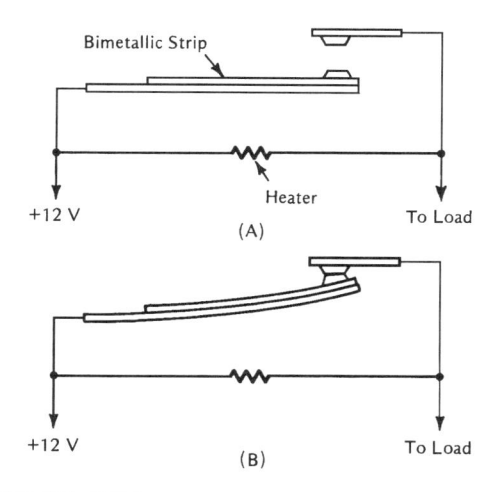

FIGURE 15–26 Schematic of hazard flasher. Normally open contacts (a) close (b) when bimetallic strip is heated.

circuit through the turn signal switch to provide the following:

1. If the turn signal switch is not activated (RIGHT or LEFT), all dual-filament rear lamps will burn.
2. If a right turn is selected by the turn signal switch, only the left-rear lamps will burn. The right-rear lamps will flash on and off.
3. Conversely, if a left turn is selected by the turn signal switch, only the right-rear lamps will burn. The left-rear lamps will flash on and off.

The brake light circuit runs from the battery, through the fuse block, to the switch. It employs the same fuse as the hazard circuit. When the hazard switch is closed and all lights are flashing, pressing the brake pedal will override the hazard flasher and turn on all lamps (nonflashing).

COURTESY AND CONVENIENCE LAMPS

Courtesy and convenience lamps are not necessarily required for the operation of a car. They provide illumination for added driver convenience. Included in this group is fiber optics, the latest type of lighting to make the automotive scene. Other convenience lights include map, door, dome, and dash lamps.

Dome Lamps

Dome lamps illuminate the interior of a car. They are found either overhead or on the rear-upper panels of the interior. They are turned on in three ways: by opening a door, by turning the headlamp switch fully clockwise (cw), or by activating a switch on the lamp itself or at some other designated point (Figure 15–28).

Some four-door cars are equipped with a dome lamp switch on all four doors, whereas others have a switch only on the two front doors. Most station wagons have a manually operated rear-door switch, although some nine-passenger wagons have a rear-door switch that is activated when the door is opened. Regardless of the number and location of the switches, the circuit is usually of the grounding type. Closing any or all of the switches grounds the circuit, and the lamps burn. The dome lamp circuit is usually protected by a fuse in the fuse block. This fuse often serves the cigarette lighter also.

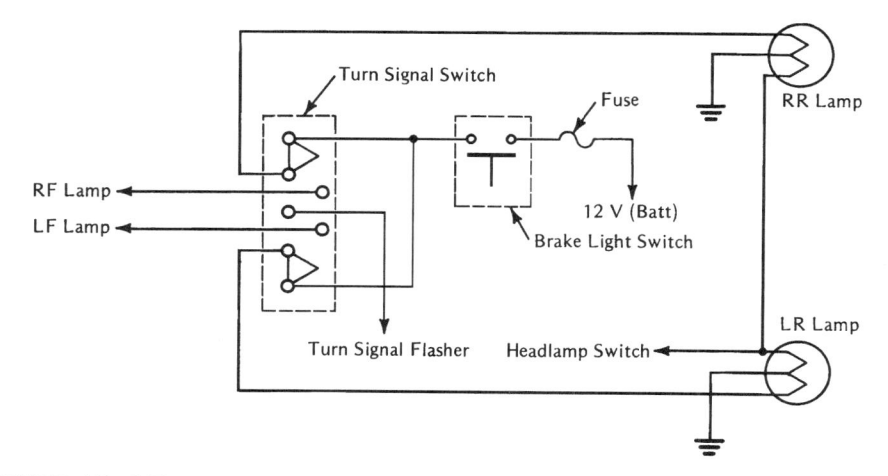

FIGURE 15–27 Typical schematic of a brake light circuit

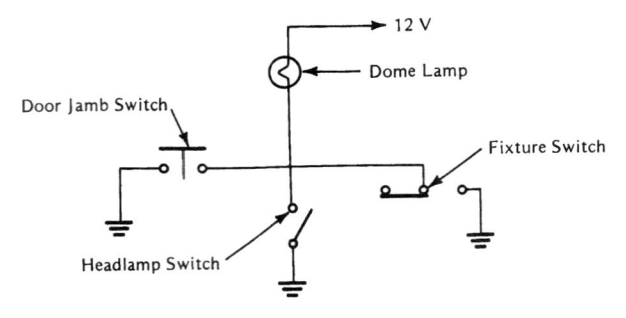

FIGURE 15–28 Typical schematic of a dome lamp circuit

Map Lights

Some cars have a map light, which is a small light usually on the passenger side of the dash panel. It is located so that it does not distract or interfere with the driver's nighttime vision. Map light circuits vary, but they can be easily traced in manufacturers' wiring diagrams and schematics.

Door Lights

Many cars are equipped with interior door lights. These lights primarily "mark" a car door to provide for safer entry and exit of the car. The circuit is usually a part of the dome light circuit and is controlled by the same switches.

Dash Lights

Dash lights are provided to illuminate the speedometer, the clock, the radio, and the other gauges and switches on the dash panel. The dash lights are turned on with either ON position of the headlamp switch, and their brilliance is controlled by the rheostat of the headlight switch. Turning the knob in a clockwise (cw) direction brightens them; turning in the counterclockwise (ccw) direction dims them.

Rheostats

A *rheostat* consists of a calibrated coil of wire and a sliding contact. Rheostats are used in lighting control to change the brilliance of lamps, such as in the dash light circuit. The greater the amount of resistance wire in the circuit, the dimmer the lamps. The total circuit resistance impressed by the rheostat depends on the position of the movable contact (Figure 15–29).

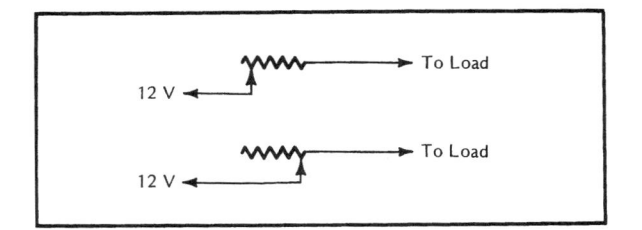

FIGURE 15–29 Schematic symbol for a rheostat: (a) greatest resistance; (b) least resistance

FIBER OPTICS

A *fiber optic* is an assembly of strands of transparent glass fibers, known as polymethyl-methacrylate, bonded together parallel to each other. The bundle of strands is sheathed by a transparent polymer and enclosed in a flexible, opaque sleeve. This bundle of fibers can transmit light from one of its surfaces to the other, around curves, and into otherwise inaccessible places (Figure 15–30). It can transmit light over a reasonable distance with an extremely low loss by the process of total reflection.

The light intensity at the viewing end of the optic is dependent on two factors: the intensity of the light source and the number of fibers in the optic strand. In some nonautomotive applications, fiber optics (also called optical fibers) are used to transmit pictures and sound. In automotive applications, fiber optics can provide several lighted areas from a single lamp. Fiber optics can also be used to determine whether exterior lights are burning by picking up light from a lamp and transmitting it to the dash panel. The current cost of fiber optics, however, limits this application. Fiber optics, used to illuminate front door-lock cylinders, are available as optional equipment on some cars. The circuit in Figure 15–31 becomes a part of the interior lamp circuit. When the door handle push button is pressed, in-line bulbs located inside the front doors and the interior lighting come on. Magnified light is transmitted from the in-line

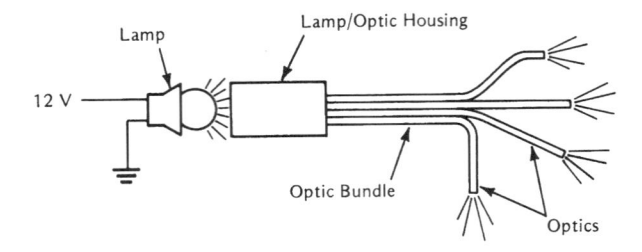

FIGURE 15–30 Illustration of fiber optic system

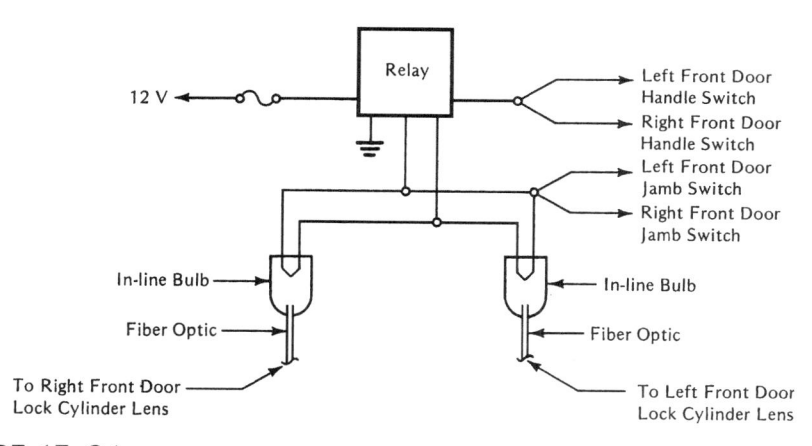

FIGURE 15–31 Schematic of fiber optic door lock illumination system

bulbs through a fiber optic harness to a lens located above the door lock cylinders. The door lock and interior lighting, when activated by the door handle push button, are controlled by a time delay relay. This allows the lights to remain on for 20–30 seconds, then automatically turn off.

SUMMARY

- Lighting is a very important feature in modern automotive operations.
- Both round and rectangular headlamps are currently in use.
- Integral (built-in) circuit breakers are commonly found in headlamp switches.
- Solid-state components provide for automatic control of headlamps.

PRACTICAL EXERCISE

This exercise is for classroom and shop participation and discussion.

An automobile and the appropriate manufacturer's electrical service manual are required for this exercise.

1. Inspect the vehicle exterior and try to determine the number of lamps used:

    a. For primary illumination. _____

    b. For secondary illumination. _____

2. Where are the switches and controls for these lamps located? _____

3. Inspect the vehicle interior and try to determine the number of lamps. How many do you find? _____

4. Where are the switches and controls for these lamps located? _____

5. Consult the manufacturer's service manual to locate lamps not observed in Steps 1 and 3. How many more did you locate? _____

6. Consult the manufacturer's service manual to locate control devices not observed in Steps 2 and 4. How many more did you locate? _____

NOTE: Some of the lamps in the system are controlled by more than one device. Try to identify them and locate all the controls for each lamp in the exercise.

Unit 15 LIGHTING

REVIEW

(T) (F) 1. Dome lamps are controlled by door switches only.

(T) (F) 2. All cars are equipped with map lights.

(T) (F) 3. Dash light brilliance is controlled by the dimmer switch.

(T) (F) 4. Fiber optics are commonly used in automotive lighting circuits.

(T) (F) 5. A rheostat is used to vary total circuit resistance.

6. Federal legislation of 1940 mandated the use of
   a. sealed-beam headlamps
   b. a two-lamp headlamp system
   c. a four-lamp headlamp system
   d. all of the above

7. Halogen sealed-beam headlamps provide about
   a. 75,000 cp
   b. 150,000 cp
   c. 200,000 cp
   d. 7500 cp

8. Current overload protection for the headlamp circuit is found in or on
   a. the fuse panel
   b. an in-line fuse holder
   c. the headlamp switch
   d. the wiring harness

9. The headlamp switch has no function in the following circuit.
   a. parking lamps
   b. tail lamps
   c. side marker lamps
   d. brake lamps

10. Which of the following should be considered most important when repairing the electrical system?
    a. Disconnect the battery.
    b. Wear eye protection.
    c. Work with adequate lighting.
    d. All are considered important.

11. There are at least _____ double-filament lamps in a turn signal circuit.
    a. two
    b. four
    c. six
    d. eight

12. Front side marker lamps, when a part of the turn signal circuit, _____ when the turn signal switch is activated with the parking lamps OFF.
    a. flash simultaneously with the turn signals
    b. flash alternately with the turn signal circuit
    c. will burn continuously at all times
    d. won't burn or flash at all

13. Turn signal flasher contacts are
    a. normally open (NO)
    b. normally closed (NC)
    c. adjustable
    d. none of the above

14. The hazard lamp circuit is integrated with
    a. the brake lamp circuit
    b. the turn signal lamp circuit
    c. the brake lamp and turn signal lamp circuits
    d. neither the brake lamp nor the turn signal circuits

15. The most important signaling device is the
    a. brake lamps
    b. turn signals
    c. hazard lamps
    d. side markers

16. What is the purpose of cornering lamps? _____

17. What does the term 4 cp mean? _____

18. What is the second filament of a double-filament park lamp used for? _____

19. How are backup lights turned on? _____

◆ UNIT 16 ◆

SAFETY SYSTEMS

OBJECTIVES

On completion of this unit you will be able to:

- Understand the function, purpose, and operation of the safety systems of the modern automobile, such as:
  - windshield wipers
  - windshield washers
  - rear window wiper/washer
  - rear window defogger
  - seat belt warning

There are many safety systems in the modern automobile; some are covered in other units of this text. This unit covers those systems having to do with clear vision: windshield wipers, windshield washers, defoggers, and deicers. The seat belt warning system is also covered in this unit.

WINDSHIELD WIPERS

There are two basic types of windshield wiper systems: the nondepressed park system and depressed park system. The nondepressed park system "parks" the wiper blades above the hood line; the depressed park system below the hood line. Gear ratios, from the motor shaft to the wiper output shaft (Figure 16–1), are from 30 to 1 to as high as 50 to 1 (30:1 to 50:1) to provide the high torque required to drive the wiper blades across the windshield.

Many standard equipment wiper motors are two-speed (HI-LO) or three-speed (HI-MED-LO). However, pulse or intermittent wiper systems employ a three-speed (HI-MED-LO plus PULSE) motor. Also, in two-speed intermittent wiper systems, the low speed is slower and the high speed is faster than in the standard two-speed system. Typical schematics for two-pulse wiper systems are shown in Figures 16–2 and 16–3. The pulse, or intermittent, system provides a delay of 1 to 15 seconds between sweeps of the wiper blades. The time delay between sweeps is controlled manually by the driver, depending on conditions.

Nondepressed Park Systems

Typical schematics for the nondepressed park (often called nonconcealed) wiper system are shown in Figures 16–4 through 16–6. The General Motors schematic will be used to trace the circuits for LO, HI, and PARK positions of the dash-mounted control switch (Figures 16–7 through 16–9). In actual operation, it is important that the wiper motor and dash switch be securely grounded to body metal.

Turning the wiper control switch in the system in Figure 16–6 to the LO position completes circuits from wiper terminals 1 and 3 to ground. Current then flows from the battery circuit through wiper terminal 2 and then through the series field circuit, where it divides. One circuit is then through the armature to ground through wiper terminal 1 and the switch.

The other circuit passes through the shunt field to ground through wiper terminal 3 and the wiper switch, as shown in Figure 16–7. Turning the wiper switch to the HI position (Figure 16–8) opens the

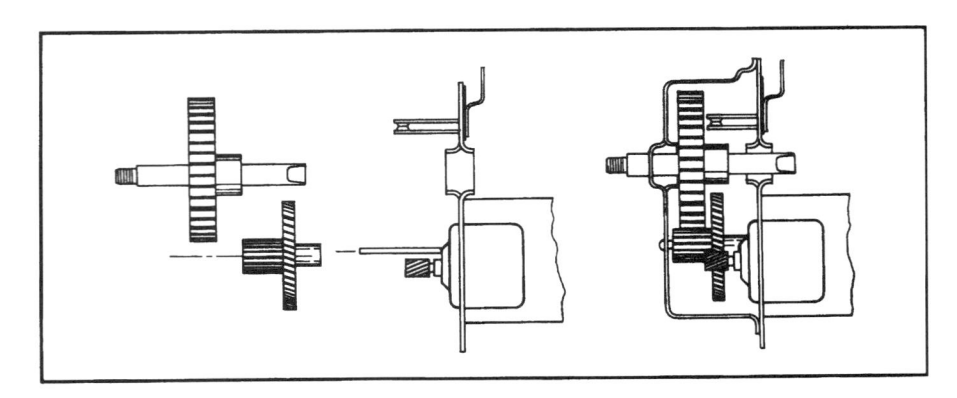

FIGURE 16–1 Gear reduction increases torque of wiper motor

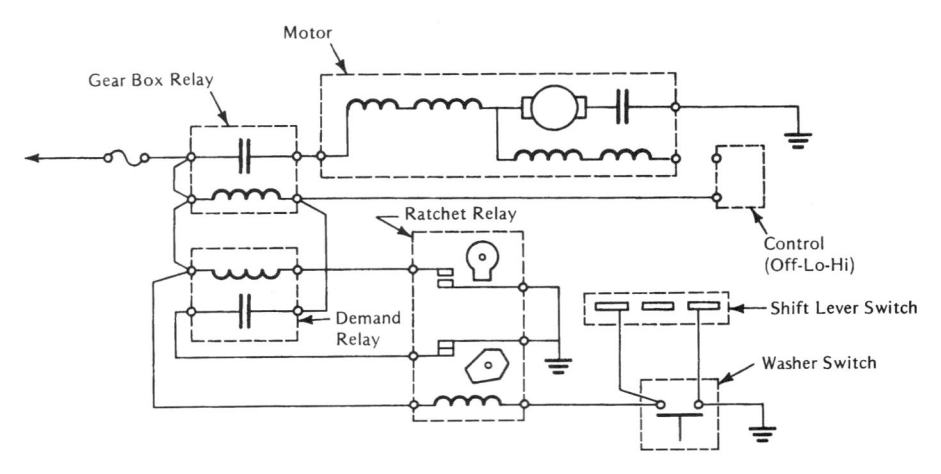

FIGURE 16–2 Schematic for early pulse-type wiper system

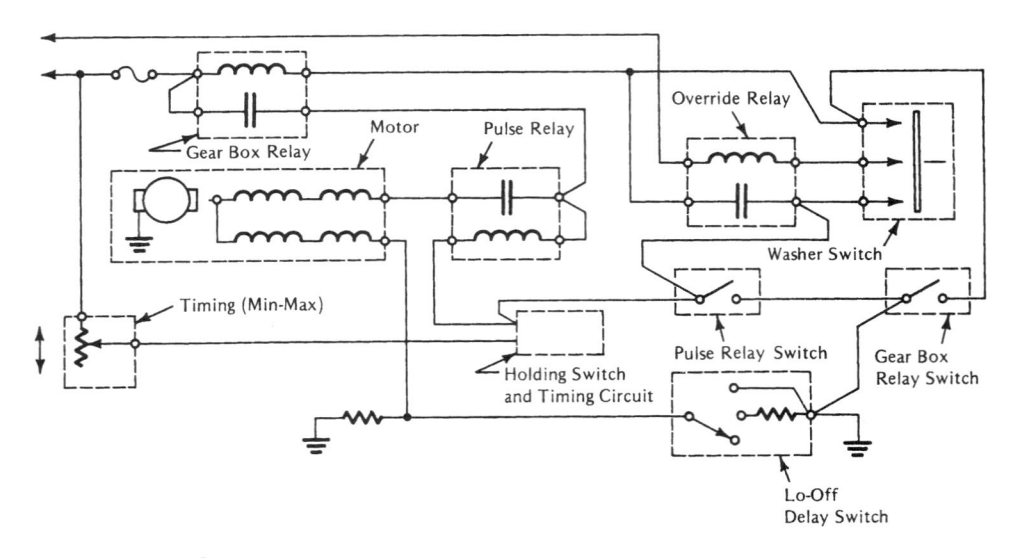

FIGURE 16–3 Schematic for late-model pulse-type wiper system

WINDSHIELD WIPERS 179

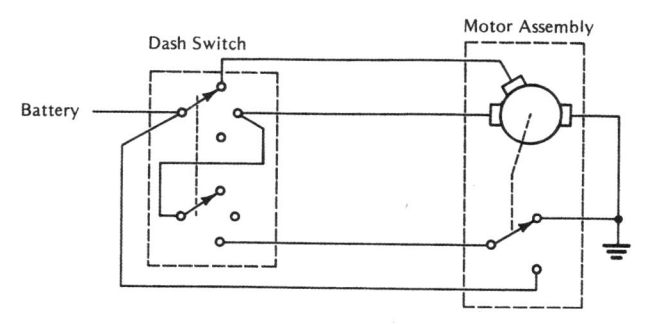

FIGURE 16-4 Chrysler nondepressed windshield wiper system schematic

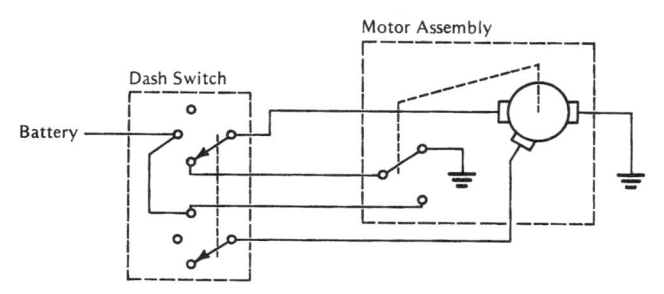

FIGURE 16-5 Ford nondepressed windshield wiper system schematic

shunt field circuit to ground at the switch. The shunt field is, however, connected to a resistor across terminals 1 and 3. Shunt field current then flows through terminal 3 through the resistor to terminal 1. The circuit is complete to ground through the switch.

Assuming that the wiper blades have not reached the park position when the control switch is turned to the OFF or PARK position (Figure 16-9), the wiper continues to rotate until the output gear is turned to a position where its cam opens the park switch.

When the control switch is turned to OFF or PARK, the shunt field circuit is completed from terminal 3 through the wiper switch to terminal 1 and on through the cam-operated park switch to ground.

Since the shunt field is connected to ground bypassing the resistor, the wiper motor operates at LO speed during the park operation. When the output gear cam opens the park switch contacts, the wiper blades are in their park position, and the wiper motor is turned OFF. Tracing the circuits in the other schematics is accomplished in the same manner. The schematics may be redrawn with the control switch in each position to trace the circuits from the battery supply through the components to ground.

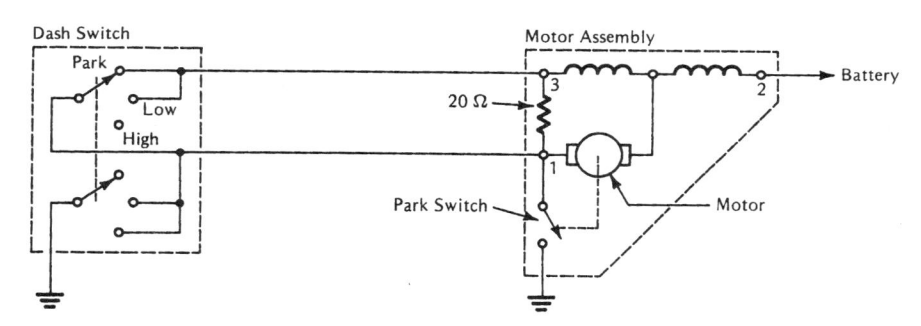

FIGURE 16-6 Typical schematic for General Motors nondepressed wiper system

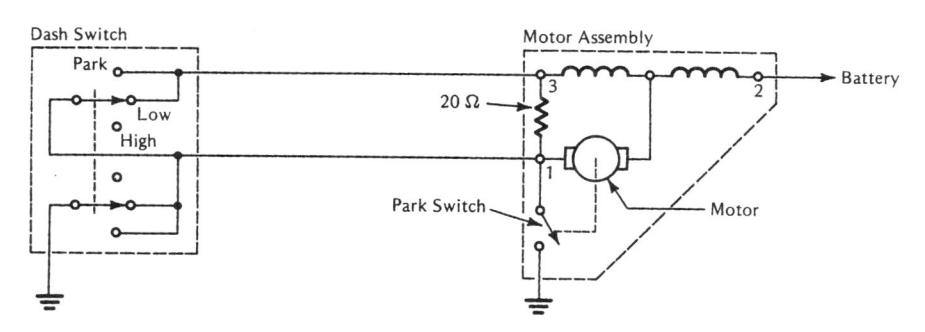

FIGURE 16-7 Nondepressed wiper system in low-speed operation.

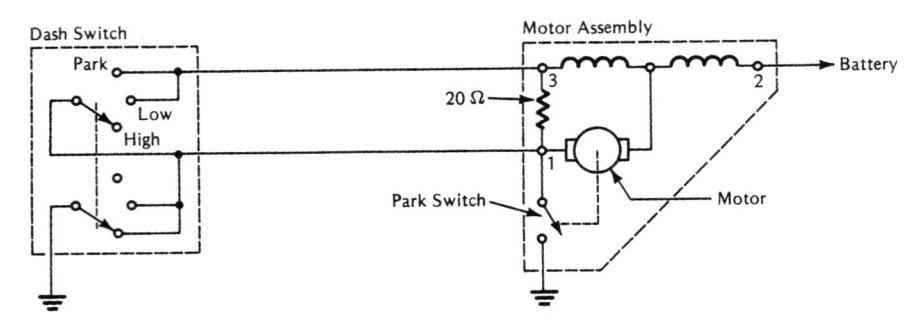

FIGURE 16–8 Nondepressed wiper system in high-speed operation

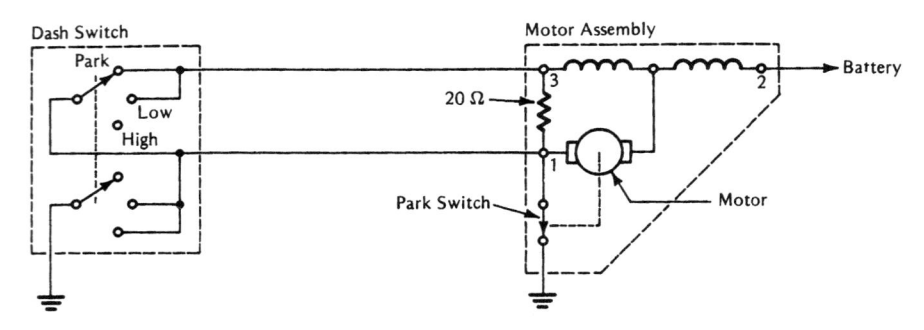

FIGURE 16–9 Nondepressed wiper system in PARK position

Depressed Park Systems

Typical schematics of the depressed park (also called concealed) wiper system are shown in Figures 16–10 through 16–12. The Chrysler schematic will be used to trace the circuits for LO, HI, and PARK positions of the dash-mounted control switch (Figures 16–13 through 16–16). In actual operation, it is important that the wiper motor and dash switch be securely grounded. Turning the control switch to the HI position completes circuits from the battery feed terminal to terminal 2 of the control switch.

This circuit is to terminal 5 of the motor, through the motor, and out on terminal 8. The circuit is then completed to ground through terminal 1 of the control switch, as illustrated in Figure 16–13.

Turning the wiper switch to the LO position (Figure 16–14) completes circuits from the battery feed ter-

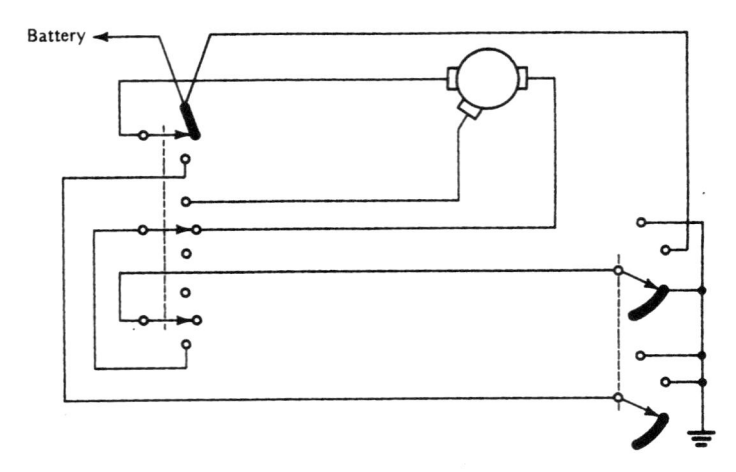

FIGURE 16–10 Typical schematic for Ford depressed park windshield wiper system.

WINDSHIELD WIPERS 181

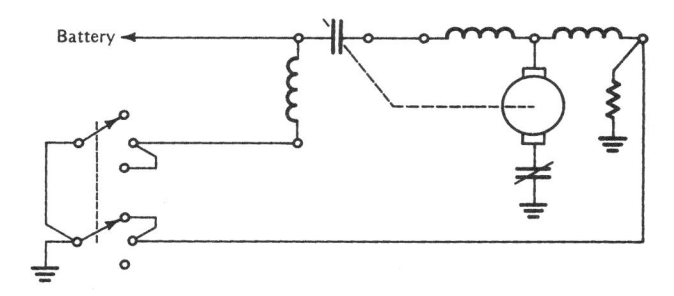

FIGURE 16–11 Typical schematic of General Motors depressed park wiper system

minal to terminal 3 of the control switch. This circuit is then to terminal 6 of the motor, through the low-speed torque-limiting resistor, and then through the motor. The circuit is completed to ground through motor terminal 8 and control switch terminal 1.

The wiper blades do not reach the park position during normal operation. When the control switch is turned to OFF or PARK (Figure 16–15), the motor's direction of rotation is reversed to park the blades.

This circuit is from the battery feed B to terminal 4 of the control switch and into the motor at terminal 7. The circuit is completed through the park switch and

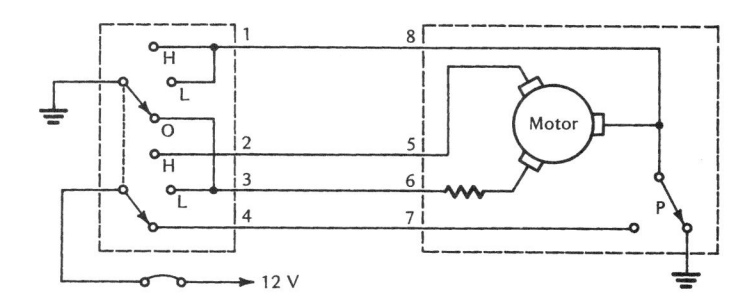

FIGURE 16–12 Typical schematic for Chrysler depressed park windshield wiper system

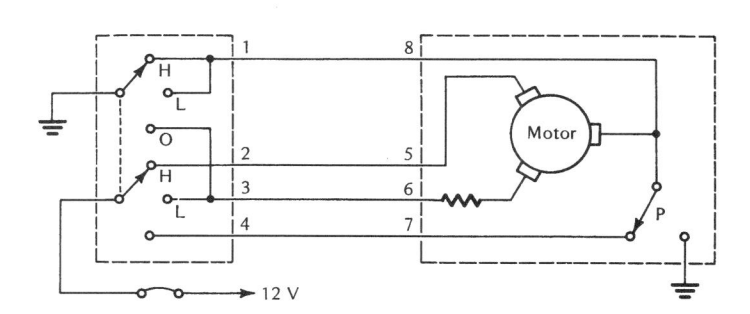

FIGURE 16–13 Schematic showing high-speed wiper operation

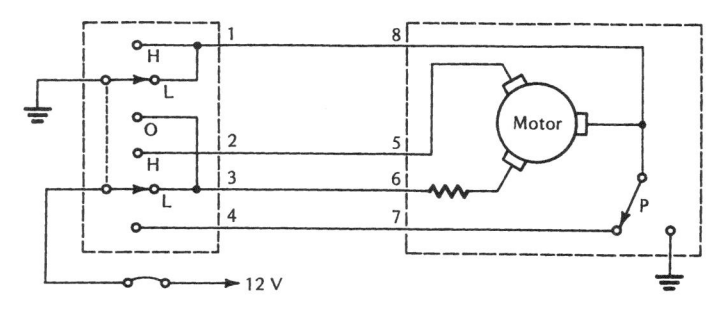

FIGURE 16–14 Schematic showing low-speed wiper operation

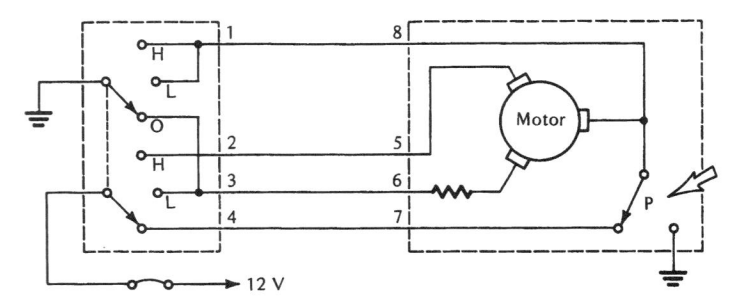

FIGURE 16–15 Schematic showing parking operation of park switch

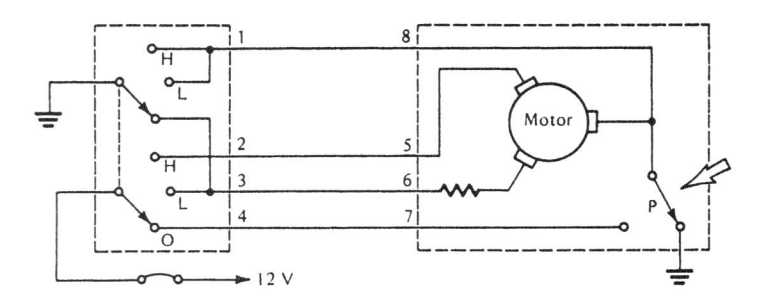

FIGURE 16–16 Schematic showing PARK position of park switch

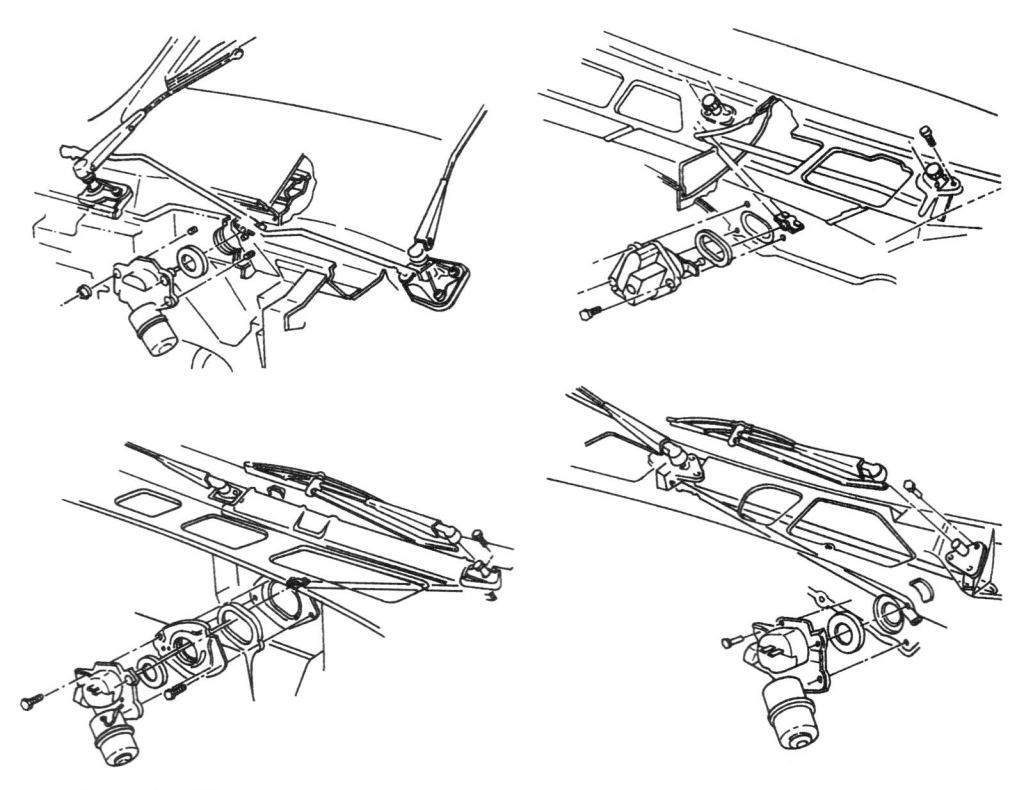

FIGURE 16–17 Location of wiper motor and methods of mounting

motor low-speed winding to ground via the resistor, terminal 6, and terminal 3 of the control switch. When the gear cam opens the park switch (Figure 16–16), the wiper blades are in their park position, and the wiper motor is turned OFF.

Wiper motors are not interchangeable. Their electrical circuits, mounting configurations, and wiper blade linkage provisions vary greatly. Several on-car installations are shown in Figure 16–17. Note the different mounting locations, as well as the configurations of the motor and gear box assemblies. Specific wiper motor check-out procedures are given in manufacturers' service manuals.

WINDSHIELD WASHERS

There are two types of pumps used in automotive windshield washer systems. One type is electrically operated, and the pump assembly mounts directly to the fluid reservoir. A permanently lubricated and sealed motor is coupled to a rotary pump, and gravity-fed fluid is forced through small rubber hoses to nozzles that direct it to the windshield. Another type, used by General Motors, is mechanically operated (Figure 16–18), and is mounted directly to the windshield wiper motor assembly. Fluid is pulled from the reservoir and then forced through small rubber hoses to nozzles that direct it to the windshield.

Electrically Operated Washer Systems

The schematic of an electrically operated windshield-washer system is simple, as shown in Figure 16–19. Pushing the Wash button completes the washer motor circuit to ground and turns the motor ON, causing the pump to operate. At the same time, a mechanical linkage within the switch turns the wiper motor to LO speed.

The washer motor operates only as long as the Wash button is held in. This allows direct control over the amount of fluid delivered to the windshield. When the Wash button is released, the washer motor is turned OFF. The wipers continue to operate, however, until the dash control switch is turned to the OFF or PARK position.

Low Fluid-Level Warning Systems

A low washer-fluid level warning system is an option on some windshield-washer systems. It may be a simple circuit (Figure 16–20) that lights a dash lamp when the Wash button is pressed if the fluid level is near empty. A more complex solid-state circuit design (Figure 16–21) lights a dash lamp when the fluid is below a predetermined level when the ignition switch is in the ON position.

The electrically operated washer motor usually cannot be field repaired. If a problem is found with either the motor or pump, it must be replaced as an assembly.

The nozzles that direct fluid to the windshield are located either in openings in the car hood (Figure 16–22) or in the windshield-wiper arm/blade assembly (Figure 16–23). Nozzles through the hood can be bent slightly to direct fluid to the proper place on the windshield. Nozzles on the arm/blade assembly usually require no adjustment.

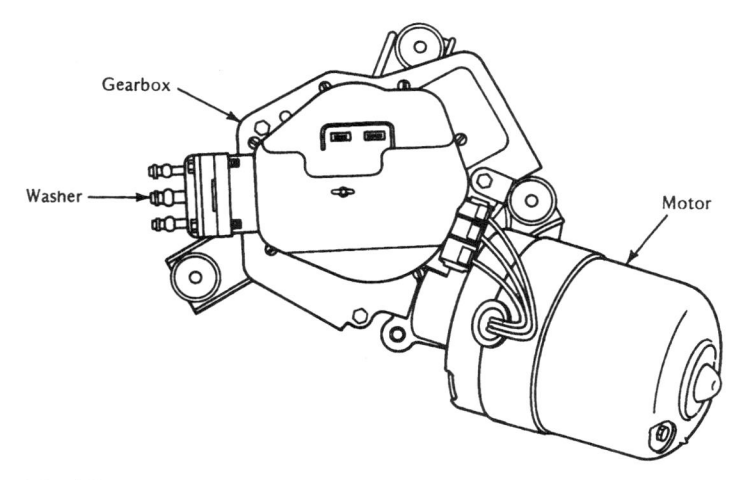

FIGURE 16–18 General Motors–type wiper-mounted, mechanically operated windshield washer pump

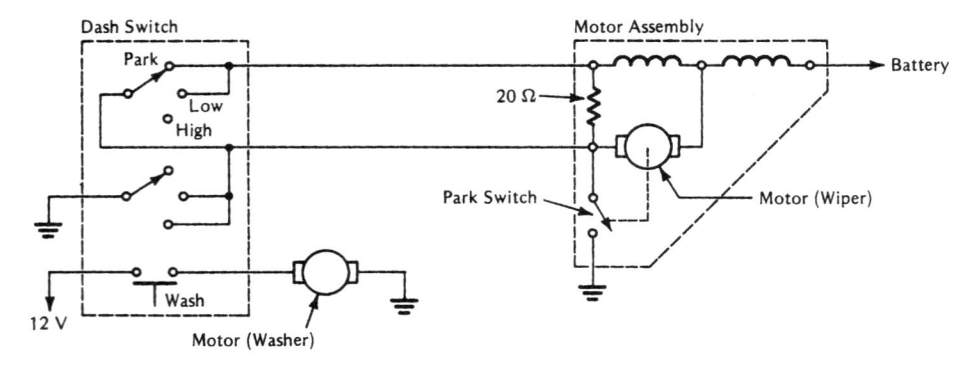

FIGURE 16–19 Wiper system schematic with added electric washer pump circuit

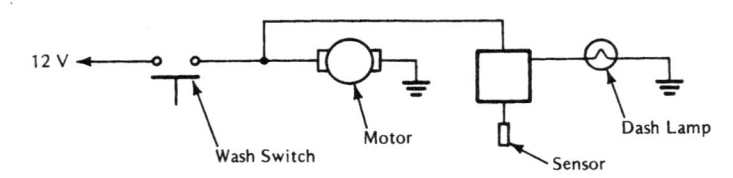

FIGURE 16–20 Schematic for low fluid-level warning system

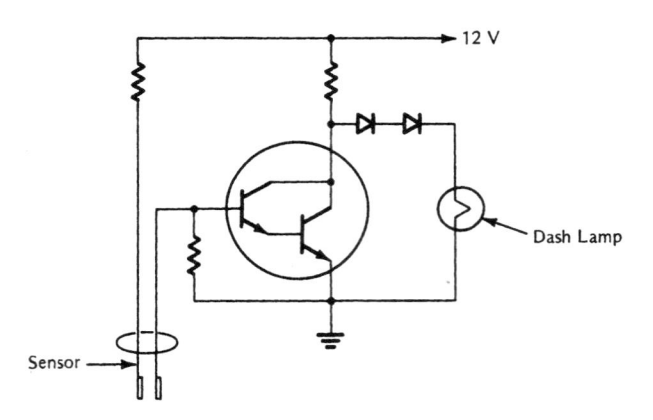

FIGURE 16–21 Typical schematic for a solid-state low fluid-level warning system

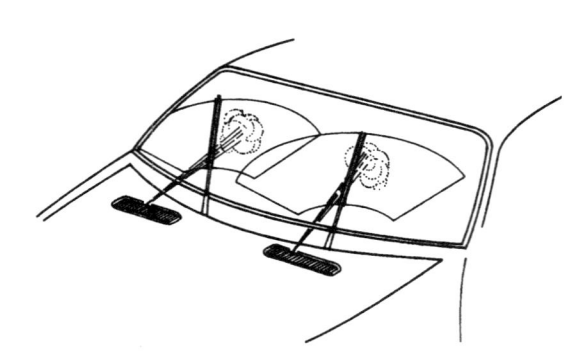

FIGURE 16–22 Washer nozzles spray from hood openings

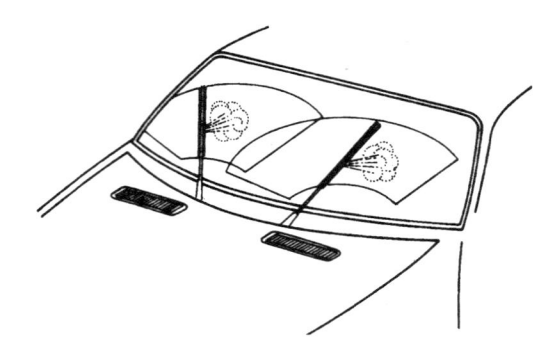

FIGURE 16–23 Washer nozzles spray from wiper arms

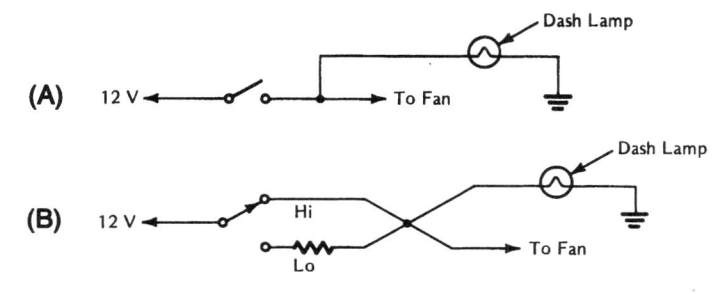

FIGURE 16–24 Schematics for (a) one-speed and (b) two-speed rear-window defogger

REAR-WINDOW WIPERS

Rear-window wipers are available as optional equipment on some cars. The wiper motor, generally one or two speed, operates in the same manner as the wipers previously discussed. Rear-window wiper motors are of the nondepressed type and, unlike the front-window wiper motors, drive only one wiper blade.

REAR-WINDOW DEFOGGERS

Rear-window defoggers are available as optional equipment on most car lines. The rear-window defogger consists of a one- or two-speed motor, a motor housing, a fan or blower, a wiring harness, and a control switch. The motor, housing, and blower assembly is an integral unit and is located in the approximate center of the package tray behind the rear seat. Inside the car air is drawn into the blower and is directed to the rear window. The control switch is dash mounted for the driver's convenience. Schematics for one- and two-speed defoggers are shown in Figure 16–24.

REAR-WINDOW DEICERS

An optional rear-window deicer system is available on most car lines. The system consists of a glass, usually tinted, with a number of horizontal ceramic–silver compound element lines that are attached to two vertical bus bars. This grid (Figure 16–25) is baked onto the inside surface of the glass during the forming process. The deicer system operates on 12 volts with a current draw of about 20 amperes. An instrument panel–mounted control switch with indicator lamp is used for primary control. It is a three-position (OFF-NORM-ON) switch that is spring loaded from ON or OFF to its center NORM position. When the switch is moved to its ON position, a latching relay coil is energized that closes the normally open (NO) contacts of the relay, which provides power to the rear-window grid and dash-panel indicator lamp. The system, illustrated in the schematic of Figure 16–26, operates continuously after being turned on until the control switch or ignition switch is turned to OFF.

Other deicer systems (Figure 16–27) employ an automatic timer that allows the rear-window grid to

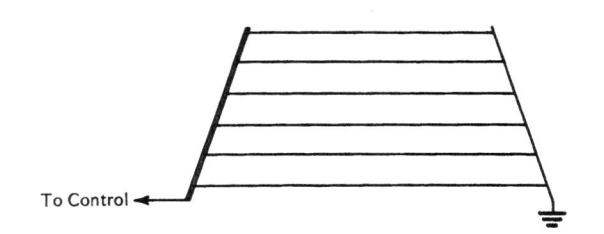

FIGURE 16–25 Rear-window grid system

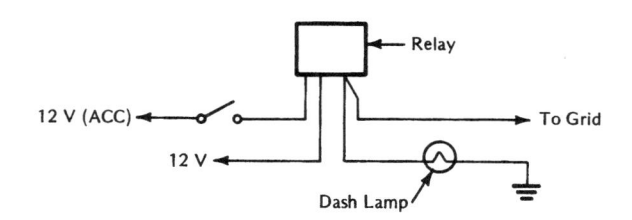

FIGURE 16–26 Typical rear-window deicer schematic

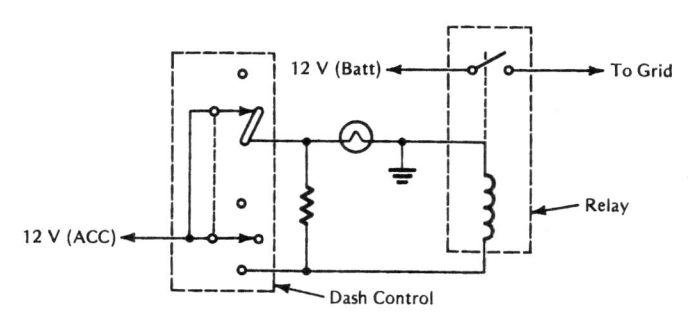

FIGURE 16–27 Typical rear-window deicer schematic with timer

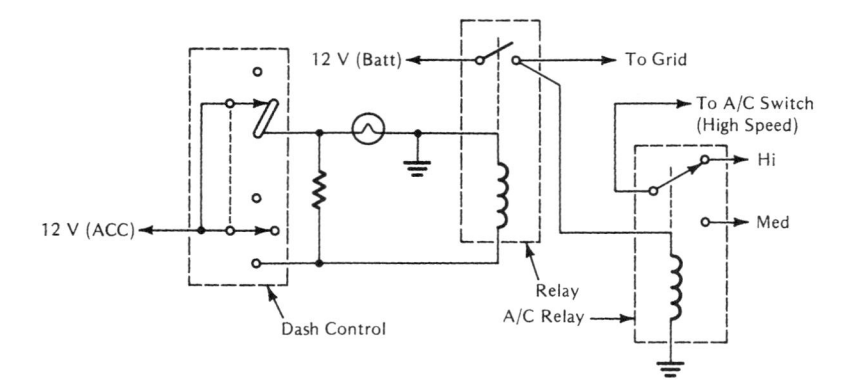

FIGURE 16–28 Schematic for rear-window deicer system with A/C relay

operate for a predetermined time period, usually 10 to 15 minutes. After this time, the system automatically turns off. The system can also be turned off at any time either by using the control switch or by turning the ignition switch to OFF.

Some deicer systems use a relay in conjunction with the heating and air conditioning circuit to regulate blower motor speed when the deicer system is in operation. This circuit (Figure 16–28) reduces blower motor speed to the MED position if it is being operated at HI when the deicer is turned on.

Standard procedures can be followed for testing and repairing deicer system wiring and components. However, manufacturers' manuals should be consulted for testing and repairing the rear-window grid.

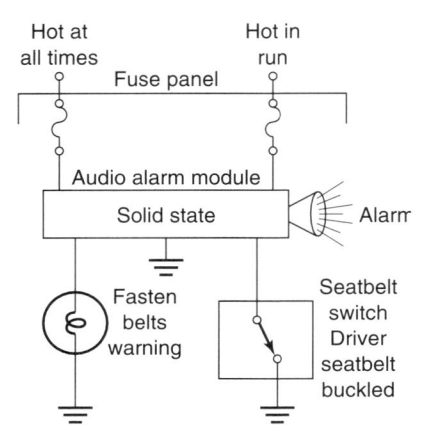

FIGURE 16–29 Schematic for seat belt warning system

SEAT BELT WARNING SYSTEMS

The seat belt warning system has an audible and visual signal consisting of a buzzer and a red lamp on the instrument panel. When the ignition switch is turned to the ON position, the seat belt warning lamp is illuminated for 4 to 8 seconds, even if the seat belt(s) is (are) fastened. The buzzer sounds during the same period only if the driver has not fastened the seat belt. If the driver has fastened the seat belt, the buzzer will not sound. Either way, seat belts fastened or not, both lamp and buzzer will stop after a 4- to 8-second interval. Rear passenger seat belts are not connected to the system. The schematic shown in Figure 16–29 is typical of the seat belt warning systems in all car lines. The system consists of a timed buzzer-relay module, a dash lamp, a seat-belt buckle switch, and a wiring harness.

SUMMARY

- There are two basic types of windshield wiper systems, the nondepressed park system and the depressed park system.
- There are many variations of the basic windshield wiper system.
- Most pumps found on the modern automobile are electrically operated.
- Seat belt warning systems are standard equipment in today's automobile.
- Two types of in-car warning signals are used: visual (flashing light) and audible (chime or buzzer).

PRACTICAL EXERCISE

A late-model vehicle with appropriate electrical service manual and a voltmeter are needed.

1. Raise the hood. DO NOT start the engine.
2. Locate the horn(s) and horn relay. Consult the service manual, if necessary.
3. Refer to the manual and note the color of the wire(s) to/from:
   a. The relay. Color(s) are: _____
   b. The horn(s). Color(s) are: _____
4. Locate the fuse or circuit breaker protecting the horn circuit. It is located _____
5. Does this fuse or circuit breaker protect any other circuit(s)? _____ If yes, what circuit(s)? _____
6. Are the colors of the wires the same as noted in Step 3?
   a. relay? _____
   b. horn? _____
7. Using the voltmeter, determine which wire on the horn relay, if any:
   a. Is "hot." _____
   b. Goes to the horn. _____
   c. Goes to the horn button/ring. _____
   d. Goes to ground. _____
8. Why, in your opinion, is a relay necessary in a horn circuit? _____
9. What would happen if the fuse or circuit breaker (Step 4) were removed from the circuit? _____
10. Remove the voltmeter and leads, connect any wires that may have been disconnected, and close the hood.

NOTES

1. There is generally no ground wire (Step 7(d)) on the horn relay. Grounding is accomplished through the horn ring/button.
2. High current draw of the horn necessitates a relay in the circuit.
3. The grounding circuit requires very little current to energize the horn relay.
4. The color(s) of the wire(s) (Step 6) should be the same as indicated in the wiring diagram.
5. If the circuit breaker or fuse is removed (Step 9) the horn will not sound.

REVIEW

1. What are the two basic types of windshield wiper systems?
   a. _____
   b. _____

2. In which wiper system would a three-speed motor most likely be found? _____

3. What would occur if the park switch in Figure 16–9 did not open when the cam reached the PARK position? _____

Unit 16 SAFETY SYSTEMS

4. What are the motor-shaft-to-output-shaft gear ratios for a windshield wiper motor assembly? _____

5. What does the term *nonconcealed windshield wiper system* mean? _____

6. The electrically operated washer (can) (cannot) be field repaired.

7. The electrically operated pump is a (rotary) (piston) type.

8. Pressing the Wash button puts the wiper motor in the (LO) (HI)-speed position.

9. The automatic timer allows the rear-window deicer to operate for about _____ minutes.

10. The rear-window deicer operates on a current draw of about _____ amperes.

11. The rear-window defogger blows (inside) (outside) air onto the rear window.

12. What is the purpose of the relay tie-in of the deicer system to the heating and air conditioning system? _____

13. How are the grids of the rear-window deicer attached to the glass? _____

14. After the Wash button is pressed on an electrically operated system, the wipers continue to operate until
    a. the Wash button is released.
    b. the dash control switch is turned to OFF or PARK.
    c. Either a or b is correct.
    d. Neither a nor b is correct.

15. The wash cycle is terminated in the electrically operated system
    a. when the Wash button is released.
    b. on completion of a full 360° cycle after the Wash button is released.
    c. when the dash control switch is turned to OFF or PARK.

16. Gear ratio from the motor shaft to the wiper output shaft may be as high as:
    a. 20:1   c. 40:1
    b. 30:1   d. 50:1

17. An intermittent wiper system provides for a delay of:
    a. 1 to 15 seconds   c. 5 to 20 seconds
    b. 5 to 15 seconds   d. 1 to 20 seconds

18. If the wiper will not park when turned off, what may be the problem?
    a. defective fuse or circuit breaker
    b. defective ground connection
    c. Either a or b may be the cause.
    d. Neither a nor b is the cause.

19. The nozzles to direct fluid to the windshield are
    a. located on the wiper arm.
    b. located in openings in the hood or cowling.
    c. Either a or b is correct.
    d. Neither a nor b is correct.

20. A rear-window defogger operates on
    a. inside-car ambient air.
    b. air heated by a resistance heater.
    c. a grid attached to the rear window.
    d. Any of these answers may be correct.

◆ UNIT 17 ◆

PANEL INSTRUMENTS AND WARNING DEVICES

OBJECTIVES

On completion of this unit you will be able to:

- Understand the function of and the schematic diagrams for
  - fuel-level and low fuel indicators.
  - coolant temperature and over-temperature indicators.
  - oil pressure and low pressure-/oil-level indicators.
  - battery over-/undercharging and discharging warnings.
  - theft deterrent systems.
  - speed minder and over-speed alarms.
  - key-in warning.
  - headlamp on warning.
  - reverse gear select warning.
  - door ajar/open warning.
  - hood or trunk open warning.

Warning devices are provided in today's modern vehicle to alert the driver of pending hazards in order to protect the driver, passengers, and vehicle from harm. In the car of tomorrow some of these devices may not rely on driver intervention, but will automatically take appropriate evasive action, thereby eliminating the time lag required for a driver's reflex and reaction. Many of the warning devices found on the modern vehicle are covered in this unit.

DASH AND DISPLAY PANEL INSTRUMENTS

An important display device used in modern vehicles is the digital indicator. There are two main types of indicator devices: the light-emitting diode (LED) and liquid crystal display (LCD). Regardless of type, the computation and determination of what is to be indicated takes place in the on-board computer. The display is just a read-out device.

There are two types of instrument gauges used: the thermoelectric gauge and the electromagnetic gauge. The thermoelectric gauge operates from a supply voltage of about 5 volts, and the electromagnetic gauge operates from a standard 12-volt supply. The 5 volts required by the thermoelectric gauge is provided by a voltage-dropping device known as a *voltage limiter*. Solid-state voltage regulators are now available that can supply a stable 5-volt output.

VOLTAGE LIMITERS

The voltage limiter ensures that an almost constant 5 volts is available to provide gauge accuracy regardless of voltage variations in the battery and charging

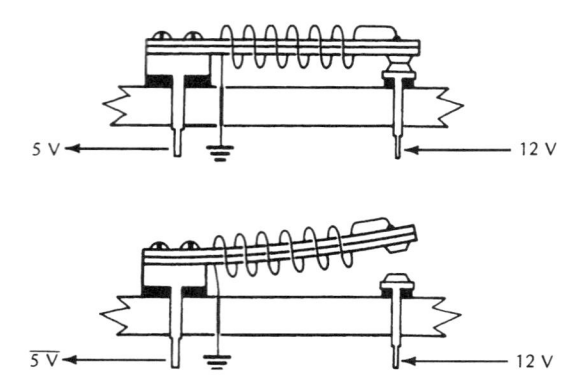

FIGURE 17-1 The opening and closing of the voltage limiter points regulates output voltage to about 5 V

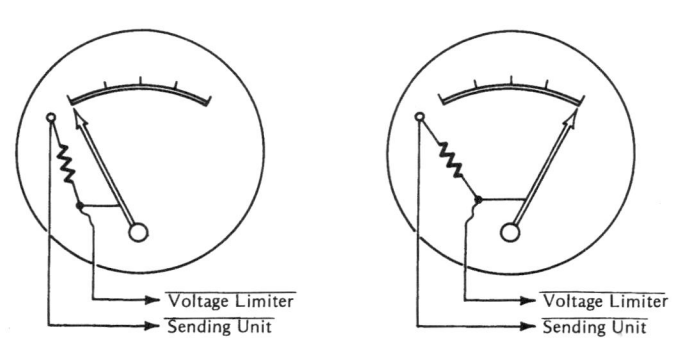

FIGURE 17-2 The thermoelectric gauge reads empty (E) or full (F) depending on the resistance of the sending unit

system. The voltage limiter operates on a principle similar to that of the bimetallic circuit breaker. With a coil of fine wire wound around a movable arm, its operation is similar to that of the hazard flasher. However, its operation, opening and closing, is much faster than that of the hazard flasher. Figure 17-1 illustrates the action of a typical voltage limiter. Whenever 12 volts is applied to the limiter, the coil is immediately heated. The calibration of the bimetallic strip is such that it quickly bends, opening the points. Since not much heat is generated because of the short heating period, the bimetallic strip quickly cools and the points close. This opening and closing action is rapid and produces a pulsating current of about 5 volts. When the input voltage is high, the point action is faster. The lower the input voltage, the slower the point action. Although the output can be considered to be pulsating dc, the action is so rapid that it is not detectable by gauge movement.

THERMOELECTRIC GAUGES

The simplest gauge is the thermoelectric gauge shown in Figure 17-2; it transfers the heat of electrical energy into a mechanical movement. The signal from the voltage limiter, at a constant 5 volts, heats a thin wire coil of a bimetallic movable arm of the gauge. This causes a pointer to move to an angle proportional to the heat generated in the coil. Actually, the 5-volt signal varies from about 1 to 5 volts, depending on the resistance of the sending unit (Figure 17-3).

ELECTROMAGNETIC GAUGES

The electromagnetic gauge (Figure 17-4) is operational through a variable voltage range of about 2 to 12 volts.

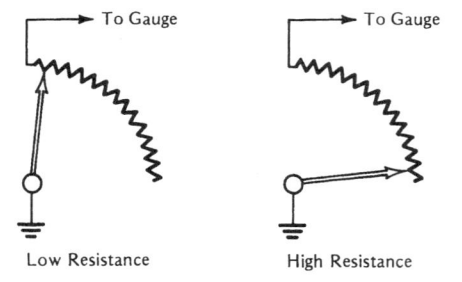

FIGURE 17-3 The resistance of the thermoelectric gauge circuit to ground is controlled by the sending unit

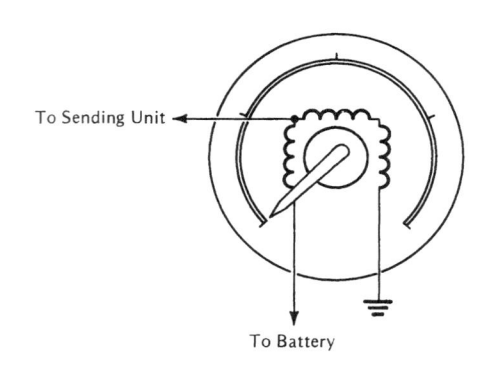

FIGURE 17-4 Electromagnetic gauge

The voltage level of the signal depends on the internal resistance of the sending unit (Figure 17-5). In the gauge, a magnetic field is produced by three coils wound on a nylon or plastic frame. The frame is constructed to allow the coils to surround a circular permanent magnet, which is free to rotate on sleeve bearings. One bearing shaft extends through the frame for attachment of a pointer. The pointer moves through its normal arc dependent on the resultant strength of the magnetic field set up by the interaction of the coils. A

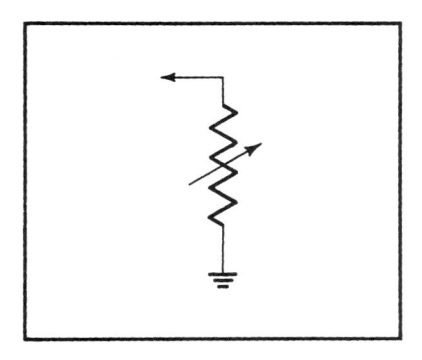

FIGURE 17-5 Voltage signal will be between 2 and 6 volts depending on the resistance of the sending unit

special silicone grease is used on the bearing surfaces to damp out pointer fluctuations caused by road shock or slight voltage variations.

SENDING UNITS

Although sending units are usually not interchangeable, the principle of operation is about the same for both thermoelectric and electromagnetic gauge applications—both operate on the principle of variable resistance. Sending units vary in appearance and application, and are matched to a particular dash gauge system. Each type of sending unit—fuel level, oil pressure, and coolant temperature—is illustrated in Figures 17-6 through 17-8. The fuel-level sending unit is located in the fuel tank and is generally accessible from either the trunk space or the underside of the car (Figure 17-9). The oil pressure and coolant temperature sending units are located on the engine at a point where average values can be obtained. Special testers are available for determining the accuracy and condition of sending units. Follow the instructions included with these testers to ensure proper use.

FUEL-LEVEL SYSTEM

The fuel-level system, depending on type, consists of a sending unit, a wiring harness, and a dash gauge, or a sending unit, a wiring harness, a voltage limiter, a choke, and a dash gauge. Regardless of the type of system, a float is attached to a movable arm that varies the

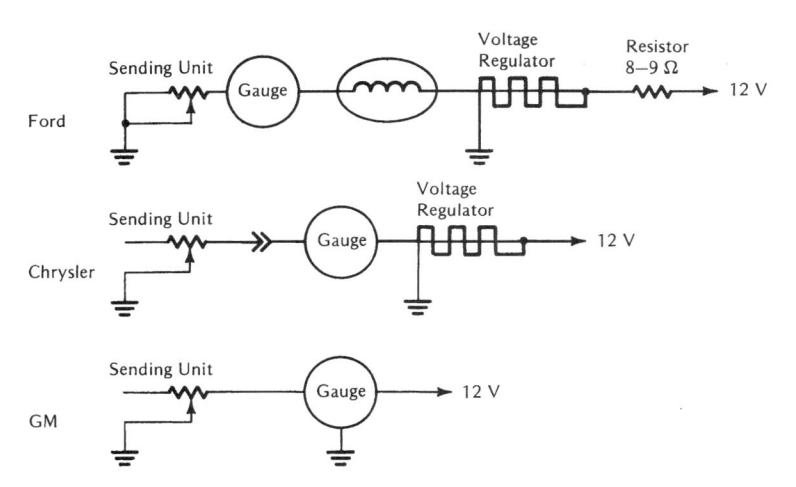

FIGURE 17-6 Typical fuel gauge schematics

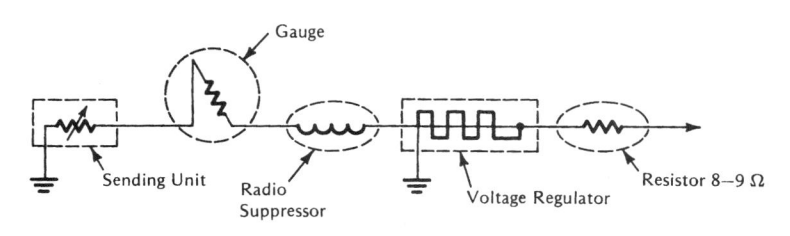

FIGURE 17-7 Typical oil pressure gauge system schematic (Ford)

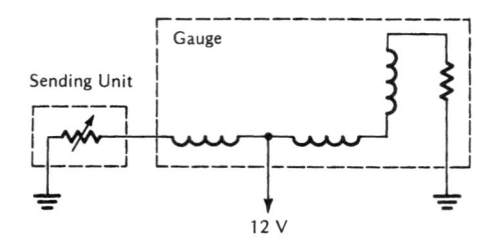

192 Unit 17 PANEL INSTRUMENTS AND WARNING DEVICES

FIGURE 17-8 Typical temperature gauge system schematic (GM)

resistance of the sending unit depending on float level (Figure 17-10).

A damaged fuel tank (Figure 17-11) can give an erroneous fuel-level indication at the dash gauge, although the components of the system are in good working order. The flow chart in Figure 17-12 shows the testing procedure for an inoperative or inaccurate fuel-level system. Schematics of two typical fuel-level systems are shown in Figures 17-13 and 17-14.

Many fuel-level systems have, as an option, a low fuel warning lamp. One of the systems uses a light-emitting diode (LED) as an integral part of the fuel gauge assembly. When the fuel level is at ⅛ of a tank or below, the LED emits a visual warning of the low fuel condition. The other system uses a relay, a 12-volt lamp, and a thermistor. The thermistor is attached to the fuel pickup tube in the fuel tank at a level that will cause it to be uncovered by fuel when the fuel level is down to about a quarter tank. The thermistor's temperature will increase without the cooling effect of the fuel, and it will trigger the sensitive relay to turn on the dash-mounted warning lamp. A simple schematic of this system is shown in Figure 17-15. Note that this circuit operates on 12 volts and is totally independent of the regulated 5-volt gauge circuit.

COOLANT TEMPERATURE SYSTEM

There are two types of dash indicators used on the modern vehicle to warn of improper coolant temperature: gauges and lamps.

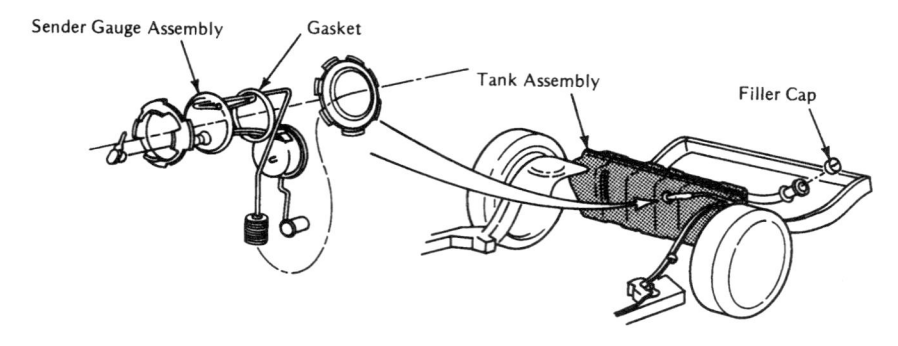

FIGURE 17-9 Location details of fuel sending units

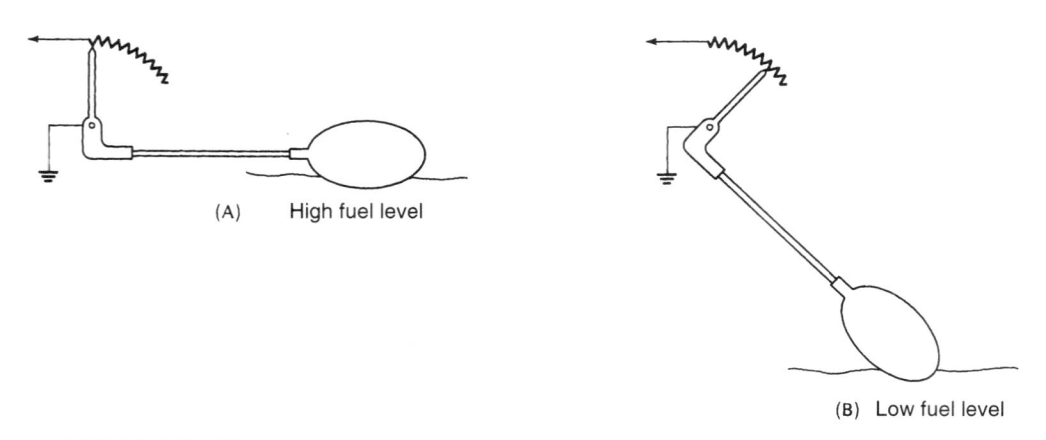

FIGURE 17-10 Sending unit provides low resistance when fuel tank is full (a) and high resistance when the fuel tank is empty (b)

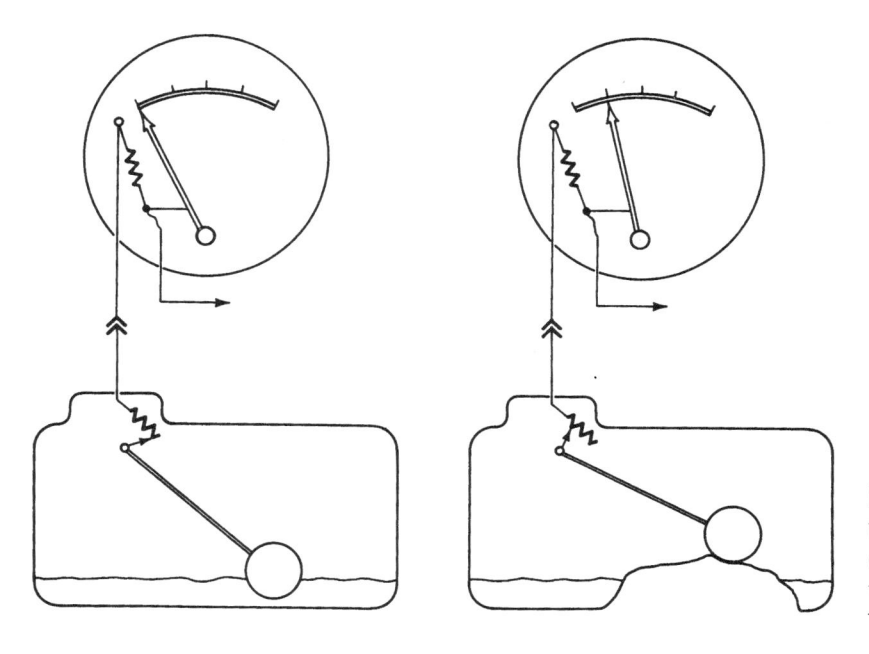

FIGURE 17-11 A damaged (bent) fuel tank can cause an improper dash reading even though the system components and wiring are undamaged

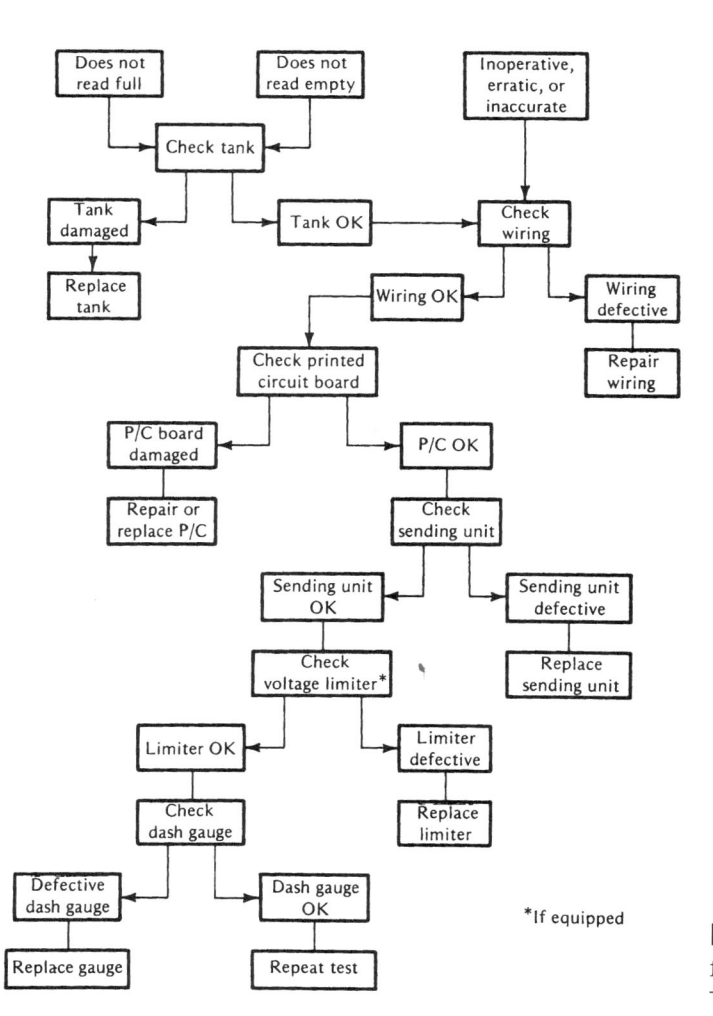

*If equipped

FIGURE 17-12 The order of testing procedures for a fuel gauge system

193

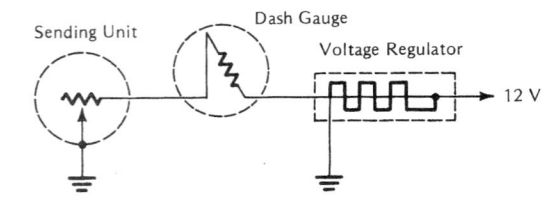

FIGURE 17-13 Typical thermoelectric fuel gauge system

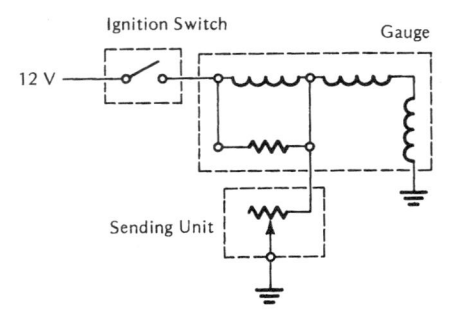

FIGURE 17-14 Typical electromagnetic fuel gauge system

Gauges

The coolant temperature warning system consists primarily of the same components as the fuel-level system, although the physical appearance and structure of the components differ. A schematic of the two types of systems is shown in Figure 17-16. The flow chart in Figure 17-17 shows the testing procedure for an inoperative or inaccurate system.

Some Chrysler temperature gauges have as an option a "hot" warning lamp as an integral part of the gauge unit. When the dash gauge indicator moves to a position representing a temperature of approximately 250°F (121°C), the warning lamp, a light-emitting diode (LED), will illuminate to warn the driver of a pending overheating condition.

The design operating temperature for most car engines is above the normal boiling point of water (H_2O) at sea-level atmospheric pressure, 14.7 psia (101.4 kPa absolute). At this pressure the boiling point of water is 212°F (100°C). Increasing the boiling point temperature of the coolant is made possible by an antifreeze additive and/or by pressurizing the cooling system. It is not uncommon for a cooling system to be pressurized to 16 psia (110.3 kPa).

Lamps

Instead of a gauge system, many cars are equipped with a lamp to indicate when the coolant temperature is near the critical point. Some cars are equipped with a second lamp to indicate that the coolant temperature has not reached the normal operating temperature. Both one- and two-lamp systems operate in basically the same manner. Figure 17-18 shows the schematic for a one-lamp system, and Figure 17-19 shows the schematic for a two-lamp system. The sending unit, unlike that used with the gauge system, is equipped with a bimetallic sensing element. The "cold" contacts are normally closed (NC) whenever the coolant temperature is below 165° to 180°F (73.9° to 82.2°C), depending on engine design. The normally open (NO) "hot" contacts close if the coolant temperature reaches 235° to 250°F (112.8° to 121.1°C), again depending on engine design.

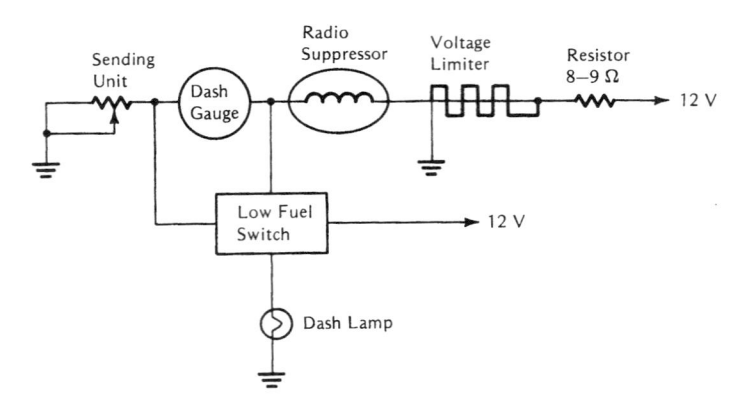

FIGURE 17-15 Low fuel warning system schematic

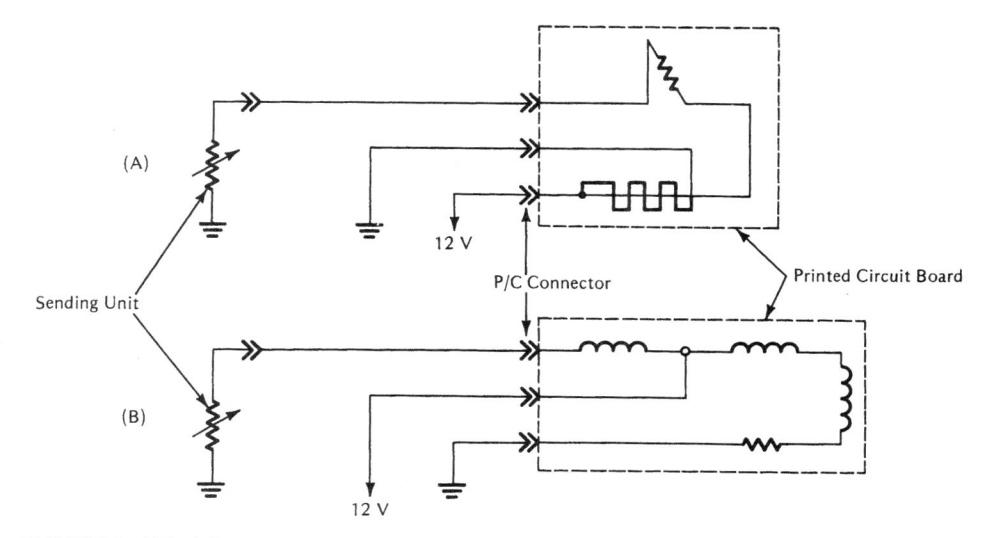

FIGURE 17-16 Typical schematics for (a) thermoelectric and (b) electromagnetic engine coolant temperature gauge systems

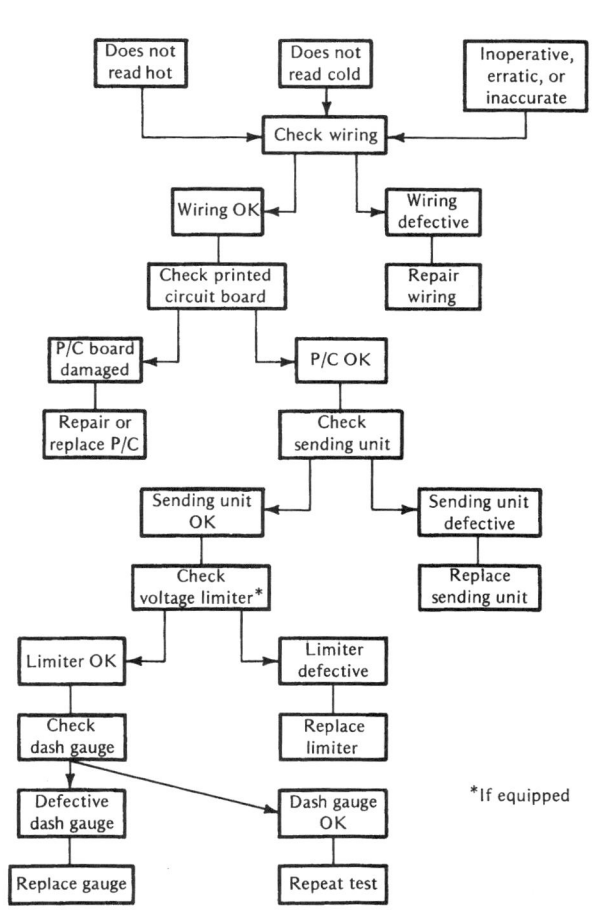

FIGURE 17-17 Order of testing procedures for temperature gauge system

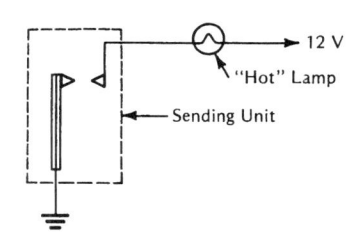

FIGURE 17-18 Schematic for one-lamp coolant temperature warning system

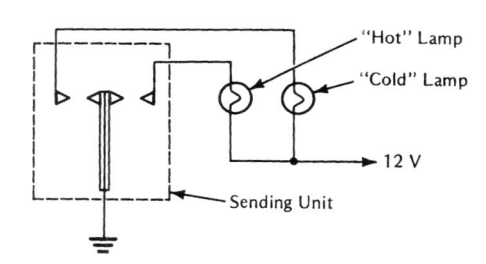

FIGURE 17-19 Schematic for two-lamp coolant temperature warning system

OIL PRESSURE SYSTEM

Gauges

The oil pressure warning gauge system consists of the same components as the coolant temperature gauge system. A schematic of two types of oil pressure

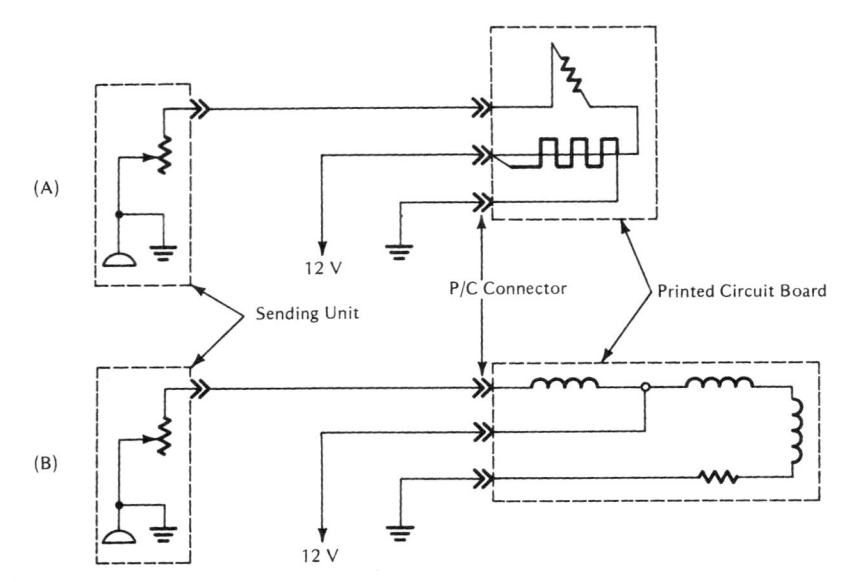

FIGURE 17-20 Typical schematics for (a) thermoelectric and (b) electromagnetic engine oil pressure gauge systems

systems is shown in Figure 17-20. The primary difference in the two systems is that the variable resistance of the sending unit in the thermoelectric system decreases with high pressure, whereas in the electromagnetic system the resistance increases with high pressure. Also, the sending unit in the oil pressure system is activated by pressure and not by temperature as in the coolant sending units. A chart showing testing procedures is in Figure 17-21.

Some cars are equipped with only a lamp to indicate when the engine oil pressure has dropped to a critically low level. This system (Figure 17-22) operates on the same principle as the engine coolant "cold" lamp circuit except that it is triggered by a pressure—not temperature—activated sending unit.

Gauges and Lamps

The oil pressure warning system is part of the electric fuel pump circuit on some cars. The sending unit is used to break the electrical circuit of the fuel pump under conditions of low oil pressure. The electric fuel pump and its circuit is covered in Unit 18.

CHARGING SYSTEM

Gauges

An ammeter and a wiring harness make up the gauge-type charging system indicator circuit. A typical circuit

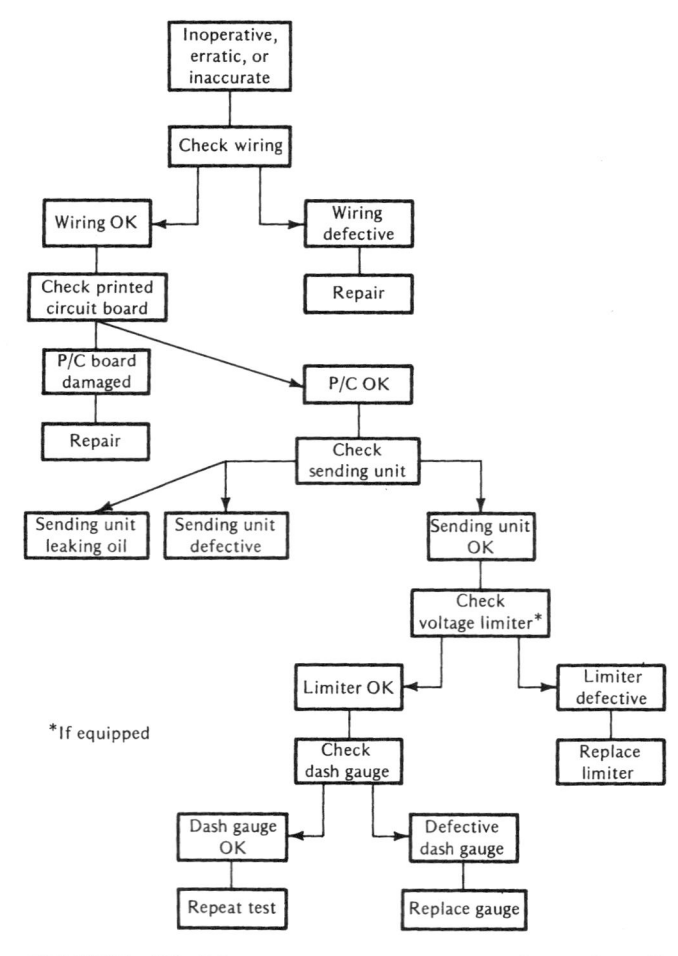

FIGURE 17-21 Order of testing procedures for oil pressure gauge system

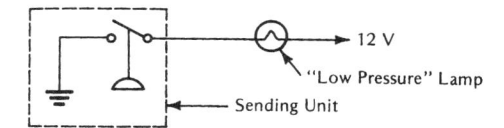

FIGURE 17-22 Schematic for oil pressure warning lamp system

schematic is shown in Figure 17-23. The current (or part of the current) to and from the battery passes through the ammeter. The direction of current flow sets up a magnetic field either clockwise or counterclockwise around the current-carrying conductor, following the left-hand rule. This field causes the ammeter (Figure 17-24) to move to the left (discharge) if current flow is out of the battery, or to the right (charge) if current flow is into the battery. The amount of current flow, in amperes, determines the degree of pointer deflection from the zero center line toward discharge (−) or charge (+).

Lamps

In many applications, the dash ammeter gauge has been replaced with an indicator lamp. Typically, the lamp is illuminated if the system is discharging and is not illuminated if the system is charging.

Two typical system schematics are shown in Figure 17-25. In this system, when the ignition switch is turned on before the engine is started, the indicator lamp should be illuminated. If it is not, the bulb or wiring is defective and should be repaired. After the engine is started, the lamp should go out. If it does not, the charging system is defective and should be repaired. The charging system was covered in Unit 14.

Gauges and Lamps

Some car lines have as an option a gauge and a lamp. The lamp, a light-emitting diode (LED), is located on the gauge (ammeter), but is independent of the gauge circuit. When the system voltage drops to a predetermined level, the LED illuminates to alert the driver of pending problems with the charging system.

HORNS

Horns are provided on all cars as an audible warning and signaling device. Small cars are usually equipped with one horn, midclass cars with two horns, and luxury-class cars with three horns. Some horn electrical circuits have no relay (Figure 17-26), whereas others do have a relay (Figure 17-27).

The horn is a vibrating device consisting of a case, a diaphragm, an electromagnetic field coil, a resistor, and a set of points (Figure 17-28). The points, normally open (NO), provide an electrical path to ground for the field coil. Whenever voltage is applied to the horn, current flows, setting up a magnetic field. The magnetic field pulls in the diaphragm and, at the same time, opens the points. When this occurs, the magnetic field collapses, and the diaphragm relaxes to its original position.

The points again close. This action is repeated as long as the horn button is pressed. The resistor in the circuit helps to prevent heavy arcing of the points while "making and breaking."

Single-horn systems have a low-note horn; dual-horn systems have both a low-note horn and a high-note horn. Three-horn systems have a low-, a medium-, and a high-note horn. The tone or pitch of a horn can be changed by turning a self-locking adjusting screw located on the horn case. There is a minimum and a maximum range, however, over which the horn may be tuned.

Horn Circuit Without Relay

The horn circuit without a relay is shown in Figure 17-26. In this circuit total current must pass through the horn-button switch. The switch, mounted in or atop the steering wheel, completes a circuit to the horn. This

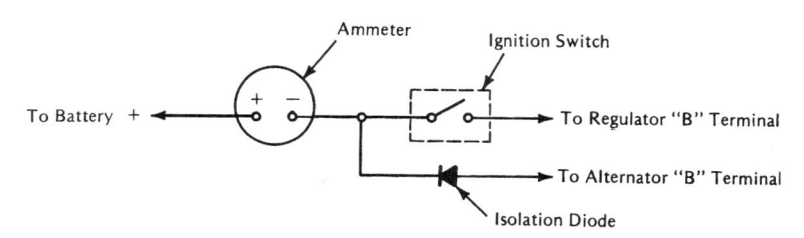

FIGURE 17-23 Schematic for typical ammeter circuit

FIGURE 17-24 Ammeter is often marked ALT for "alternator"

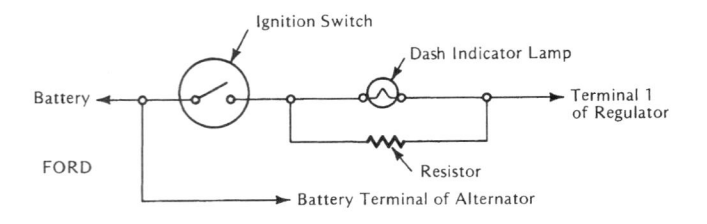

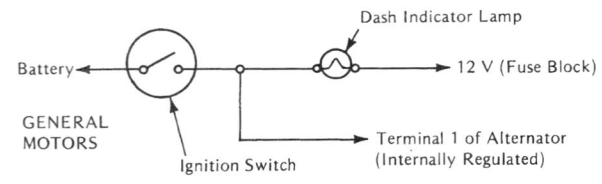

FIGURE 17-25 Typical schematics for Ford and General Motors dash lamp charging system warning indicators

FIGURE 17-26 Horn circuit with no relay

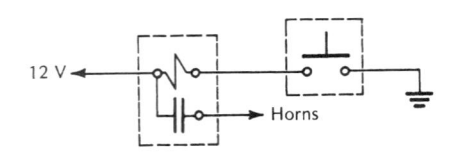

FIGURE 17-27 Horn circuit with relay

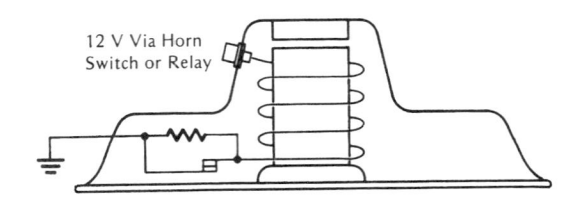

FIGURE 17-28 Sectional view of typical horn

circuit is usually fed from the headlamp switch or fuse block.

Horn Circuit with Relay

The horn circuit with a relay is shown in Figure 17-27. In this circuit only "pilot" current sufficient to operate the relay must pass through the horn-button switch. Current for the horn circuit passes through the contacts of the relay. This circuit is usually fed from the fuse block.

Horn Buttons

Although the horn button, or switch, is usually located in the center of the steering wheel, Figure 17-29, it can be a pressure-sensitive pad covering the center spoke of the steering wheel (Figure 17-30). Some car models are equipped with an optional rim-blow switch instead of the conventional button or pad. The rim-blow switch consists of a plastic insert containing two thin copper strips that are inserted around the inner diameter of the steering wheel (Figure 17-31). Pressing the plastic insert at any point around the steering wheel brings the copper strips into contact and completes the horn circuit.

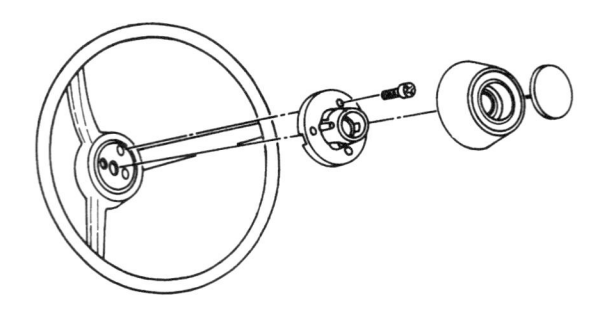

FIGURE 17-29 Horn button (switch) located at center of wheel

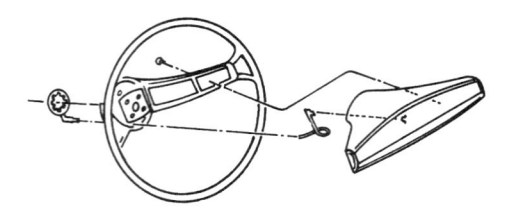

FIGURE 17-30 Pressure-sensitive horn switch located in center spoke of steering wheel

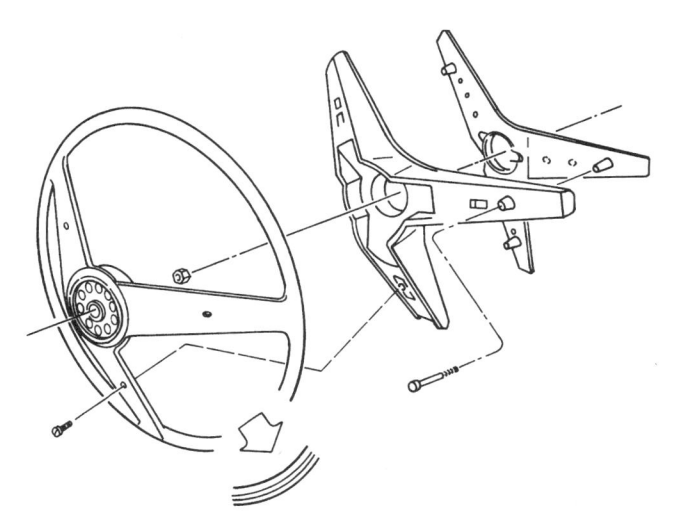

FIGURE 17–31 Details of rim-blow horn

Horn Relays

The horn relay is an electromagnetic switch. When power is applied to the relay coil, an electromagnetic field is set up that pulls the contact armature in to close the normally open (NO) set of points. Consequently, the closed points complete the horn circuit (Figure 17–32).

ANTITHEFT ALARM SYSTEMS

There are many types of antitheft alarm systems (often called burglar alarms) available. Many of them are aftermarket add-on types. They are also offered as new-car options. The system shown in Figure 17–33 is available as an option on late-model motor vehicles. This system must be armed before it can be triggered. The solid-state controller is activated by a manual selector switch located in the glove compartment. The two positions of the selector switch are labeled "arm enable" and "arm prevent." Once the switch is placed in the "arm enable" position, the system is activated. When, for example, it is moved to the "arm prevent" position after being armed, the alarm will immediately sound. When the selector switch is placed in the "arm enable" position, the system is automatically armed 90 seconds

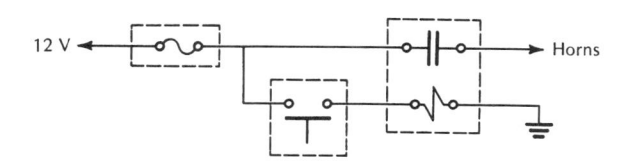

FIGURE 17–32 Typical horn circuit with relay control

after the ignition switch is turned off. If a door is held open after the ignition switch is turned off for more than 90 seconds, the system will be armed immediately after the door is closed.

Once armed, the system can be disarmed only by turning the ignition switch to the ON or ACC (accessory) position. If this is not done within 20 seconds after the door is opened, the alarm will activate. Once activated, the horn sounds, and the parking, marker, license plate, and tail lamps flash at 50 cycles per minute for 5 minutes. After 5 minutes, the system shuts off and automatically rearms. Further tampering will again activate the system.

A simpler system (Figure 17–34) also used on late-model vehicles is typical of many aftermarket devices that are available as well. (Operation and installation instructions are included with the add-on systems.) The control switch is key activated after all doors are closed and locked. Opening a door, the hood, or the trunk will activate the alarm if the control switch is closed.

SPEEDMINDERS

The speedometer speedminder or speed alert system is an option on many car lines. Its purpose is to quickly alert the driver that a predetermined speed is being exceeded. If, for example, the maximum speed permitted (or desired) were 60 mph (96.5 km/h), the driver could set the speedminder pointer at 62 mph (99.7 km/h). Then, at the preset speed, the speedminder circuit (Figure 17–35), would energize and sound an alert. The alert is usually a buzzer, but it can be a buzzer and light. The speedminder (Figure 17–36), can usually be manually set by the driver at any speed between 30 mph (48.2 km/h) and the maximum calibration of the speedometer, usually 90 to 120 mph (144.8 to 193.0 km/h).

SPEEDOMETER CALIBRATION CHECK

A speedometer can be checked for proper calibration with some degree of reliability if the odometer is accurate. The odometer can be checked against milepost markers alongside most interstate highways. It is recommended that a distance of 10 miles (18.5 km) be used to check the odometer accuracy. If the odometer is in the metric measure, fifteen kilometers (9.3 miles) should be sufficient.

Inaccuracies of a mechanical speedometer can usually be corrected by changing the drive gear located in the transmission. Increasing the number of teeth in

Unit 17 PANEL INSTRUMENTS AND WARNING DEVICES

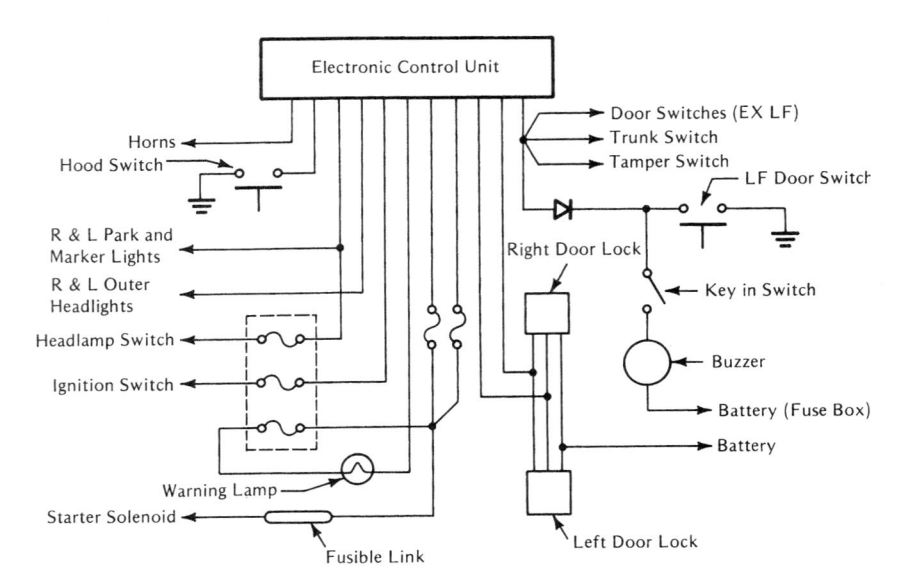

FIGURE 17-33 Schematic for Cadillac's antitheft alarm system

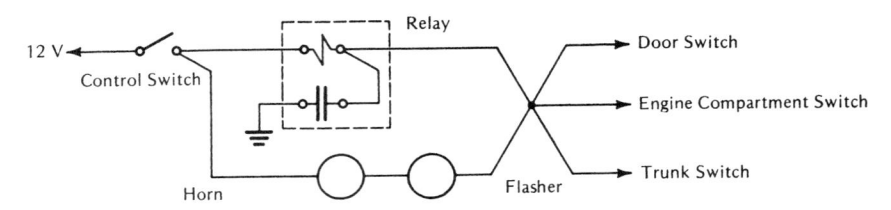

FIGURE 17-34 Schematic for simple antitheft alarm system

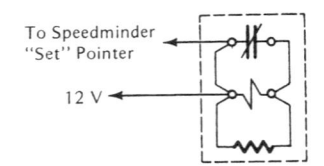

FIGURE 17-35 Speedminder buzzer circuit

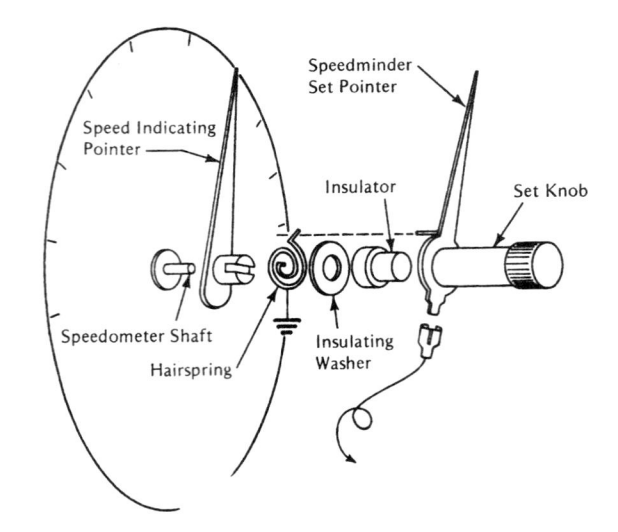

FIGURE 17-36 Exploded view of speedminder portion of a typical speedometer

the gear decreases indicated miles (kilometers) covered. Decreasing the number of teeth increases the indicated distance covered.

If the odometer is accurate, the speedometer may be easily checked for accuracy. A few miles of roadway that permits a speed of 60 mph (96.5 km/h), and a stopwatch or a watch with a sweep-second hand is required. At a constant speed of 60 mph (96.5 km/h), record the amount of time, in seconds, required to travel 1 mile (1.6 km) as indicated by the odometer. If less than 60 seconds is required, the speedometer calibration is low, and actual speed is faster. If, on the other hand, longer than 60 seconds is required, the speedometer calibration is high, and actual speed is slower. Tables 17-1 and 17-2 give the actual speeds for covering a mile (1.6 km) in 55 to 65 seconds.

TABLE 17-1 Conversion table: time vs. mph and km/hr to nearest 1/100th of a mile or kilometer

Seconds	mph	Km/hr
55	65.45	105.30
56	64.29	103.44
57	63.16	101.62
58	62.02	99.79
59	61.02	98.18
60	60.00	96.54
61	59.02	94.96
62	58.06	93.41
63	57.14	91.93
64	56.25	90.50
65	55.38	89.10

TABLE 17-2 Conversion table: time vs. km/hr and mph to the nearest 1/10th of a kilometer or mile

Seconds	Km/hr	mph
65	89.1	55.4
64	90.5	56.3
63	91.9	57.1
62	93.4	58.1
61	95.0	59.0
60	96.5	60.0
59	98.2	61.0
58	99.8	62.1
57	101.6	63.2
56	103.4	64.3
55	105.3	65.5

KEY-IN WARNING SYSTEMS

The key-in warning system provides an audible warning (buzzer or chime) when the driver's door is opened with the key in the ignition switch. On some systems, the alarm may sound if the passenger's door is opened as well. The key-in warning system serves as a reminder not to lock the keys in the car.

Key-in warning systems have not changed appreciably over the years. There are at present four

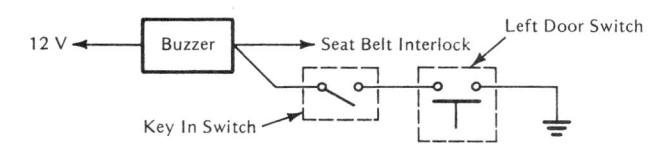

FIGURE 17-37 Typical key-in warning system integrated with seat belt interlock schematic

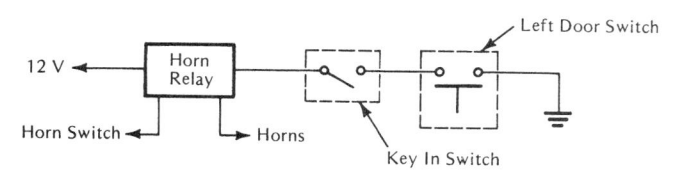

FIGURE 17-38 Typical key-in warning system integrated with horn relay schematic

systems. Figure 17-37 shows the key-in warning integrated with the seat belt interlock system. Another system is integrated with the horn relay (Figure 17-38). A third and fourth system (Figures 17-39 and 17-40) appear at first glance to be the same. However, one system is electrically grounded through the key-in switch, whereas the other is grounded through the doorjamb switch.

Quick Check

A quick check of the key-in warning system (after first checking fuse condition) consists of opening the driver's door with the key in the ignition switch. The dome lamp or courtesy lamps should illuminate. If the lamp(s) do not illuminate, the problem may be in the

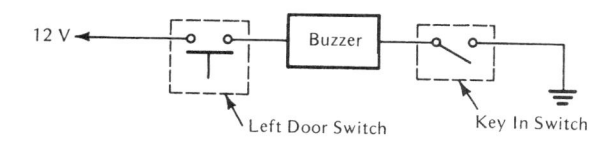

FIGURE 17-39 Typical key-in warning system grounded through the ignition switch schematic

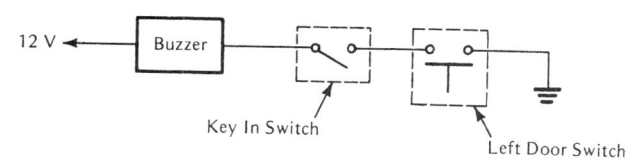

FIGURE 17-40 Typical key-in warning system grounded through the doorjamb switch schematic

doorjamb switch or related wiring. If the lamps do illuminate when the door is opened, replace the buzzer with one known to be functioning. Then, if the buzzer does not sound with the door open and the key in the ignition switch, the problem is in the wiring. Reinstall the original buzzer and trace the wiring circuit. The problem—probably an open circuit—must be repaired to restore the system.

HEADLAMP-ON WARNING SYSTEMS

The headlamp-on warning system provides an audible sound to warn the driver that the headlamps or parking lamps are turned on when he or she is exiting the car. As with other warning systems, the audible warning is provided by a buzzer. Schematics of typical headlamp-on warning systems are given in Figure 17–41. Specific wiring information is found in manufacturers' shop manuals and wiring schematics.

Like the key-in warning system, the headlamp-on warning system uses the left-front doorjamb switch to complete an electrical circuit that activates the buzzer. American Motors, Chrysler, and General Motors use doorjamb switches of the grounding type, and Ford uses nongrounding types (Figure 17–42).

Quick Check

A quick check of the headlamp-on warning system can be done by noting whether the dome or courtesy lamps illuminate when the driver's door is opened. If the lamps do not illuminate, the problem may be in the doorjamb switch or associated wiring. If the lamp(s) do illuminate, replace the buzzer with one known to be working. Then, if the buzzer does not sound when the headlamp switch is in either ON position, the problem is in the wiring. Reinstall the original buzzer and locate and correct the wiring as necessary.

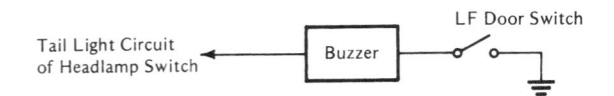

REVERSE WARNING SYSTEMS

Many vehicles are equipped with a warning signal that sounds when the transmission is placed in reverse gear. This system consists of a buzzer (internal) and/or a horn (external) that is electrically connected to the backup light system (Figure 17–43). Consisting of a 12-volt buzzer and/or horn with mounting hardware and two short pieces of wire, this feature can be easily added to any vehicle.

DOOR-AJAR WARNING SYSTEMS

The door-ajar warning system is similar to the dome and courtesy lamp circuits covered in Unit 6. A separate doorjamb switch is used to illuminate a dash lamp if a door is not completely closed. The doorjamb switch may be either grounding or nongrounding, as shown in Figure 17–44.

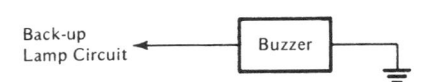

FIGURE 17–43 Reverse warning circuit

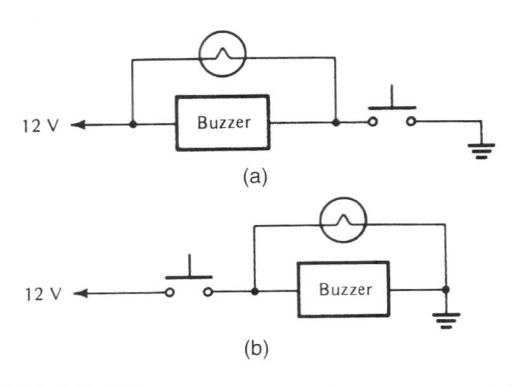

FIGURE 17–41 Headlamp-on warning system

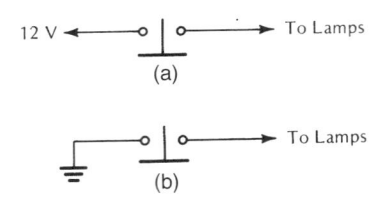

FIGURE 17–42 (a) Nongrounding doorjamb switch and (b) Grounding doorjamb switch.

FIGURE 17–44 Door-ajar warning systems; (a) typical of all American manufacturers except Ford, (b) typical Ford system

HOOD- AND TRUNK LID-OPEN WARNING SYSTEMS

The hood and trunk lid warning system circuit is similar to the door-ajar warning circuit. A grounding or nongrounding jamb switch is used to illuminate a dash lamp if the hood or trunk lid is not completely closed.

SUMMARY

Modern motor vehicles include many warning devices that were not available in earlier production years. These displays and audible warning devices are generally electrically controlled. The modern technician must be familiar with the warning display devices, as well as the sensing devices that provide the warning indicators.

PRACTICAL EXERCISE

The following exercise is intended for classroom and shop participation and discussion. A late-model vehicle and appropriate electrical wiring shop manual are required.

1. Raise the hood of the vehicle. DO NOT start the engine.
2. Locate the temperature sending unit on the engine.
3. Locate the temperature sending unit in the appropriate wiring diagram.
4. Are the wire(s) color coded in the wiring diagram the same as they are on the sending unit? _____
5. Locate where the wires terminate in the wiring diagram.
    a. _____ Color _____
    b. _____ Color _____
    c. _____ Color _____
6. Locate the temperature indicating device in the wiring diagram. Is the device a gauge or a lamp? _____
7. Repeat Steps 4, 5, and 6 for the
    a. Oil pressure sending unit. _____ Color _____
    b. Low oil-level warning sending unit. _____ Color _____
    c. Low coolant-level sending unit. _____ Color _____
    d. Door-ajar warning system. _____ Color _____
    e. Other warning devices. _____
8. After completion of this exercise reconnect any wires that may have been disconnected and close the hood.

REVIEW

1. The speedminder buzzer will (continue) (cease) to sound at speeds greater than the speed at which it is set.
2. Ford uses a (grounding) (nongrounding) door-jamb switch.
3. Chrysler uses a (grounding) (nongrounding) door-jamb switch.
4. The door-ajar system activates a (dash lamp) (buzzer) (dash lamp *and* buzzer) if a door is opened.
5. Grounding and nongrounding doorjamb switches (are) (are not) interchangeable.
6. Cars are equipped with _____ horn(s).
    a. one
    b. two
    c. three
    d. Any of the above.
7. There is (are) _____ type(s) of horn switches.
    a. one
    b. two
    c. three
    d. None of the above.
8. Properly tuned, a horn should draw _____ amperes.
    a. 2 to 3
    b. 3 to 5
    c. 4 to 5
    d. 5 to 6
9. Total current must pass through the horn switch if the circuit (is) (is not) equipped with a relay.

10. Single-horn systems are equipped with a (low-) (medium-) (high-) note horn.

11. Briefly describe the operation of a fuel-level system that uses an electromagnetic dash gauge. _____

12. Draw a simple schematic of a coolant temperature system that uses a thermoelectric dash gauge.

13. Draw a simple schematic of an oil pressure lamp system.

14. Briefly describe the operation of Ford's low fuel-level warning system. _____

15. Can a damaged fuel tank give an erroneous fuel-level indication? How/why? _____

16. The thermoelectric gauge system is used by _____.

17. The electromagnetic gauge system is used by _____.

18. A device called a(n) _____ is used to regulate voltage to the thermoelectric system gauges.

19. The resistance required for proper gauge reading is provided by a device called the _____.

20. A special _____ is used to damp out pointer "flutter" of the electromagnetic gauge.

◆ UNIT 18 ◆

OPTIONAL AND CONVENIENCE SYSTEMS

OBJECTIVES

On completion of this unit you will be able to:

- Explain the operation of convenience systems including
- mechanical and power seats.
- power windows.
- power tailgate windows.
- sun-/moon roof systems.
- door lock systems.
- keyless entry systems.
- trunk closing systems.
- electronic speed control.
- radio and antenna systems.

There are many devices and systems available in the modern vehicle that are provided primarily for the comfort and convenience of the driver and passengers. Seats, of course, are not an option. The manner in which seats are adjusted, however, is an option. This unit also covers other options such as clocks, radio and antenna systems, rear-view mirrors, and speed control. Convenience systems include power windows and door locks, keyless entry systems, trunk closing and lock systems, and so on. Though a particular car line is occasionally referenced, the systems covered in this unit should be considered typical and not representative of any car line. For specific information about the particular system manufacturer's specifications must be consulted.

FRONT SEATS

There are many types of front seats available in today's automobile (Figure 18–1). In addition to the standard full-width bench seats, there are 50–50, 60–40, and 40–40 seats. Bucket seats (Figure 18–2) are available with various options. All mechanically adjustable seats, regardless of their options or types, have two-way adjustment, fore and aft, as illustrated in Figure 18–3.

Power Front Seats

A popular option is the power adjustable seat. Power seats have two-way, four-way, or six-way adjustment, as illustrated in Figure 18–4. Because of the vast number of seat options, it is necessary to consult the manufacturer's wiring schematics and manuals for any particular type of seat. This section will cover typical power seats and their related electrical circuits.

Two-way Power Seats. The two-way power seat used by Ford and General Motors is adjustable forward or backward in a horizontal plane and is driven by a single electric motor. Flexible shafts are connected from the left and right ends of the motor to a transmission

206 Unit 18 OPTIONAL AND CONVENIENCE SYSTEMS

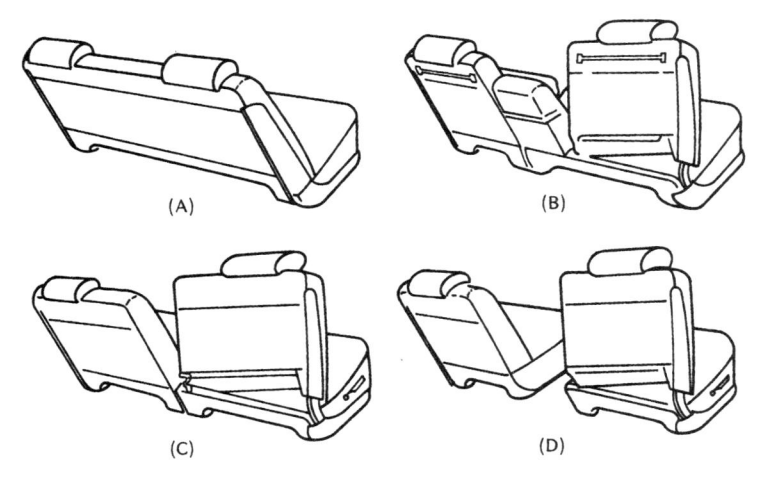

FIGURE 18–1 (a) Full bench; (b) split 60–40; (c) split 50–50; (d) 40–40 seat

and horizontal rack on left and right tracks. The motor is reversible and is actuated by a control switch to move the seat to the front or rear. Typical schematics for the Ford and General Motors two-way seat control are shown in Figure 18–5.

Four-way Power Seats. The four-way power seat used by Ford provides fore and aft movement of the seat and also provides an up and down movement of the front edge of the seat, resulting in a tilting action. The rear edge of the seat does not move up or down. The four-way seat is driven by a dual electric motor through two gearboxes and two flexible cables (Figure 18–6). When actuated by a control switch, the reversible motors move the seat to the front or rear and the front edge up or down. A typical schematic of the four-way seat control is shown in Figure 18–7.

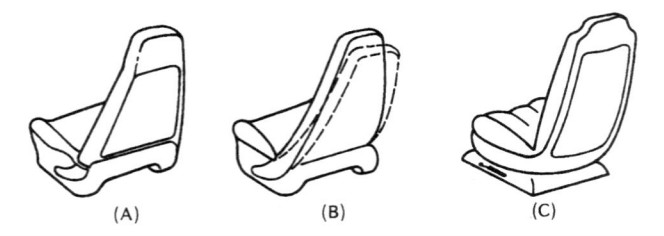

FIGURE 18–2 Typical bucket seats: (a) standard, (b) tilt-back, and (c) swivel

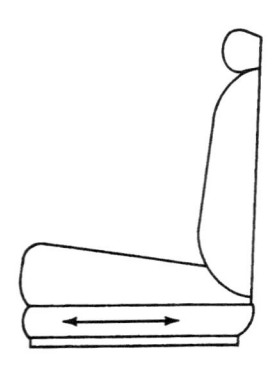

FIGURE 18–3 Two-way power seat adjusts fore and aft

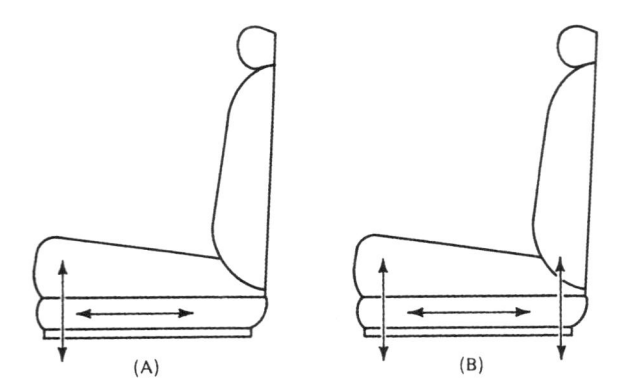

FIGURE 18–4 (a) Four-way power seat adjusts fore and aft and front edge vertical tilt; (b) six-way power seat adjusts fore and aft and vertical tilt of front and rear edges

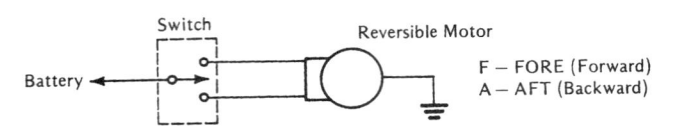

FIGURE 18–5 Typical Ford and General Motors two-way seat control schematic

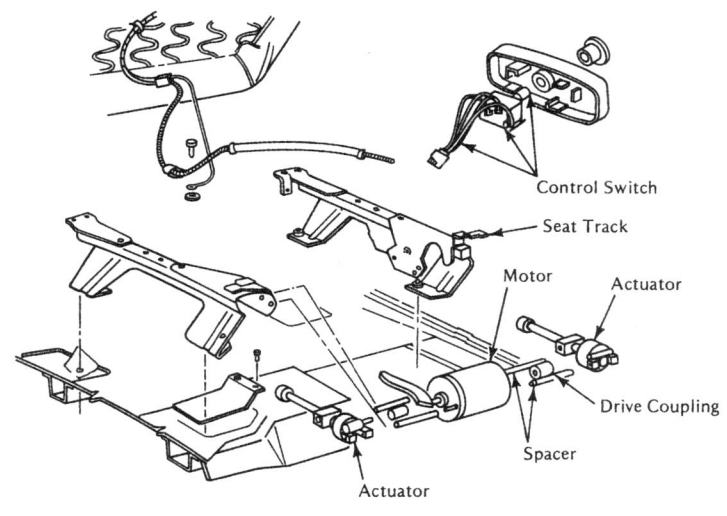

FIGURE 18–6 Typical mechanical section of four-way seat control

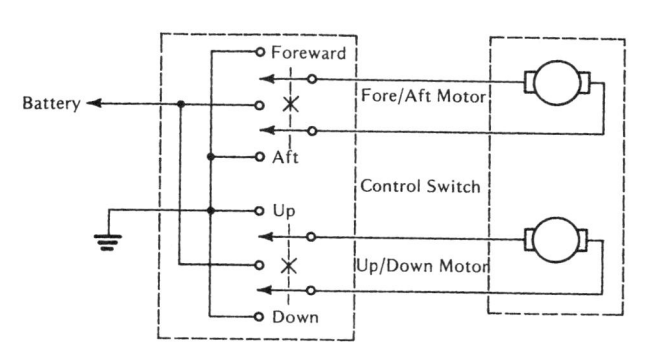

FIGURE 18–7 Typical schematic of four-way seat control

Six-way Power Seats. The six-way power seat is by far the most popular of the power seat options. It provides horizontal (fore and aft), vertical (up and down), and vertical tilt (front and rear) adjustments. The six-way power seat is driven by a three-armature reversible motor through vertical worm gear drives and horizontal rack and pinion drives (Figure 18–8). The horizontal drive mechanism of the seat consists of a rack and pinion on each track. The pinion, housing, and motor are attached to a movable section of track. The seat is moved either forward or backward by pinion gears traveling in a rack in the lower track sections. Worm gear mechanisms are used for the vertical drive portion of the seat. When a switch is actuated, center and rear armatures are energized, and the vertical drive units are activated. The seat is then moved upward or downward to the desired position by the worm gears.

Activating the front tilt switch energizes the center motor armature. This drives the front vertical worm gear to move the seat to the desired position. When the rear tilt switch is activated, the rear armature is energized, which drives the rear vertical worm gear. This description of operation, although typical, is of the Ford six-way power-operated seat system. Typical schematics for Chrysler, Ford, and General Motors six-way seat controls are shown in Figure 18–9.

Seat Back Locks

An optional seat back lock-release system is available on some two-door car lines equipped with electric door locks. The seat back lock system utilizes a solenoid on each of the front seat backs. These solenoids are operational through independent electrical circuits that are completed through associated doorjamb switches. As shown in Figure 18–10, some car lines use grounding doorjamb switches, whereas others use insulated switches.

The solenoids have both an "unlock" and a "hold-in" coil that are wound in tandem around a single plunger. The unlock coil draws about 18 amperes, and the hold-in coil draws less than one ampere. Both are energized at the same time but are deenergized individually. For example, when a door is opened, the associated seat back lock-release solenoid is energized. The unlock and hold-in coils pull in the release plunger. When the plunger reaches its full inward travel, it simultaneously trips an internal switch to open the circuit for the unlock coil. The circuit of the hold-in coil remains complete to hold the plunger in the unlock position. When a door is closed, the solenoid is deenergized to allow the seat back lock to return to its normal

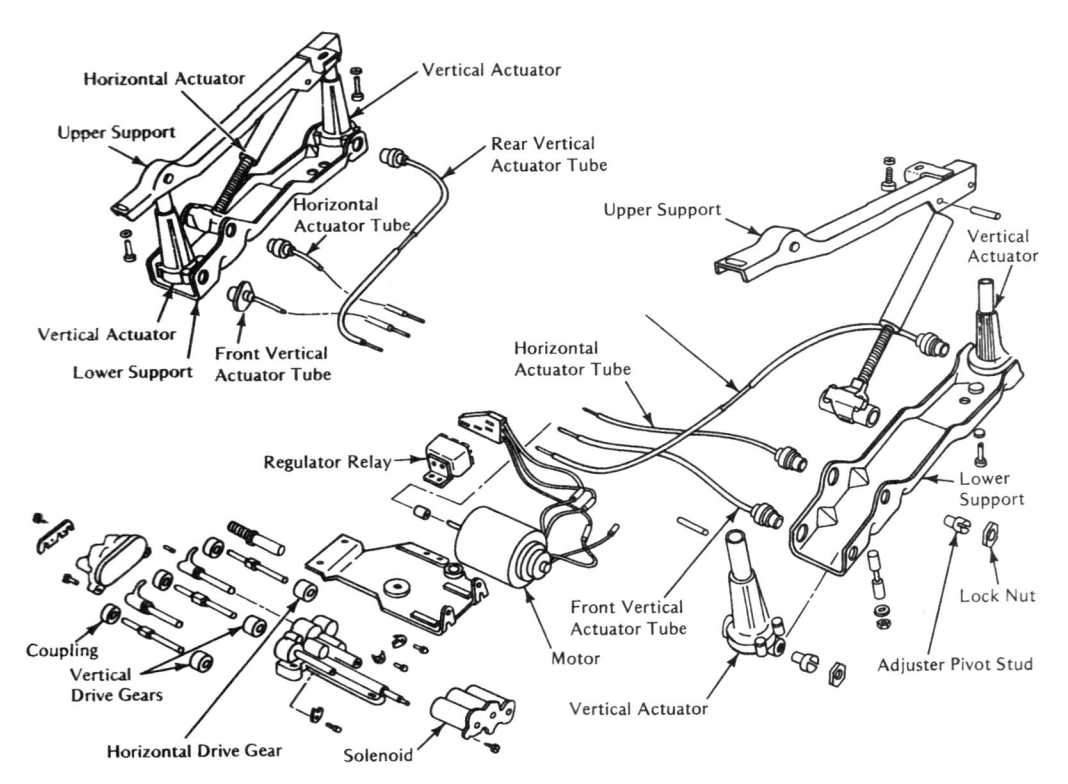

FIGURE 18–8 Exploded view of a typical six-way power seat mechanism

"lock" position. A manual override is provided for seat back release in the event of an electrical system failure.

Power Reclining Seat Backs

Some car lines offer an optional power-operated reclining seat back. A flexible drive cable connects a permanent-magnet reversible motor to a reclining actuator that, when activated, reclines the seat back about 20° rearward of its normal upright position. The motor is operated by a separate control, usually located on the outboard side of the seat cushion. This option, available for the passenger seat only, can be used with either manually or power-operated seat adjusters. A schematic of a simple seat recliner is shown in Figure 18–11.

POWER WINDOWS

Power windows, or power-assist windows, as they are often called, are a popular option on most car lines. Each window is equipped with an electric motor controlled by an individual, independent switch in each door. The driver's door control contains four switches, for the control of all four windows. The driver's control, or master window switch panel, is often equipped with a fifth switch, a lock that cuts out all other switches to allow the driver to control all windows.

Electric window motors are of the 12-volt reversible type and are attached to the window regulator mechanism. Most are mounted directly to the window regulator (Figure 18–12), but some are equipped with a nylon or plastic and rubber coupling between the motor output and the gear and pinion input shafts. Two types of motors are used: a permanent-magnet type and a split-series wire-wound type. Ford and Chrysler generally use the permanent-magnet type, and General Motors generally uses the split-series wire-wound type. Because of variations in systems, manufacturers' wiring schematics and diagrams must be followed for circuit tracing and troubleshooting.

Safety Precautions

Window lift regulator mechanisms are usually spring loaded in the down position (Figure 18–13). This means that whenever the window is in the down position, the mechanism is under heavy spring tension. Because of the scissorslike arrangement of the regulator mechanism, extreme care must be exercised when adjusting

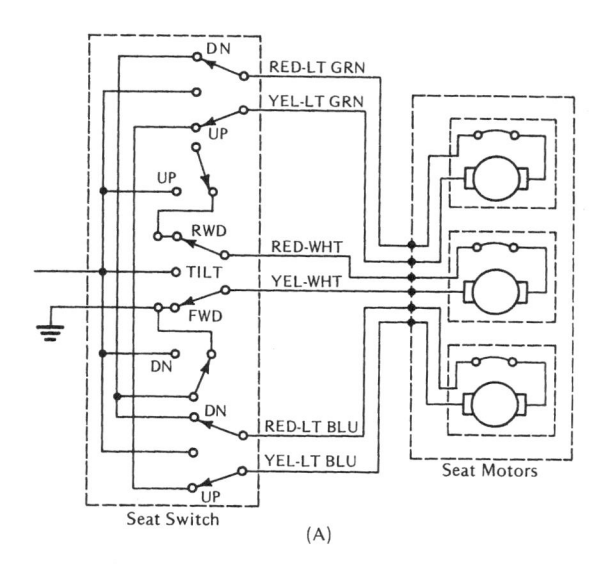

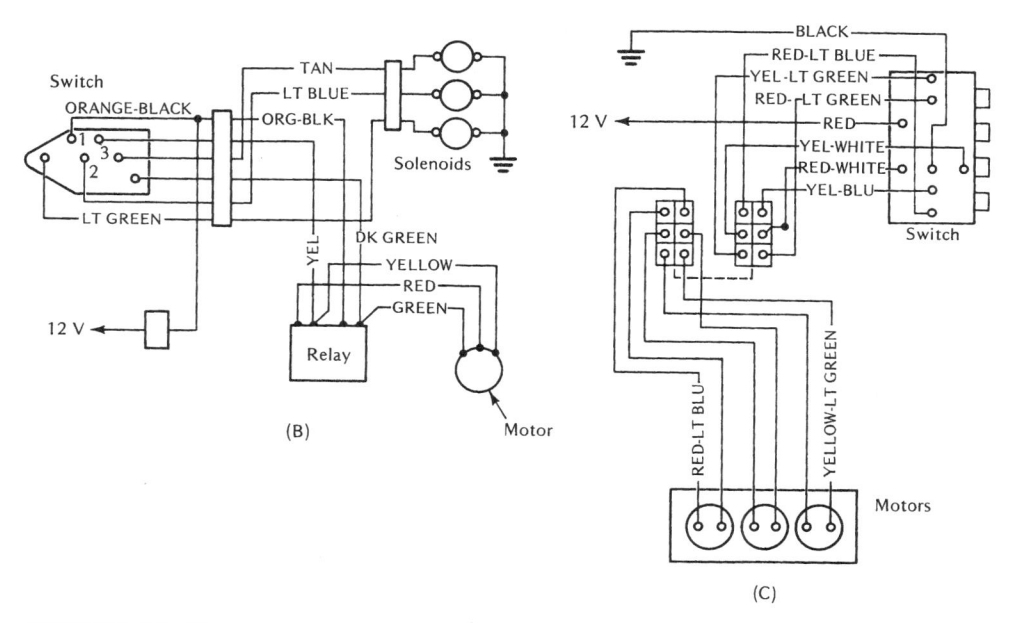

FIGURE 18-9 Six-way seat schematics: (a) Ford (b) General Motors (c) Chrysler

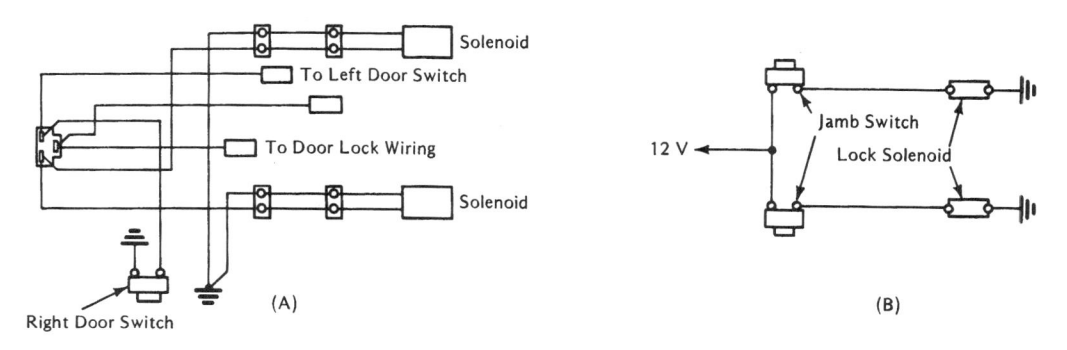

FIGURE 18-10 Schematics for seat back lock system: (a) grounding doorjamb switch and (b) GM's nongrounding doorjamb switch

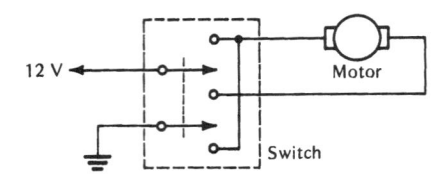

FIGURE 18–11 Schematic for a power seat-back recliner

or repairing electric window units. This is especially important if the glass is not attached to the regulator mechanism or if the glass is broken.

When removing the motor for service or replacement, the window should be in the up position. If the motor failed with the window in the down position, the regulator mechanism must be clamped or blocked to prevent it from moving (Figure 18–14). The motor drive pinion is usually the only part of the system that holds the window-regulator mechanism in the down position. If not clamped or blocked when the motor is removed, the window may quickly and immediately move to the up position. Keep hands and fingers clear of all internal parts and work with extreme caution. This is especially important when removing a motor while the window is in the down position.

Power Tailgate Windows

Tailgate window motors and mechanisms used on station wagons are basically the same as those used for doors. The tailgate window is operated by a control switch that is usually located on the dash panel and by a key-lock switch in the rear door. Some 9-passenger station wagons have a third switch that is accessible to the rear passengers.

Some tailgate window circuits include interlock switches to prevent window operation while the tailgate is opened. This is necessary to prevent glass damage. Typical schematics of tailgate window circuits are shown in Figure 18–15.

Power Window Adjustment

All power windows have a provision for adjustment of the mechanism to provide for freedom of up-and-down

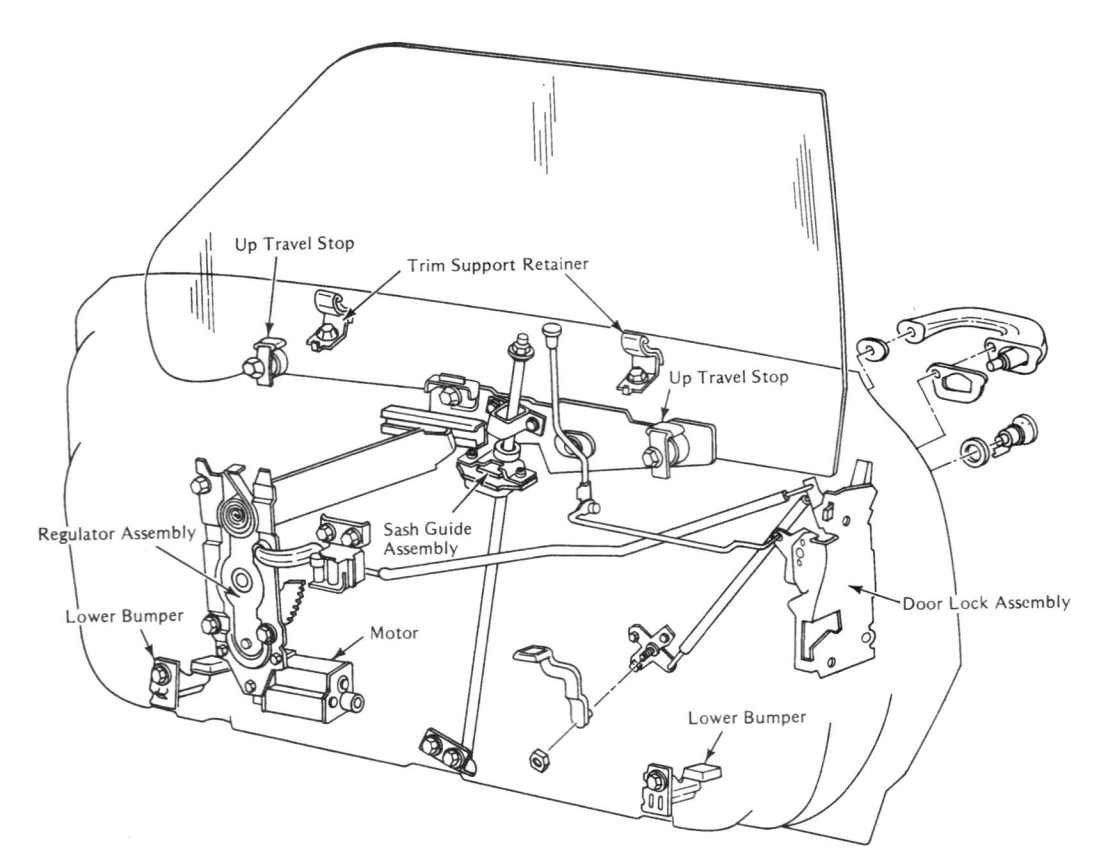

FIGURE 18–12 Typical window-regulator mechanism for power windows

POWER WINDOWS 211

Power Window Motor Testing

Power window motors cannot be tested on a car because of various factors such as glass and regulator movement, which can affect motor current draw. Therefore, window motors must be bench tested for "no load" current draw. For this test, the window motor is connected to a battery with an ammeter connected in series. The motor should draw no more current than that specified by the manufacturer, usually 5 to 6 amperes. Also, the motor should run quietly and smoothly. It should draw the same amount of current when running in either direction.

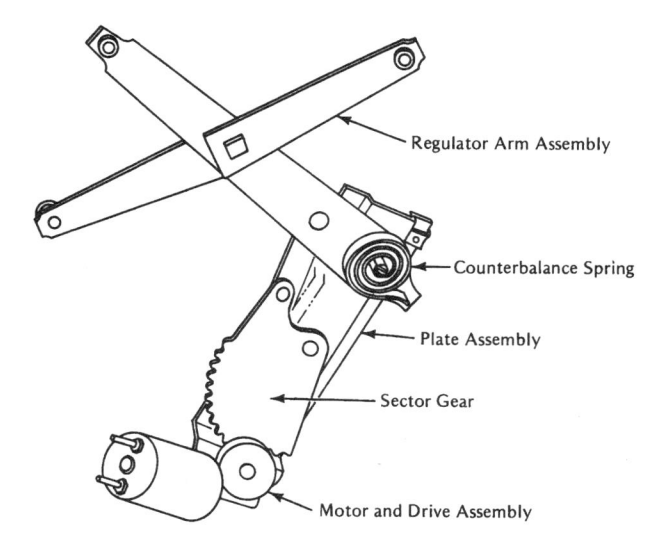

FIGURE 18-13 Counterbalance spring loads the regulator arm assembly in the down position. Rapid unloading, such as by removing the motor and drive assembly, could cause personal injury if care is not taken.

movement and to provide for proper "seating" of the glass when fully raised or lowered. These adjustments include fore and aft, in-and-out tilt, upper and lower stop, equalizer arm, and stabilizer.

Additionally, some windows have separate vent adjustments that include up-down or in-out, upper or lower stop, and catch block. No two adjustment procedures are the same, and proper procedures are given in shop manuals for each year and model car.

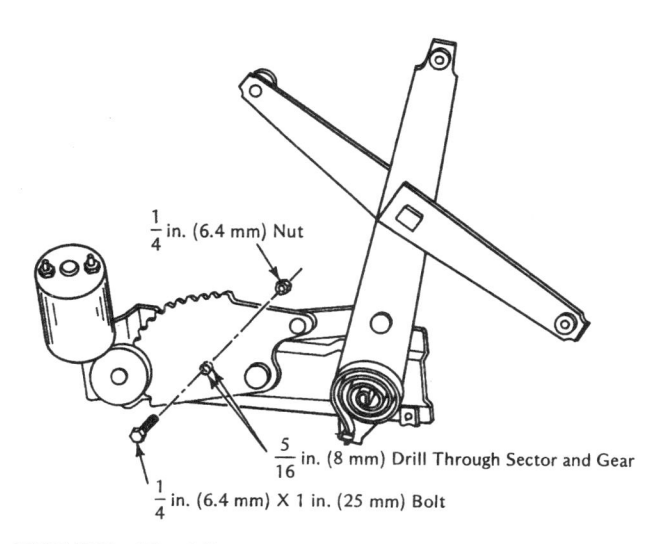

FIGURE 18-14 Regulator mechanism "locked" with bolt and nut

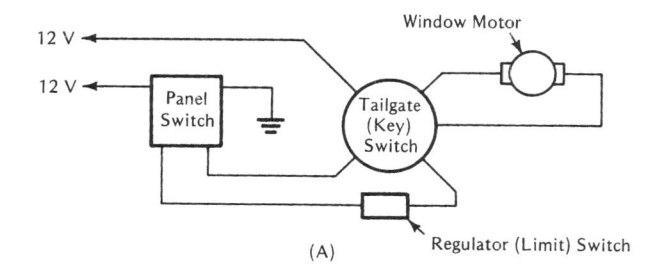

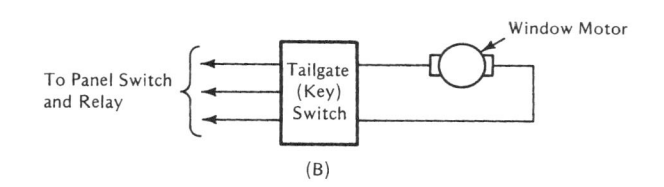

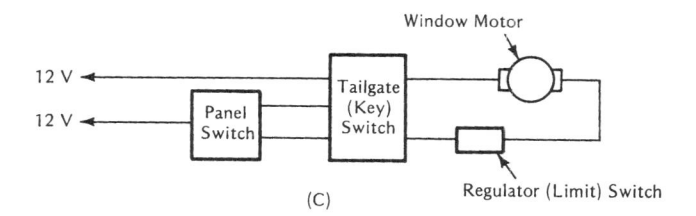

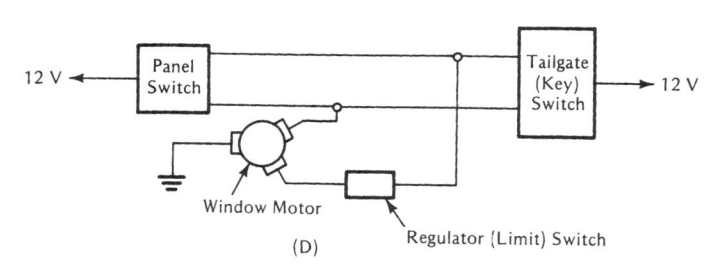

FIGURE 18-15 Schematics for typical tailgate window circuits: (a) Chrysler, (b) Ford, and (c) General Motors

If the motor draws an excessive amount of current, inspect the worm gear and pinion for damage or binding. If the motor can be disassembled, inspect it for damage. While disassembled, it should also be cleaned and lubricated. Petroleum jelly is a good lubricant for electric window motors; a light coat is sufficient. If, after reassembly, the current draw is still excessive, the motor must be replaced.

Power Window Electrical Testing

Switches and electrical circuits can be tested by following the appropriate electrical schematic. Typical schematics for both types of window lift systems are shown in Figures 18–16 and 18–17. Switch testing using two jumper wires is a common practice. An ohmmeter can also be used to check switch continuity.

Power Window Relays and Interlocks

Most electric window circuits, except for the tailgate circuit, are equipped with a normally open (NO) relay that is energized only when the ignition switch is in the ACC or ON position. This relay (Figure 18–18) prevents window operation when the ignition switch is in the OFF position.

Most tailgate electric window circuits are equipped with an interlock normally closed (NC) switch that opens when the tailgate is opened. This prevents window operation while the tailgate is open. Unlike the other windows, the tailgate window can be raised or lowered with a key in the lock. Also, when using the key, the window can be lowered or raised regardless of the position of the ignition switch. In most systems, however, the interior switches cannot be used unless the ignition switch is in the ACC or ON position.

Sunroofs

The power-operated sunroof (or moon roof) was first offered as an option by Chrysler and General Motors more than 20 years ago.

The sliding roof panel is operated by a reversible motor controlled by a normally open (NO), three-position (OPEN-OFF-CLOSE) switch. This switch is usually located in the area of the windshield header safety pad. The roof panel can be fully opened, closed, or left partially open in either direction of travel.

Although the roof panel is normally operated by a reversible electric motor, it can be closed manually with a crank or, in some cases, an Allen wrench,

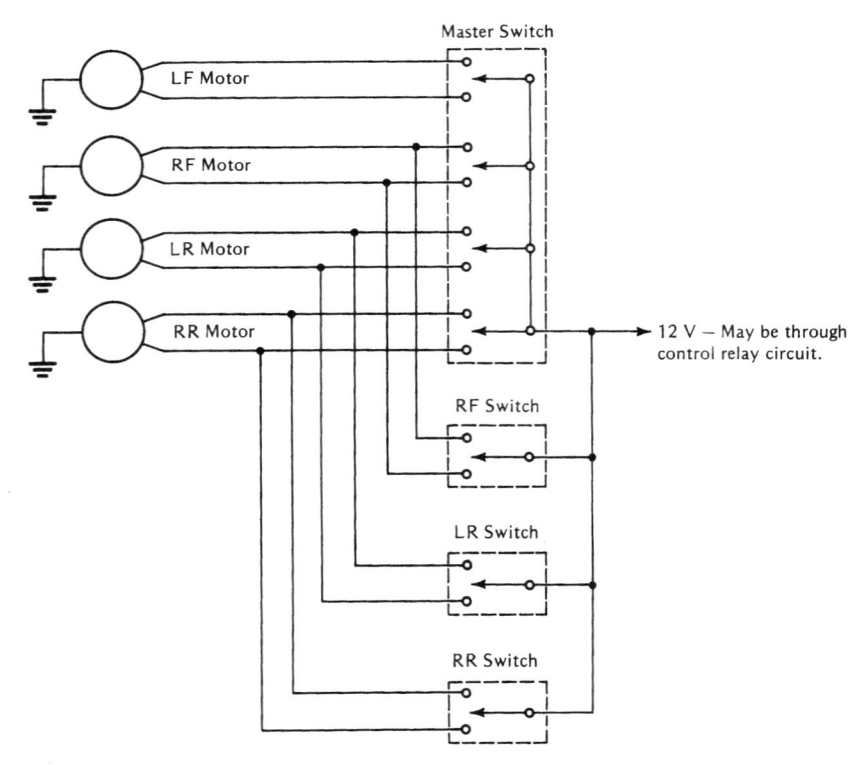

FIGURE 18–16 Typical internally grounded type of window motor circuit

if necessary. Electrical schematics for typical sunroof circuits are given in Figure 18–19. Sometimes panel adjustment is critical to ensure weathertight sealing when the panel is closed. If adjustment is required, specific

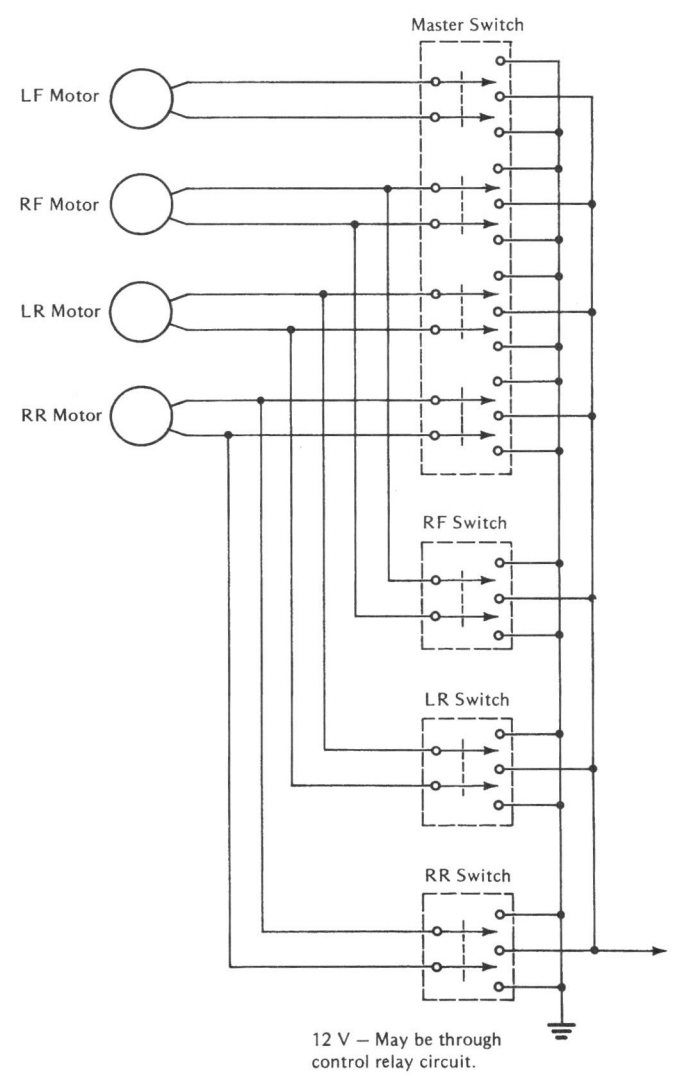

FIGURE 18–17 Typical externally grounded type of window motor circuit

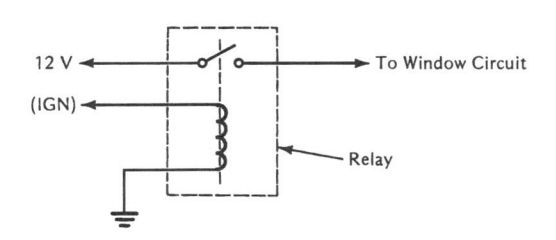

FIGURE 18–18 Schematic of ignition switch-controlled window circuit relay

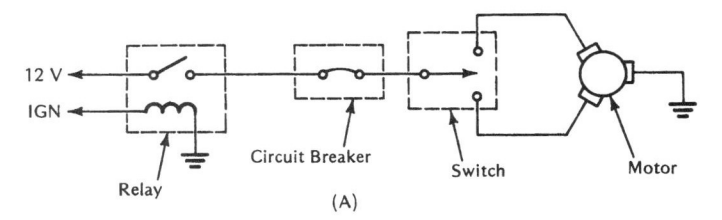

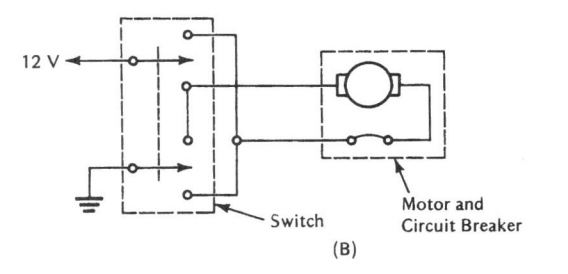

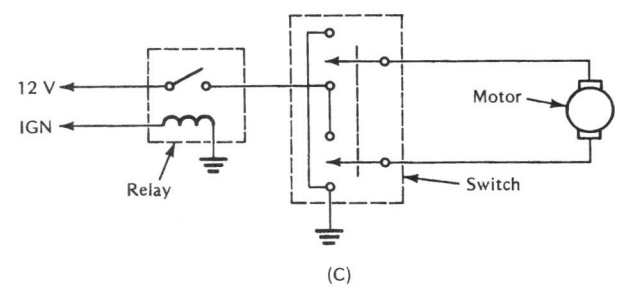

FIGURE 18–19 Typical sunroof circuits: (a) Chrysler, (b) Ford, and (c) General Motors

procedures and specifications should be followed to ensure proper alignment and closing.

ELECTRIC DOOR LOCKS

The electric/electronic door locks system is another popular option on all car lines. There are two basic types of door lock systems: solenoid operated or motor operated, depending on the car line. Many systems include a "keyless entry" as part of the system.

Like many other automotive options, the electric door lock systems and schematics vary considerably from one year's model to the next. The 1994 Taurus/Sable electronic lock system, for example, requires more than 36 pages of manufacturer's service manual instructions to outline diagnosis and troubleshooting procedures. Therefore, it is important that manufacturers' specifications, wiring schematics, and diagrams be consulted for specific circuit tracing and troubleshooting. The following is an overview of typical system operation.

Solenoid-Operated Door Locks

Unlike the solenoids previously covered, the door lock solenoid is a push-pull type. Instead of the usual one-winding coil, the door lock solenoid coil has two windings. When one coil is energized, the core is pulled into the solenoid; when the other coil is energized, the core is pulled out of the solenoid. The core, or plunger, is attached by mechanical linkage to the door lock mechanism, and action of the solenoid either locks or unlocks the door. Typical schematics are shown in Figure 18–20.

Motor-Operated Door Locks

Motors used in power door locks are of the permanent-magnet type and are operated through a relay by conventional switches. The reversible permanent-magnet motor, which is serviced as an assembly, is controlled by a double-pole, double-throw (DPDT), double-coil relay that is externally grounded. A clockwise rotation of the motor output shaft extends the shaft to unlock the door. When polarity is reversed, the output shaft rotates counterclockwise, retracting the shaft to lock the door. A typical schematic of a motor-operated power door lock system is shown in Figure 18–21.

AUTOMATIC DOOR LOCKS

The automatic door lock system provides for the automatic locking of all doors under the following circumstances:

1. All doors are closed with courtesy lamps OFF.
2. The ignition switch is ON.
3. The driver is seated.
4. The transmission selector is in DRIVE.

When the transmission gear-selector lever is moved to PARK, NEUTRAL, or REVERSE, all doors will automatically unlock if an inside door handle is pulled. Also, if the car is stopped and the gear selector is in DRIVE, an individual door can be unlocked by pulling its inside door handle, or all doors can be unlocked from the main control switch. The door, or doors, will automatically relock when the door is closed.

The automatic door lock system (Figure 18–22) consists of an electronic logic module, a permanent-magnet motor lock actuator relay, a backup lamp switch, right- and left-front door lock remote-control handle switches, and a driver's seat sensor switch. This system is interconnected with the regular electric door lock harness and the permanent-magnet motor lock actuators and switches.

Keyless Entry System

In 1980 Ford first introduced an electronic door lock system (Figure 18–23) that permits the driver to unlock the car door and/or release the deck lid without a key. This feature, now available as an option on most car lines, consists of five touch buttons located on the driver's door belt molding. Each button has two numbers (1–2, 3–4, 5–6, 7–8, 9–10), and all are connected to a microcomputer for "code" and "sequence" identification.

Two five-digit codes are available: one code is programmed at the factory and cannot be changed; the other code is for owner programming and can be changed at any time. Entry into the car is gained by pressing the proper buttons in the proper sequence.

For nighttime use, pushing any of the five buttons will illuminate all the buttons, the interior car lighting system, and the key entry lock light. The driver can then unlock the door by use of a key, in the conventional manner, or by using either of the two codes—factory or owner programmed.

Tailgate Locks

Station wagon tailgate locking systems operate on the same principle as door locking systems. On station wagons not equipped with electric door locks, the tailgate lock system does not include a relay. A typical tailgate locking system schematic is shown in Figure 18–24.

The tailgate lock system is not necessarily a part of the door lock system. A car with an electrically or mechanically operated door lock system can have an electric or mechanical tailgate lock.

Trunk Lock-Release Systems

The trunk compartment lid-release system (Figure 18–25) consists of a lock-release solenoid, a wiring

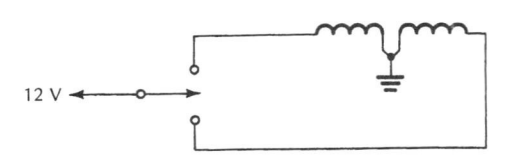

FIGURE 18–20 A simple two-winding solenoid door lock circuit

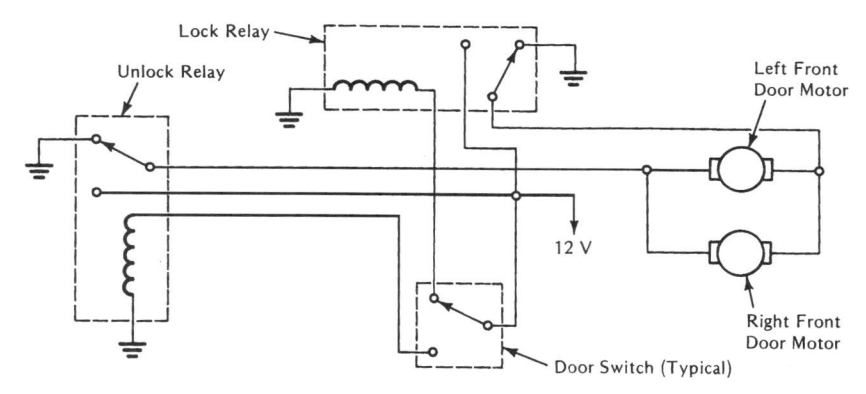

FIGURE 18–21 Schematic of a motor-operated door lock system

harness, and a switch. The solenoid is externally grounded, and its circuit is complete through the wiring harness from a normally open (NO) push button switch that is usually located inside the glove compartment. When the ignition switch is ON and the release button is pressed, voltage is supplied to the solenoid. The solenoid is energized and the plunger retracts, releasing the lock hook from the trunk latch striker.

Trunk Lid-Closing Systems

The trunk lid-closing system consists of a permanent-magnet motor with a gear-reduction drive, a relay, a

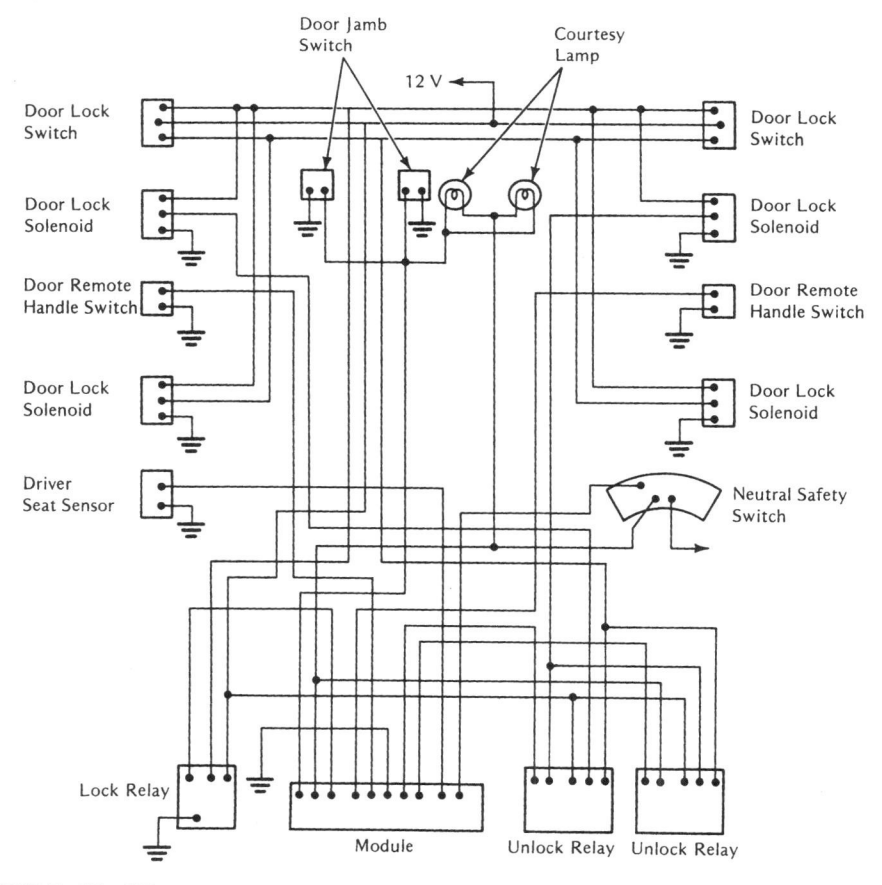

FIGURE 18–22 Schematic of an automatic door lock system

216 Unit 18 OPTIONAL AND CONVENIENCE SYSTEMS

FIGURE 18–23 Keyless electronic entry system (Courtesy of BET, Inc.)

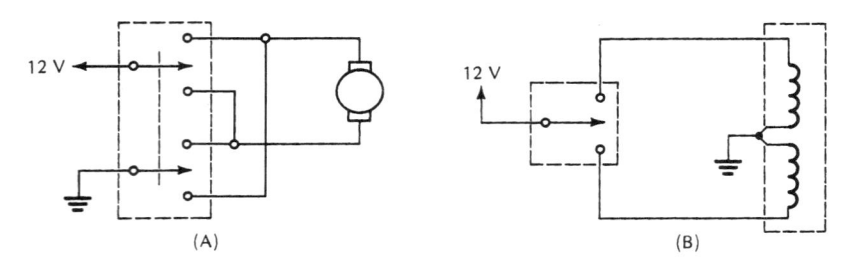

FIGURE 18–24 Typical door lock circuits: (a) motorized and (b) solenoid

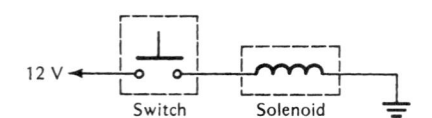

FIGURE 18–25 Schematic of solenoid trunk release circuit

flexible drive cable, and a pull-down striker assembly. These components are used with a lid-release switch, a lock solenoid, and a lock assembly. Follow the schematic shown in Figure 18–26. As the trunk lid starts upward, contacts in the unit switch assembly close to provide a circuit to the motor through normally closed (NC) contacts of the relay.

The motor circuit is complete to ground through a second set of normally closed (NC) relay contacts. As the striker reaches the limit of its up cycle, a metal tab enages the up-cycle cutout switch to break the circuit.

When the trunk lid is closed, the lock hook engages the striker, providing a ground circuit through the lock assembly to body metal.

The relay coil is energized and its closed contacts provide a circuit to the motor (polarity now reversed). The motor circuit is complete to ground through a second set of relay contacts. As the lid reaches its closed position, the lock frame depresses the plunger switch to open its contacts, thereby breaking the ground circuit to the motor.

ELECTRONIC FUEL-INJECTION SYSTEMS   217

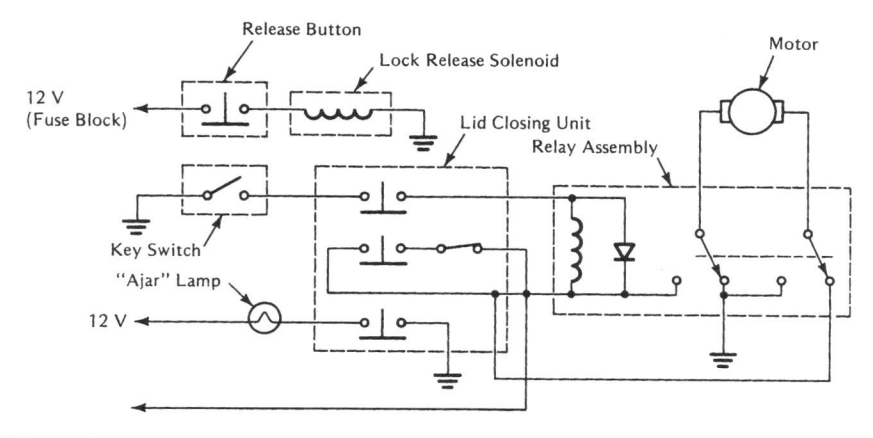

FIGURE 18-26 Schematic of typical trunk-closing system

POWER-OPERATED REAR-VIEW MIRRORS

The power-operated rear-view mirror, (Figure 18–27) is controlled by a double-pole, double-throw (DPDT), center-off switch. Moving the mirror up, down, right, or left is accomplished by reversing the polarity of two permanent-magnet motors with gear reduction units. The motors and gear reduction units are located inside the mirror housing and are serviced as a complete assembly.

ELECTRONIC FUEL-INJECTION SYSTEMS

Although the electronic fuel-injection system is not considered to be part of the automotive electrical system, some of the subsystems of the electronic fuel-injection system are common to, and a part of, automotive electricity. The electronic fuel-injection system (Figure 18–28) monitors all engine operating conditions and electronically meters fuel, controlling the air/fuel ratio necessary to meet those conditions. The electronic fuel-injection system uses an electrically operated fuel metering valve that releases a small quantity of fuel into the engine through individual injectors. The injectors, one for each cylinder, are timed to open so that the fuel is metered into each cylinder at the beginning of the intake stroke. A common fuel rail supplies fuel under high pressure to each injector. The amount of fuel injected into each cylinder depends on the amount of time the injector is held open. All of this is determined by such factors as intake manifold pressure, intake air temperature, engine speed, and engine load.

Engine conditions, such as temperature, speed, and throttle position, are monitored by sensors. Information provided by the sensors is computed by the electronic control unit (ECU) to determine the proper injector duration and interval with respect to engine firing order. The ECU is a small, preprogrammed analog computer that translates sensor signals into command signals for fuel delivery control.

The electronic fuel-injection system consists of four main subsystems: the electronic control unit

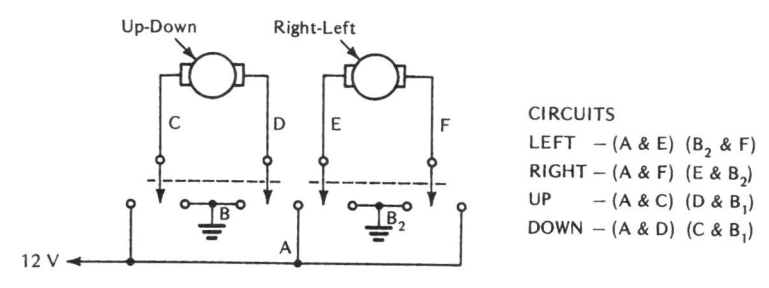

CIRCUITS
LEFT — (A & E) (B$_2$ & F)
RIGHT — (A & F) (E & B$_2$)
UP   — (A & C) (D & B$_1$)
DOWN — (A & D) (C & B$_1$)

FIGURE 18-27 Power-operated rear-view mirror circuit

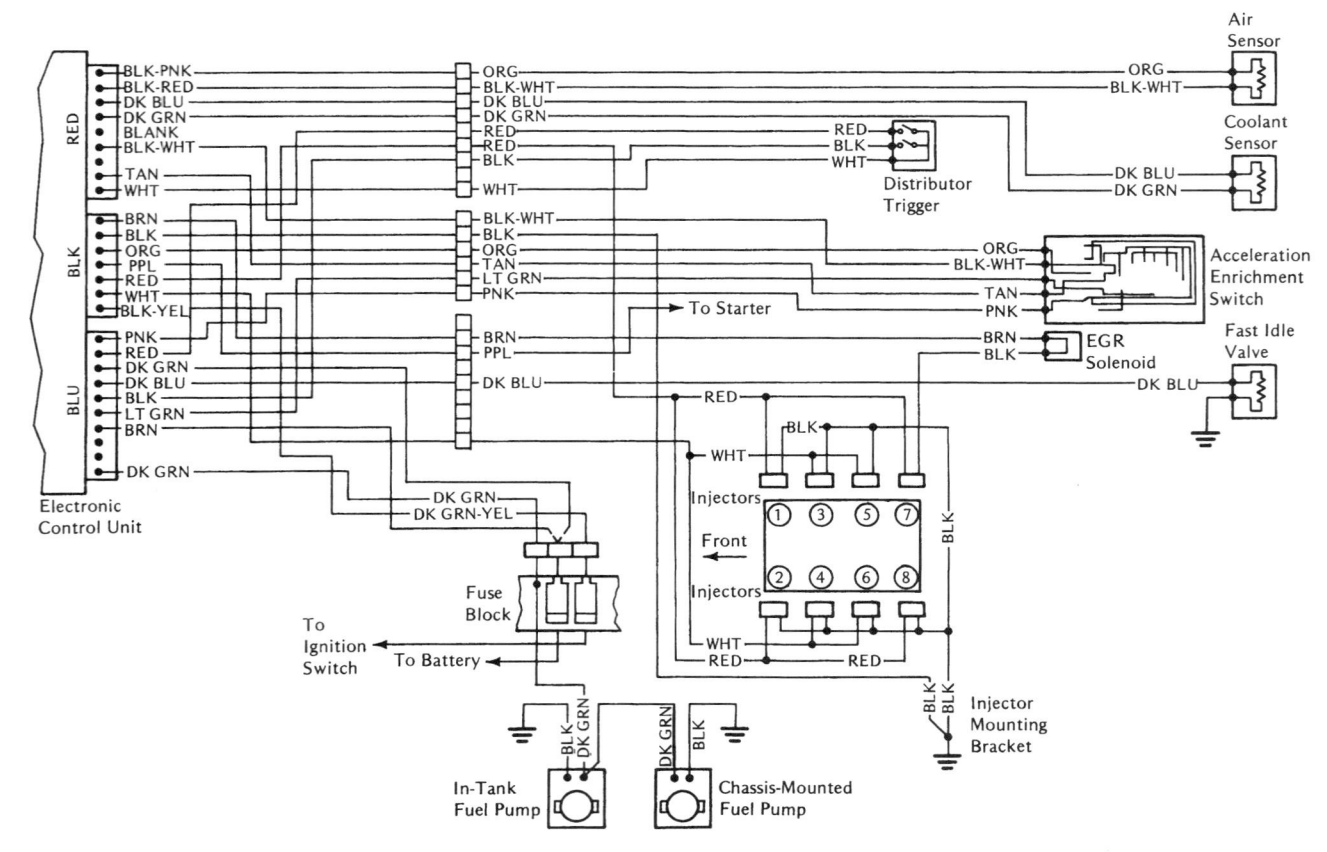

FIGURE 18-28 Schematic of typical electronic fuel-injection system

(ECU), the fuel delivery system, the air induction system, and the sensor system.

Electronic Control Unit

The ECU activates the electric fuel pumps, the fast-idle valve, the injection valves, and the exhaust-gas recirculating (EGR) solenoid. The optimum air/fuel ratios for various conditions are preprogrammed. According to sensor signals, the computer sends signals to the injectors indicating a specific time duration that each is to be open.

Fuel Delivery System

In addition to the fuel delivery and supply lines, the fuel delivery system consists of seven major components, including the fuel tank, two fuel pumps, the fuel filter, the fuel rail, the injector valves, and the fuel pressure regulator.

Fuel Tank. The fuel tank is different from conventional tanks in that it incorporates a reservoir below the pickup tube. The purpose of the reservoir is to ensure a constant supply of fuel under adverse conditions, such as low fuel or severe maneuvering.

Fuel Pumps. The system is equipped with two electric fuel pumps. An in-tank boost pump is an integral part of the fuel tank and gauge unit. It supplies fuel to the external chassis-mounted main fuel pump and prevents vapor lock in the suction side of the main pump. The main fuel pump contains a check valve to prevent bleed-back to the fuel tank when it is not in use. It also has an internal pressure relief valve that opens at 55–95 psi (379–655 kPa) to protect against excess pressure.

Fuel Filter. The fuel filter consists of a disposable element, usually paper, housed in a metal casing. The fuel filter has the appearance of a miniature version of a standard oil filter.

Fuel Rail. The fuel rail provides fuel flow passage for the pressure regulator and each of the injectors.

Injector Valves. The solenoid-operated injector valves meter fuel into each cylinder based on a pulsed electrical signal from the electronic control unit (ECU).

Pressure Regulator. The pressure regulator maintains a constant 39 psi (269 kPa) pressure differential across the injectors.

Air Induction System

The air induction system consists of either three or four components, depending on the system. In addition to a throttle body assembly, an intake manifold, and a fast-idle valve assembly, some systems have an air solenoid valve. Air for combustion enters the intake manifold through the throttle body. Primary air flow is controlled by the throttle valves, which are similar to a conventional carburetor system.

Sensor System

There are five basic sensors: a manifold-pressure sensor, a throttle position sensor, an engine-speed sensor, and two temperature sensors. Each sensor is electrically connected to the ECU but also operates independently. Individual sensor signals are analyzed by the ECU and converted into appropriate system signal commands.

FUEL PUMPS

Some car lines feature an electric fuel pump as standard equipment. The electric fuel pump is usually located inside the fuel tank and provides for dual-pressure operation depending on engine conditions. Additionally, it provides for positive fuel shutoff in the event of reduced oil pressure. Follow the schematic shown in Figure 18–29 for the operation of the electric fuel pump.

When the ignition switch is in the START position, full battery voltage is available through diode D_1 directly to the fuel pump. This produces maximum fuel pump pressure for maximum fuel delivery to the carburetor. After the engine is started, and the ignition switch is in the ON position, the normally closed (NC) manifold vacuum switch is opened, placing a resistor in series with the fuel pump circuit through the normally closed (NC) contacts of the relay. This resistance reduces battery voltage to the fuel pump to produce minimum fuel pump pressure sufficient for adequate fuel delivery to the carburetor under normal driving conditions.

The circuit of the relay coil is not completed because the normally closed (NC) oil pressure switch contacts open when the engine is running, as long as there is sufficient oil pressure. When the oil pressure drops to a predetermined low level, the oil pressure switch contacts close to energize the relay. This causes the relay contacts to open, breaking the electrical circuit to the fuel pump motor. This action cuts off all fuel supply to the engine.

At full throttle, the engine manifold vacuum is reduced to a predetermined level that closes the manifold vacuum switch. This action bypasses the resistor to provide full battery voltage through the relay contacts to the fuel pump. This produces full fuel pump pressure to provide maximum fuel delivery to the carburetor.

The engine coolant "hot" warning lamp is also included in the circuit. The lamp is illuminated when the normally open (NO) contacts of the temperature switch close. However, this does not affect the action of the fuel pump. Two diodes, D_1 and D_2, are provided to prevent unwanted feedback or sneak circuits.

ELECTRIC OVERDRIVES

An *overdrive* unit is a second transmission, available as optional equipment on some car lines equipped with manual three-speed transmissions. Essentially, an overdrive unit is a two-speed transmission that is attached to the rear of a standard three-speed transmission. This arrangement provides four forward speeds instead of three. At any road speed, an overdrive unit can reduce engine speed by about 30 percent. Many people think that overdrive units, which were once a popular option, will make a comeback because of the high cost of fossil fuel.

The basic components of an electrically controlled overdrive unit consist of a solenoid, a relay, a speed-sensitive governor switch, a manual kick-down switch, and a wiring harness. A schematic of a typical electrically operated overdrive system is shown in Figure 18–30.

SPEED CONTROLS

Automatic speed control (cruise control) has been an option for many years. The speed control system is a device used to keep a car at an even, preselected speed by semiautomatically taking over manual throttle control. The speed control can maintain car speed uphill or downhill as well as on level terrain.

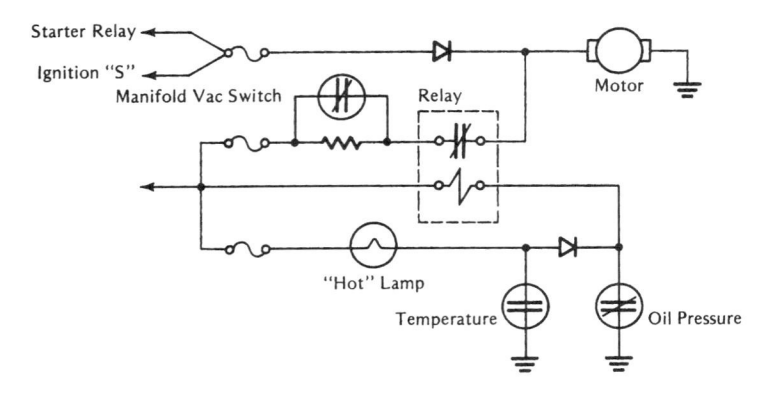

FIGURE 18–29 Schematic of typical electric fuel pump circuit

The speed control is engaged by first reaching the desired speed and then pressing an ON-SET button. It is disengaged by pressing an OFF button. It is also automatically disengaged by activating several release switches or, depending on the system, by applying the brakes, actuating the turn signals, or blowing the horn.

There are two types of systems in current use: electropneumatic and electrical. The electropneumatic system operates on battery voltage and manifold vacuum. The electrical system operates on battery voltage only. A brief description of both types of systems follows.

Electropneumatic Speed Controls

The electropneumatic speed control system consists of four major subsystems: an engaging switch, a regulator assembly, a vacuum servo, and a release switch. A typical electropneumatic schematic is shown in Figure 18–31.

Engaging Switch. The engaging switch (usually a push button) is at the end of the turn signal switch lever, on the lever, or at some other convenient location. To engage the speed control, the driver accelerates to the desired speed and then momentarily presses the button. Once actuated, the car will hold the preselected speed even if the driver removes his or her foot from the accelerator (gas) pedal. The car speed can be adjusted upward or downward using the engaging switch in a manner outlined in the owner's manual.

Regulator Assembly. The regulator assembly contains a flyball-type governor, a valve body, and a magnet assembly, and is cable driven from the transmission. A second cable off the regulator is used to drive the speedometer. The regulator assembly is the

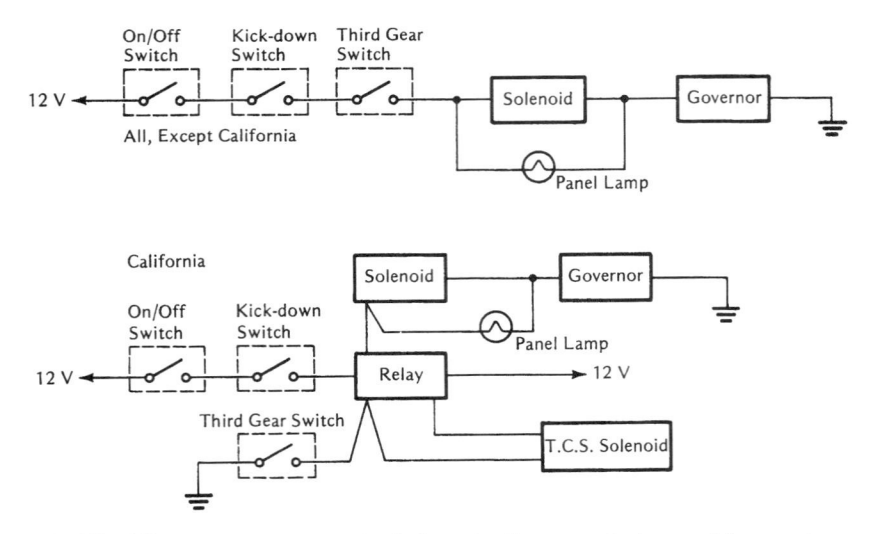

FIGURE 18–30 Schematic of typical electrically operated overdrive system.

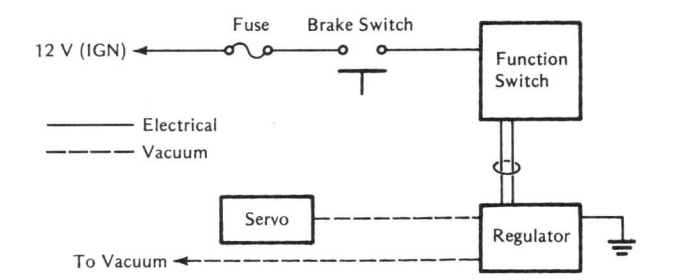

FIGURE 18-31 Typical electropneumatic cruise control circuits

speed-sensing and control device used to regulate and maintain the preselected speed.

Vacuum Servo. The vacuum servo is controlled by the regulator assembly and is connected to the throttle linkage. It controls throttle speed through the accelerator linkage to the carburetor. It is actuated by a vacuum "signal" from the regulator assembly.

Release Switch. The electropneumatic system usually has two release switches, one electric and one vacuum operated. The purpose of both switches is to release the speed control whenever the brake pedal, turn signal, or horn is actuated.

Electrical Speed Controls

The electrically operated speed control system consists of five major subsystems: an engaging switch, an amplifier assembly, a sensor, a throttle actuator, and a release switch. A typical schematic is shown in Figure 18-32.

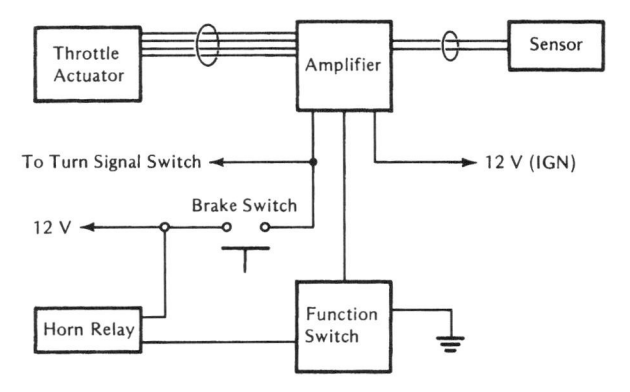

FIGURE 18-32 Typical electronic cruise control circuit

Engaging Switch. The engaging switch is similar to the engaging switch of the electropneumatic system.

Amplifier Assembly. The amplifier assembly receives an electrical signal from the sensor indicating car speed and relays the information to the throttle actuator. The amplifier assembly can be considered to be the brains of the speed control system.

Sensor. The sensor is cable driven from the transmission. A second cable off the sensor is used to drive the speedometer. Car speed is determined by small electrical signal impulses of the sensor, which are fed into the amplifier assembly.

Throttle Actuator. The throttle actuator is connected to the throttle linkage and is controlled by the amplifier assembly. It controls throttle linkage position, thereby controlling car speed. The throttle actuator position is based on an electrical signal from the amplifier assembly.

Release Switch. Release switches are provided to disengage the speed control system if the brake pedal, turn signal, or horn is actuated.

The speed control system, although automatic in operation, does not take primary control from the driver. At all times the driver has immediate override control of the speed control system for obvious safety reasons.

ELECTRIC CLOCKS

The electric clock is an inexpensive and popular automotive option. There are several types of clocks available, including analog, digital, and solid-state clocks.

Regardless of the type, testing procedures for an inoperative clock are simple. Using a test lamp or voltmeter, check to ensure that 12 volts is available to the clock and that the ground circuit is sound. If this is not the case, replace the fuse or repair the wiring as necessary. If the wiring is sound but the clock is inoperative, it must be removed and replaced. Because of the high cost of service, repairs are not generally practical. Also, repair parts and information are not usually available to the automotive electrical technician.

RADIOS AND SOUND SYSTEMS

Radios and sound-producing systems are probably the most popular type of optional equipment. The first radio to be offered to the motoring public was manufactured by Westinghouse and made available on the 1922 Chevrolet model 490 Roadster. An interesting feature was that the antenna required the entire roof. A novelty for the wealthy, the radio cost $200; the entire car, without radio, cost $510.

Radios became more popular in 1928 and 1929; in 1930 the Cadillac, Chrysler, DeSoto, LaSalle, Mormon, and Roosevelt were factory wired for radio installation. By 1934 radio controls were found on the dash, and push-button station selection was available in 1939.

There are many types of radios and sound systems available today. These include AM/FM, AM/FM stereo, AM/FM/tape, tape, and compact disc (CD). Almost any combination of these devices may be found as an integral unit. In addition there are many tuning variations available, including manual, push button, semiautomatic, and automatic. Preprogrammed memory selection for frequencies in the AM and FM bands is also available. It is not unusual for the digital clock to be a part of the radio system.

AM Radio

AM, an abbreviation for *amplitude modulation*, provides for a frequency range of 550 kHz to 1600 kHz. AM radio signals are reflected by the atmosphere, providing long-range reception. AM signals travel further at night than during the day, and many transmitting stations are required to reduce their transmitting power at dusk. The AM method of transmission is subject to noise, and most electrical disturbances can enter an AM radio. Weak stations suffer the most interference.

Quality AM circuits include an automatic gain control (AGC) that maintains the volume at a constant level. Although AM reception is not generally good when passing under bridges or near steel structures, the AGC helps to maintain a constant volume level.

FM Radio

FM, the abbreviation for *frequency modulation*, provides a frequency range of 88 mHz to 108 mHz. FM radio signals, often referred to as "line of sight" signals, are not usually reflected by the atmosphere. This limits reception distance, because the receiving antenna must "see" the transmitting antenna. A building or hill can easily block out an FM signal. In metropolitan areas where FM signals are strong, reception is possible because the radio waves can bounce off buildings and other structures.

The normal range for FM reception is about 25 miles (40 km), but there are exceptions. On flat terrain, and with a powerful transmitter, reception can be extended to 35 miles (56 km) or more. Reception under 20 miles (32 km) is reliable on most commercial stations, and flutter caused by buildings and hills should not be troublesome.

FM Stereo Radio

An FM stereo radio was first introduced by Chevrolet in 1965. Stereo, an abbreviation for *stereophonic*, consists of two audio signals broadcast simultaneously on an FM frequency. "Stereo," then, defines a sound reproduction system in which at least two channels of sound are delivered to the listener, creating an illusion of depth and sound separation (Figure 18–33).

Cassette Player

The cassette player, a long-time favorite option for motorists, accepts a thin, flat, rectangular, sealed "package" containing a length of prerecorded 1/8-inch (3.2-mm) wide magnetic tape, each end of which is attached to flangeless hubs or reels. Sound is picked up from the tape as it passes across a pickup called a play head. At the end of play for side one (generally 30 or 45 minutes), the tape may be rewound, turned over for side two play, or, in some systems, automatically reversed.

Cassette tape systems are available for both recording and playback. Those designed specifically for automotive service, however, are playback-only systems.

Compact Disc

The compact disc (CD) system, introduced in 1983, may be either dash or remote mounted. It is a sound reproduction system that uses a laser (light) to detect audio signals made by digital recording on a disc. The new term CD-ROM refers to a disc that has read-only memory.

Unlike other types of playback system, there is no physical contact between the pickup and the recording medium. Therefore, there is virtually no wear and/or surface noise. Also, unlike other types of disc, such as records, all data is recorded on the back (unlabeled) side, and is "read" from the center out. A small, low-power laser is focused on a small spot on the disc, and light reflected onto a phototransistor produces an electrical signal that corresponds to the recorded informa-

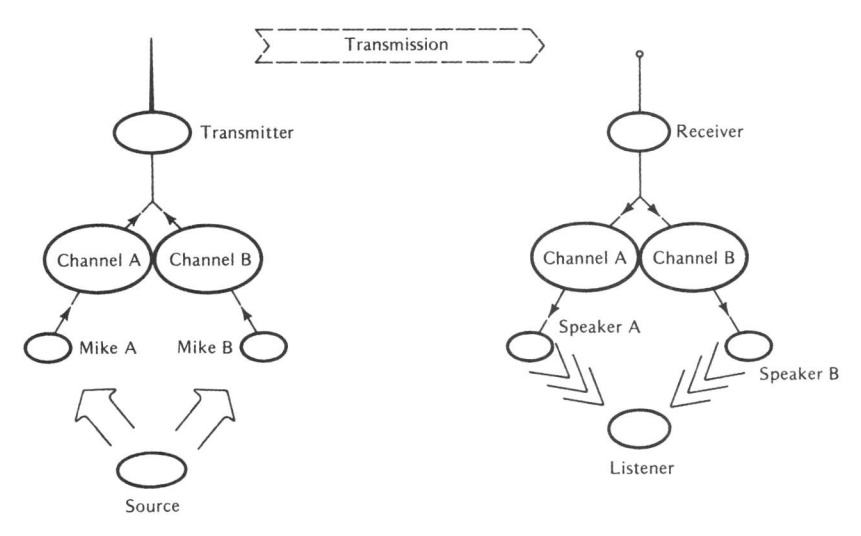

FIGURE 18–33 Stereophonic transmission and reception

tion. Though a vast amount of information can be stored on and retrieved from a CD, a standard size with a maximum of 75 minutes' playing time has been agreed on by the industry. This is sufficient for 10 to 15 full-length songs or an average symphony. To provide for longer play without interruption or repeat, CD changers are available that will accommodate up to 10 or 12 CDs, for up to 15 hours of play time.

CB Radio

The citizens band (CB) radio, once a popular accessory for two-way communication, has given way to another form of mobile communications, the cellular telephone. The CB, a truckers' standard for mobile communications for many years, came to public attention in the late 70s and early 80s with hit songs and the movie *Cannonball Run*. The CB may still be found in automotive applications.

It is interesting to note that the first two-way communication known from an automobile was in 1910, between a Chalmers-Detroit Company automobile moving in New York City's Central Park and the Terminal Building (42nd Street and Park Avenue), a variable distance of 1 to 3 miles.

Most citizens band radios operate on a 40-channel frequency range of 26.965 mHz to 27.405 mHz. Channel 9 (27.065 mHz) is reserved for emergency use and is often monitored by local police. Channel 19 (27.185 mHz) is used for highway information.

CB power output is limited by Federal Communications Commission (FCC) regulation to 4 watts. Signals are usually transmitted and received for 1 to 5 miles (1.6 to 8.0 km), with an average of about 2.5 miles (4 km). In open country and under ideal conditions, reception is often extended to 18–20 miles (29–32 km). Under rare and extremely favorable atmospheric conditions, reception of hundreds of miles may be encountered. It is illegal, however, to attempt to communicate over a distance of more than 150 miles (241 km).

Cellular Telephones

There are three basic types of cellular telephones: portable, mobile, and transportable. They all function in the same manner, and kits are available to adapt one type for use as another type. The primary differences are in size, weight, and power. Cellular telephones have been by far the most popular means of two-way communications in recent years.

Unlike the CB, there are generally no distance restrictions. A cellular phone may be used to call just about everywhere from just about anywhere. There is no need to be within range of your home cellular system to place and receive calls. When traveling, you may use a feature called "roaming" to place and receive calls in much the same manner as with a corded telephone.

New cellular systems enable motorists to send and receive calls or faxes, and to check travel information such as weather and road conditions. Some even have a built-in answering machine. Calls may be encrypted for total security, to prevent others from listening in to calls made on a digital cellular system.

Maintenance and repair of cellular telephones, other than replacing the antenna or changing the battery, are restricted by the FCC to those licensed to service transmission equipment.

Portable Phone. Many users prefer the convenience of a portable phone, which operates on rechargeable batteries, has a built-in antenna, and can be taken anywhere. These phones generally weigh 8 to 10 ounces with the battery, and are usually easily carried in the pocket. This type of phone is often carried in the glove compartment, with a fresh battery pack, and is used for on-the-road emergencies.

Mobile Phone. A mobile phone, or car phone, as it is often called, is preferred because it draws unlimited power from the car battery and can achieve superior reception from its connection to an antenna on the roof or rear window. The greater power helps connect calls and improves sound quality when in an area where reception is weak or interference is common. Original-equipment phones, those supplied by the auto manufacturer, have built-in conveniences that aftermarket phones do not have. These include in-dash, sun visor, or remote custom installation. In common with aftermarket mobile phones, most have hands-free operation. Some have voice activation, a voice recognition feature that allows the driver to dial by simply reciting the telephone number.

Car phones are designed to offer features that specifically address the needs of behind-the-wheel cellular users. A horn alert, for instance, honks the horn when an incoming call is received, a handy feature if you are within hearing range of your vehicle. A car-radio mute is available that automatically reduces the sound system volume when a call comes in. Some have a volume control for the speaker, data interface, dual mode (digital or analog), and an electronic lock to prevent unauthorized use.

Transportable Phone. A third type of cellular phone, called the transportable, weighs up to five pounds. It is a hybrid combining full power with battery operation in a carrying bag or case. Many transportable telephones have a cigarette lighter adaptor, making them suitable for automotive use, though they do not have the convenience features of a mobile phone.

ANTENNA SYSTEMS

There are several types of automobile antenna systems available, including fixed, manually retractable, and electrically retractable. There are specialized antennas and multiuse antennas, such as AM/FM, CB, AM/FM/CB, and cellular phone. Although some antennas are designed for AM/FM/CB use, no attempt should be made to transmit with a CB or cellular telephone over a standard AM/FM antenna. Doing so will damage the transmitter and/or sound system.

Antennas

Antenna performance can be tested by substituting a known good antenna. An antenna can also be checked for opens or shorts by disconnecting it from the radio and using an ohmmeter.

The power-operated antenna is extended and retracted by a coiled nylon cord actuated by a reversible electric motor. Power antennas are controlled either manually, by a dash-mounted switch, or automatically when the radio is turned on or off. The motor circuit can be electrically tested according to procedures previously given for similar motors. The antenna circuit can be tested using an ohmmeter. Most power-operated antenna problems can be avoided by frequent cleaning of the telescoping sections. In the winter months, after cleaning, wipe the sections with a cloth moistened with a light oil.

AM/FM Antennas

The AM/FM antenna can be mounted on either side of a car on the front or rear fender. A shielded cable connects the antenna to the radio. The shielding helps to eliminate electrical interference. On some car lines, the antenna is laminated into the windshield (Figure 18–34).

AM/FM Antenna Trimming. AM/FM antennas can be "trimmed" for best radio reception. Near the point where the antenna plugs into the radio, a small hole in the radio case allows access to the trimmer. To trim the antenna, extend the antenna to its normal height and turn the radio on. The ignition switch should be in the ACC position for radio operation. Manually tune the radio to the weakest signal on the AM band between 1400 kHz and 1600 kHz. Increase radio volume and set the tone control to maximum treble. Adjust the antenna by carefully turning the trimmer back and forth until the position is found that gives maximum volume (loudness).

CB Antennas

CB antennas can be mounted on the roof, the trunk lid, the rear bumper, or the fenders. There are two types of

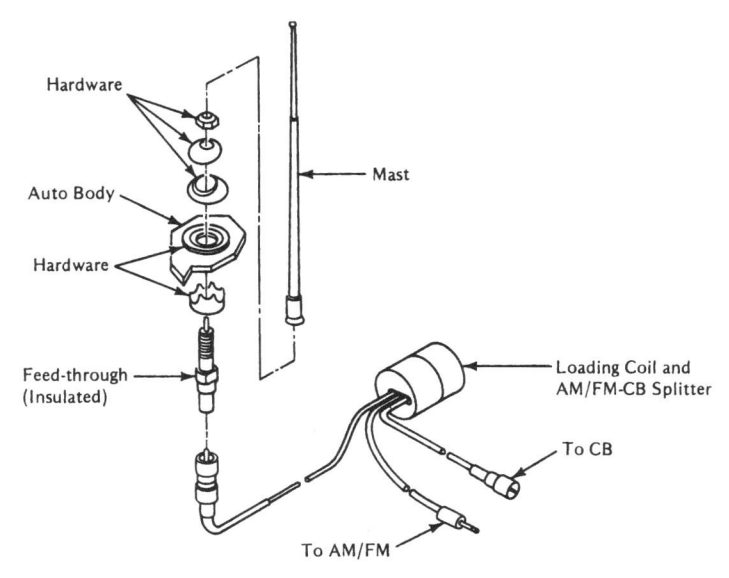

FIGURE 18–34 Typical AM/FM antenna: (a) traditional cowl mount, (b) windshield (concealed) mount

antennas: the full-length whip and the loaded whip. The full-length (quarter-wavelength) whip is too long for most mounts except the bumper because it is 102 inches (2.6 m) long. Therefore, to provide for proper electrical length, the antenna is shortened, and a loading coil is used.

AM/FM/CB Antennas

The AM/FM/CB antenna is designed to receive all AM, FM, and CB signals and to transmit CB signals. A splitter assembly, including a loading coil, divides the AM/FM and CB functions of the antenna. All functions and adjustments of this antenna are the same as previously given.

Cellular Phone Antennas

The cellular phone antenna is a special type that may be affixed to the rear window, and the signal is "fed" through the glass by capacitive coupling. It is therefore not necessary to provide a hole in the glass for a lead wire. The antenna may also be mounted on the roof, where a hole may be provided for the lead wire to feed through. Some manufacturers produce a "cellular look" CB antenna that is not intended for cellular use. For maximum performance, only an antenna specifically designed for cellular service should be used.

REPAIRS

Because of the specialized equipment and knowledge required, any sound equipment determined to be defective should be referred to an authorized repair shop for service. Although radio-receiver technicians need no special FCC license to make repairs, repair of CB transceivers and cellular telephones must be made by, or under the supervision of, a holder of an appropriate FCC license. Internal adjustments or modifications to this equipment can lead to illegal operation of a CB radio or cellular phone, as defined in FCC Rules and Regulations, Part 95. Such illegal operation can lead to serious consequences.

SUMMARY

The modern automobile includes many items that are options on the part of the purchaser. The auto-motive electronics technician should expected to consult manufacturers' specifications to perform maintenance required on a wide variety of optional devices.

PRACTICAL EXERCISE

This exercise is intended for classroom and shop participation and discussion. A late-model vehicle with the appropriate electrical service manual and assorted electrical components (fuses, circuit breakers, and switches) are required.

1. In the manufacturer's electrical service manual, follow the schematic diagrams of the components covered in this unit.
2. In the vehicle, identify the components and wiring serving the convenience systems covered in this unit.
3. What other systems are found on the vehicle that are not covered in this unit?
4. Test a switch:
    a. Using an ohmmeter.
    b. Using a jumper wire.
5. Test for a defective fuse or circuit breaker:
    a. Using an ohmmeter (out of the circuit).
    b. Using a voltmeter (in the circuit).

REVIEW

1. The two-way power seat provides seat movement
    a. fore/aft
    b. up/down
    c. fore/aft and tilt
    d. up/down and fore/aft

2. The six-way power seat does *not* provide seat movement
    a. fore/aft
    b. up/down
    c. fore/aft and tilt
    d. side to side

3. The seat back lock system utilizes a(n) _____ for operation.
    a. two-armature motor
    b. one-armature motor
    c. two-winding solenoid
    d. one-winding solenoid

4. Mechanically adjustable seats provide for _____ operation.
    a. one-way
    b. two-way
    c. four-way
    d. six-way

5. The power reclining seat back is actuated by
    a. a separate reversible motor
    b. a jack shaft from the seat adjuster mechanism
    c. a two-winding solenoid
    d. a mechanical crank

6. Briefly describe the safety precautions that must be taken when working on electric window motors and regulators. _____

7. Which car lines use the permanent-magnet motor for window lift operation? _____

8. What is the purpose of the window lock switch? _____

9. What should the window motor "no load" current draw be? _____ amperes.

10. Briefly describe the operation of a solenoid-operated door lock system. _____

11. Briefly describe the operation of a motor-operated door lock system. _____

12. How many and what type of motors are used in the power-operated rear view mirror system? _____

13. Identify the following abbreviations:
    a. cw _____
    b. ccw _____

14. Write the abbreviation for:
    a. normally open _____
    b. normally closed _____

15. The electronic fuel-injection system has (one) (two) fuel pump(s) that (is) (are) (mechanically) (electrically) operated.

16. The standard electric fuel pump system has (one) (two) fuel pumps(s) that (is) (are) (one-) (two-) speed.

17. In the event of (low) (high) engine oil pressure, the standard electric fuel pump system will break the (fuel pump) (ignition) circuit electrically to stop the engine.

18. (Sensors) (Thermistors) are used to (send) (receive) an electrical signal for subsystem operation.

19. Speed controls (automatically) (semiautomatically) take over manual throttle control, taking (primary) (secondary) control from the driver.

20. Explain the following abbreviations:

   a. AM _____

   b. CB _____

   c. FCC _____

   d. SWR _____

◆ UNIT 19 ◆

COMFORT SYSTEMS

OBJECTIVES

On completion of this unit you will be able to:

- Describe the heating and cooling systems presently available, including
  - thermostats and temperature control devices.
  - fans, blowers, and speed controls.
  - principles of air conditioning.
  - over-temperature and over-pressure devices.
  - typical electrical and mechanical system schematics.

The comfort system of today's modern automobile consists of the heater, ventilator, and air conditioner. There are three basic types of system: manual, semiautomatic, and automatic.

HEATERS

The heater was one of the first accessory options available for a car. Some of the early heaters, which are still available for air-cooled engine cars, were gasoline fired. Hot-water heaters are now commonly used.

The heater of today is relatively simple. Its controls consist of a cable-or vacuum-operated outlet door, an outside recirculate door on some models (Figure 19–1), and a blower control switch assembly. Some heaters have a cable-or vacuum-operated water control valve to prevent hot coolant circulation in the heater core when the heater is not in use.

Some heater systems are equipped with a normally open (NO) engine thermostatic switch that closes when engine coolant temperature reaches approximately 125°F (52°C). This switch is in series with the blower motor (Figure 19–2), and prevents blower operation until sufficiently heated coolant is available in the heater core.

Two to four-speed blower motor control is provided by dropping resistors. Some after-market systems have a variable speed control. With either method, the more resistance in the circuit, the slower the blower motor.

Troubleshooting heaters is relatively simple (Figure 19–3). If the blower motor is not operational, the problem is electrical: defective motor, wiring, or control. If the blower motor is operational, but no air comes from the outlet (or if air comes out but is not warm), the problem is in the mechanical (cable) or fluid (vacuum) system. If air—cool or warm—comes from the outlet when the car is in motion but the blower motor is turned off, again, the problem is in the cable or vacuum system.

Figures 19–4 through 19–6 show the schematics for Chrysler, Ford, and General Motors heater systems. These schematics are typical of car lines equipped with a heater only—not car lines equipped with heater/air conditioning combinations.

Other heaters, such as the rear-window deicer, were discussed previously in Unit 16. For cars equipped with gasoline-powered heaters, manufacturers' diagrams and troubleshooting procedures should be followed.

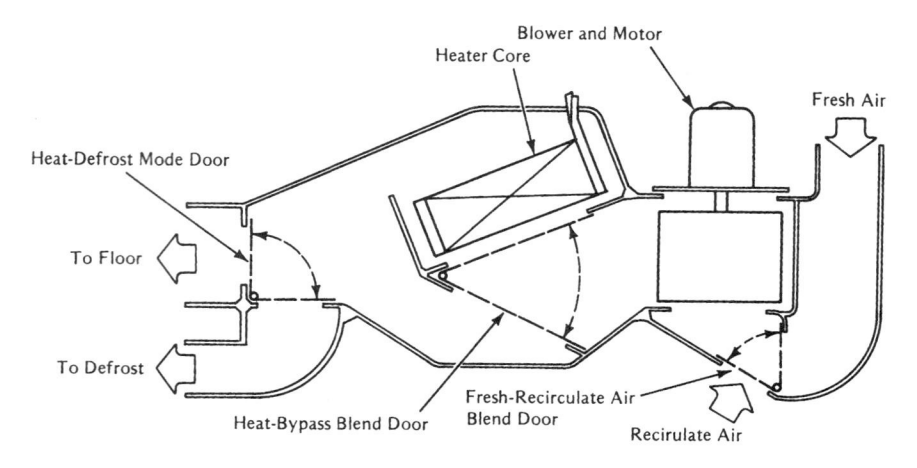

FIGURE 19-1 Heater door and duct control circuits

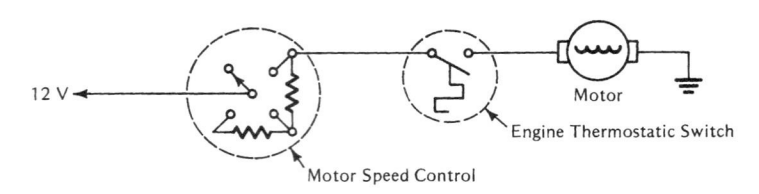

FIGURE 19-2 Simple heater motor schematic with speed control and engine thermostatic switch

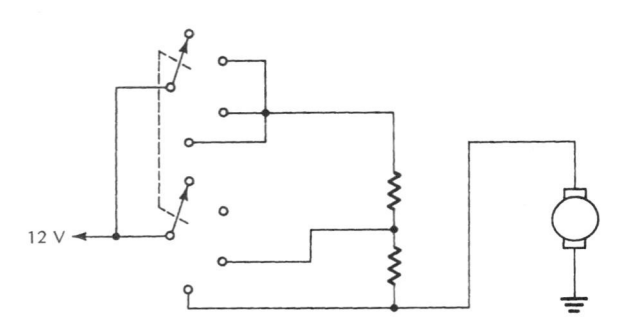

FIGURE 19-3 Typical heater schematic

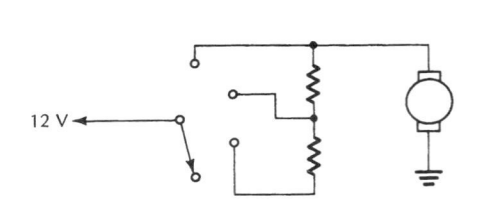

FIGURE 19-4 Typical Chrysler heater schematic

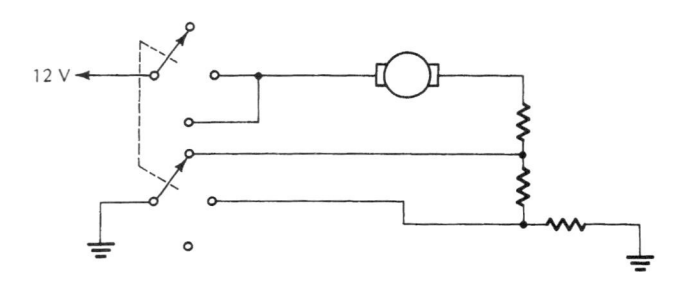

FIGURE 19-5 Typical Ford heater schematic

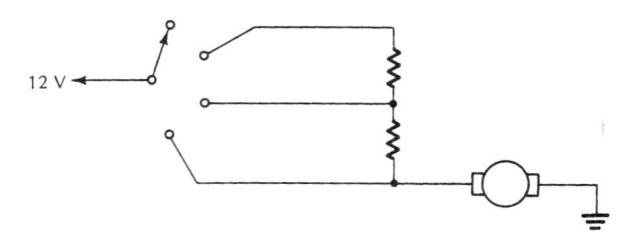

FIGURE 19-6 Typical General Motors heater schematic

AIR CONDITIONING SYSTEMS

Automotive air conditioning is, by far, the most expensive automotive accessory. It is standard equipment on most luxury car lines and an option on most other car lines. Automotive air conditioning is currently found in over 80 percent of all new cars sold. It is expected that the total percentage will increase over the next few years in spite of the high cost of fossil fuels. This means that automotive air conditioning service is an essential part of the trade for the automotive technician.

Air conditioning troubleshooting and repairs are usually accomplished by technicians with an overall knowledge of the various systems. This includes electrical and mechanical knowledge as well as an understanding of the physics associated with heat transfer. Many good textbooks have been written on this subject and are listed in the bibliography of this text. Accordingly, this text will cover only briefly the mechanics and physics of the automotive air conditioning system.

Mechanical Aspects

Aside from the cable-or vacuum-operated "mode" doors directing air flow, there is only one mechanical moving part in an automotive air conditioning system—the compressor.

The compressor is belt driven off the engine crankshaft and "pumps" refrigerant liquid and vapor through the system to provide for an exchange of heat. Compressors, depending on the model, have from one to six cylinders. Each cylinder has a suction and discharge valve. Figure 19–7 shows a cutaway drawing of a simple one-cylinder automotive compressor.

The automotive electrical technician's first task is to ascertain whether the compressor is operational. This is easily determined by observing the clutch. If the center of the clutch is turning, the compressor is operating. If it is not turning, the problem may be mechanical (defective compressor) or electrical (defective wiring or components).

Air Conditioning Physics

The physics of heat transfer dictate that heat must be picked up inside the passenger compartment and moved into the outside air. The three basic laws of physics apply:

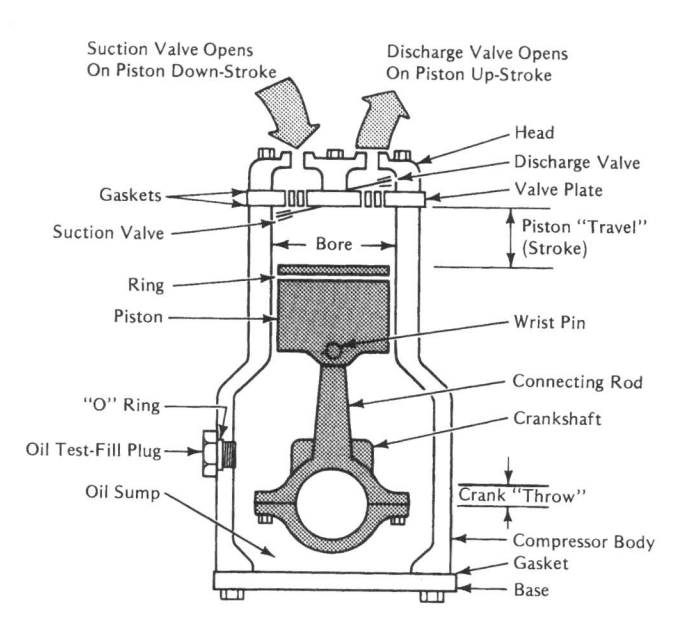

FIGURE 19–7 Single-cylinder automotive air conditioner compressor

1. Law of matter: All things are composed of matter in one of three states—gas (vapor), liquid, or solid.
2. Law of heat: All matter contains sensible, latent, and specific heat.
3. Law of pressure: The heat content (law of heat) at which a change of state (law of matter) takes place is dependent on the pressure.

In the air conditioning system shown in Figure 19–8, there are two states of matter, liquid and gas, and two pressure ranges, low and high. These two states and two pressure ranges are essential if the air conditioner is to produce a cooling effect.

The automotive electrical technician is concerned with two things: Is the compressor operational? and Is air coming out of the duct? If the compressor is operational and the blower motor is heard to be running, but no air comes out of the duct, the problem is mechanical (stuck or binding "mode" door or cable) or in the vacuum system (switch or tubing leak). If the air coming out of the duct is not cool, the problem may be mechanical (stuck or binding "blend" door), in the vacuum system (switch or tubing leak), or in the system itself (physics). If no air is coming out of the duct, and the blower motor is not running, the problem is probably electrical. Note, however, that many system protective and interlock electrical switches are functional by system temperature or pressure.

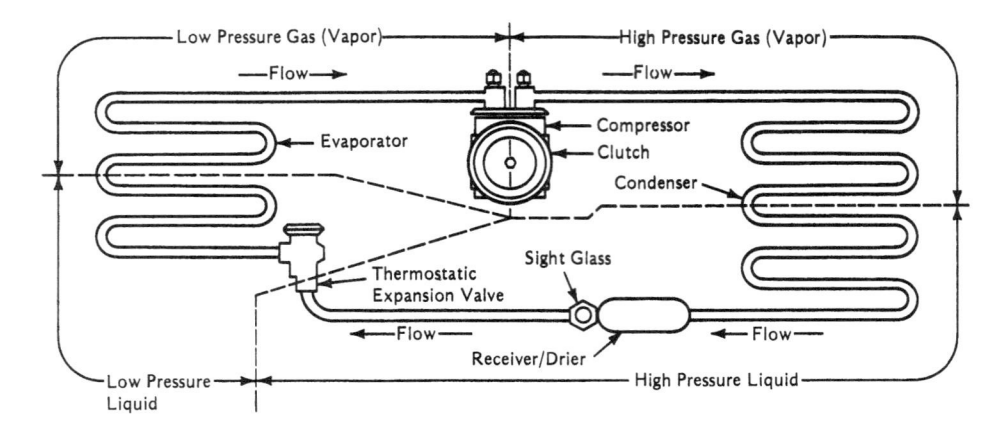

FIGURE 19–8 Simple air conditioning mechanical system circuit illustrating two states of matter (liquid and gas) and high-/low-pressure zones

Electrical Aspects

Almost every year and model's automotive air conditioning schematic differs to some extent. Schematics shown in this text are representative, and are for no particular year or model car line. Figure 19–9 is typical of an air conditioning system. This circuit is from the ACC terminal of the ignition switch or fuse block, through an in-line fuse or circuit breaker, and to the blower and thermostat switches. The circuit protection is usually 20 amperes to provide sufficient current for high blower and clutch operation and at the same time maximum circuit protection.

The blower switch shown is a four-position, OFF-HI-MED-LO switch. When in position 1, the electrical circuit to the motor is open, and it does not run. In position 2, full-battery voltage is available to the motor, and it runs in high speed. A resistor is placed into the circuit in position 3 to reduce battery voltage to the blower motor so that it operates at medium speed. Low speed is provided at position 4 with the introduction of the second resistor.

A thermostat cycles the clutch on and off to provide the temperature control selected by the driver. Most thermostats are preset for a temperature differential (td), called "delta t" (Δ_t) of 12°F (6.7°C). For example, if the thermostat were set for 50°F (10°C), its electrical contacts would close when the evaporator air temperature reached 50°F (10°C) or higher. When the evaporator air temperature drops to the set point, the thermostat contacts open. They remain open for a 12°F (6.7°C) temperature rise and close at 62°F (16.7°C).

Protective Switches

Some factory-installed air conditioning systems have one or more system protective switches. These include engine thermostatic switches, compressor discharge switches, thermal limiters, under-hood ambient switches, in-car ambient switches, ambient cutoff switches, deicing switches, and low-pressure switches. Using the same basic schematic (Figure 19–9), we will discuss the function of each of these switches.

Engine Thermostatic Switch. The engine thermostatic switch (Figure 19–10) is placed in the blower circuit to prevent blower operation until after proper engine warm-up. This normally open (NO) switch closes when engine coolant temperature reaches 95°F (35°C).

Compressor Discharge Switch. The compressor discharge switch (Figure 19–11) prevents the compressor from operating when low refrigerant pressure is sensed. This normally closed (NC) switch, located in the high-pressure side of the system, is electrically

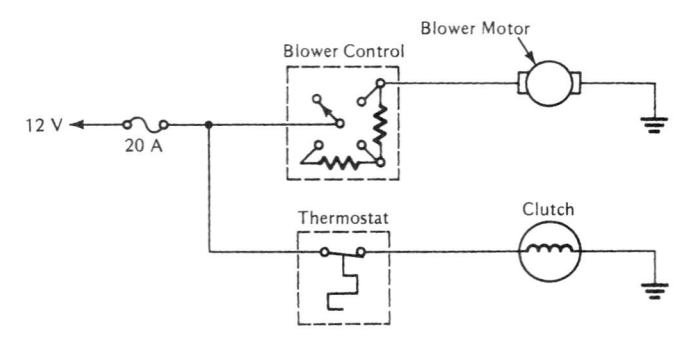

FIGURE 19–9 Schematic of an aftermarket air conditioning system

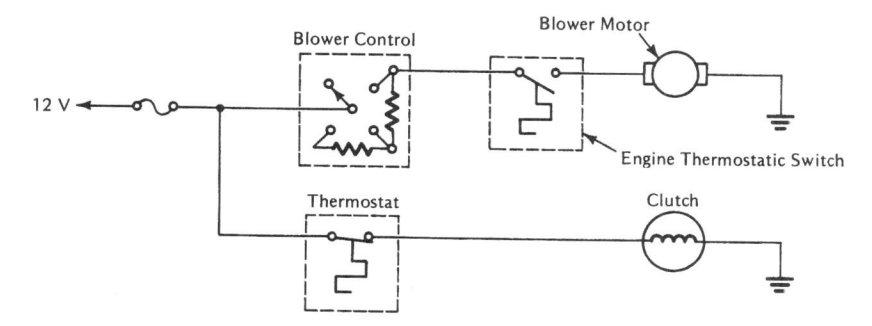

FIGURE 19–10 Schematic showing location of engine thermostatic switch

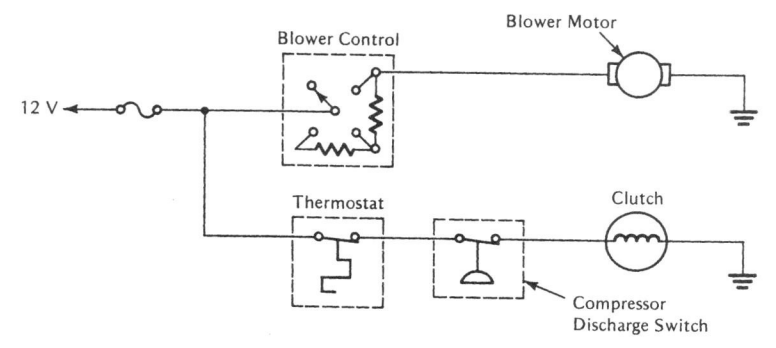

FIGURE 19–11 Schematic showing location of compressor discharge switch

connected into the clutch circuit. This switch will open if the pressure drops to 37 psi (255 kPa) or below Once open, a 5-psi (34.5-kPa) pressure increase is required to close its contacts; it closes at 42 psi (289.6 kPa).

Thermal Limiters. The thermal limiter works in conjunction with a superheat switch. This device (Figure 19–12) is a unique method of system protection used by General Motors cars and other cars equipped with six-cylinder Frigidaire compressors. An increase in system temperature at the compressor above that determined to be maximum for proper operation causes the normally open (NO) contacts of the superheat switch to close. This shorts the clutch circuit to ground through the resistor and blows the fuse.

After a high-temperature condition has been corrected, the thermal limiter must be replaced. The superheat switch automatically resets and requires no attention.

Ambient Switches. The under-hood ambient switch and the in-car ambient switch are used to control compressor cutoff at predetermined temperatures. These switches (Figure 19–13) are normally closed (NC), and open if the ambient temperature drops below a predetermined level. For example, if the in-car ambient temperature in the evaporator dropped below 32°F (0°C), the switch would open, stopping compressor action. The ambient cut-off switch used by Ford is designed to open if the outside-air ambient temperature drops below 40°F (4.4°C).

Deicing Switch. The deicing switch (Figure 19–13), is only another in-car ambient switch. Its purpose and operation is the same as previously described.

Low-Pressure Switch. The low-pressure switch used by Chrysler serves the same purpose and function as the compressor discharge switch (Figure 19–11). The only difference is that its electrical contacts open at pressures below 30 psi (207 kPa) and close at pressures above 30 psi (207 kPa).

Air Conditioning System Schematics

Complete air conditioning system electrical schematics are shown in Figures 19–14 through 19–17. These typical schematics can be used for circuit tracing to

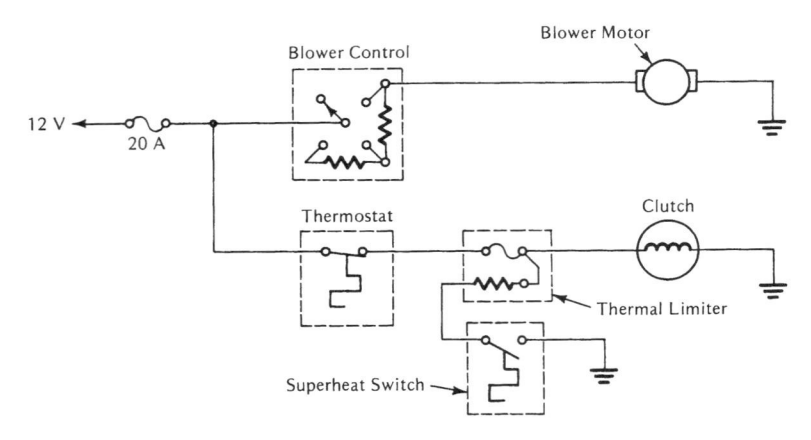

FIGURE 19-12 Schematic showing location of thermal limiter and superheat switch

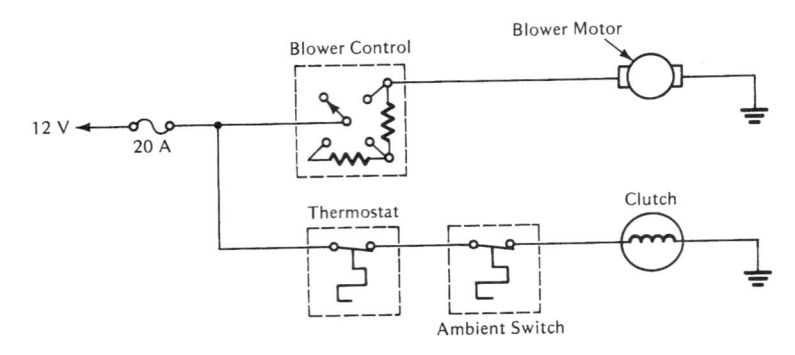

FIGURE 19-13 Schematic showing location of ambient switch

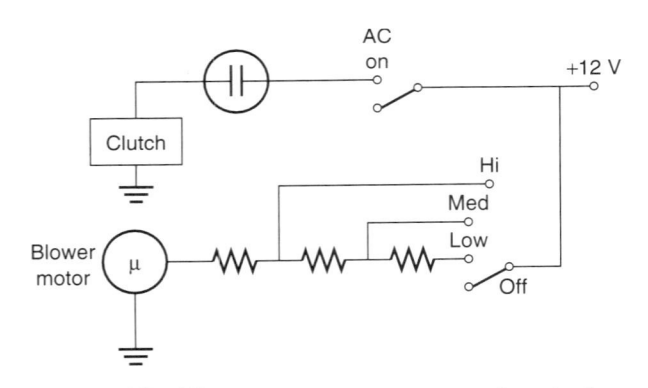

FIGURE 19-14 Typical air conditioner electrical system schematic

determine electrical system problems. However, it may be necessary to consult specific year and model schematics because of variations.

Note that most air conditioning systems depend on vacuum-operated components to be functional.

A typical vacuum system circuit schematic is shown in Figure 19–18. There are different vacuum schematics for almost every year and model car. The vacuum system controls open and close "mode" doors and, in some cases, electrical switches. Reference sources for vacuum systems are given in the bibliography of this text.

AUTOMATIC CONTROL AIR CONDITIONING SYSTEMS

The automatic control air conditioning system works in conjunction with the heater system to blend heated and cooled air to provide a preselected in-car temperature of from 65°F (18°C) to 85°F (29°C) and a relative humidity (rh) between 40 and 55 percent. The ideal comfort range has been determined to be 72°F (22°C) to 80°F (26.6°C) at a relative humidity of 45 to 50 percent.

AUTOMATIC CONTROL AIR CONDITIONING SYSTEMS 235

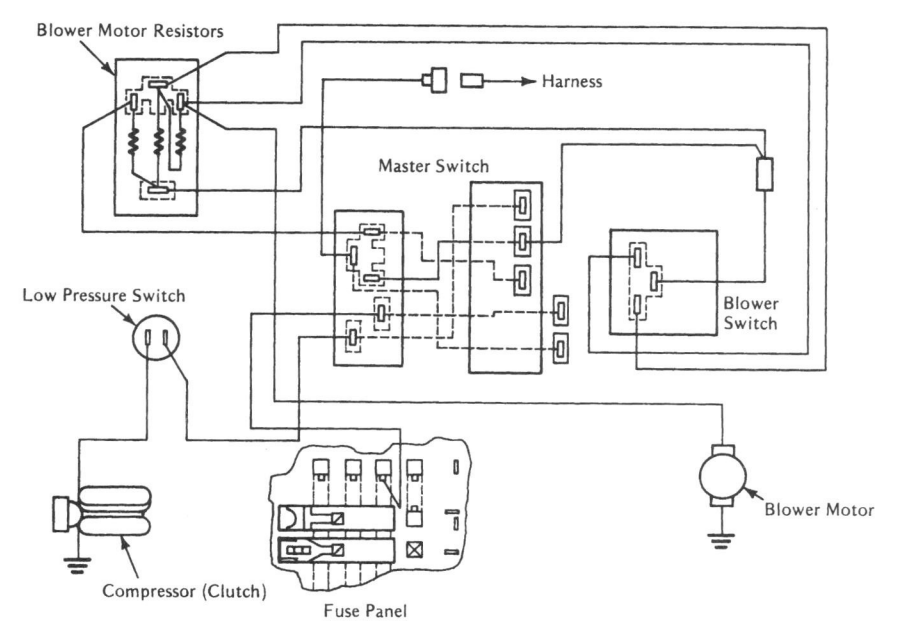

FIGURE 19-15 Typical Chrysler air conditioning electrical schematic

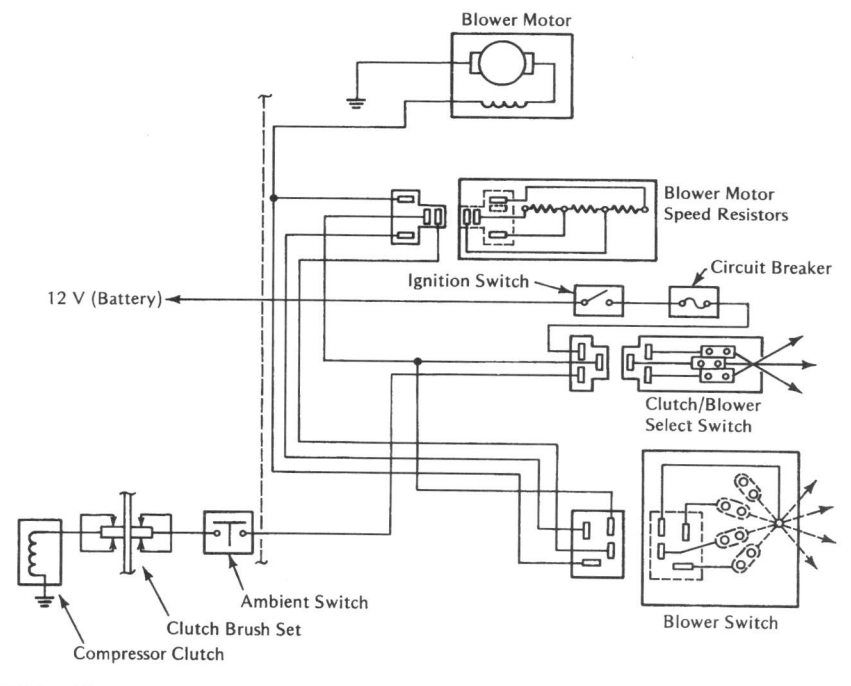

FIGURE 19-16 Typical Ford air conditioning electrical schematic

The complex electrical circuits of the automatic controls called Auto Temp or Auto Temp II by Chrysler; Automatic Temperature Control (ATC) by Ford; and Automatic Climate Control, Comfortron, or Tempmatic by General Motors are integrated with vacuum circuits, one dependent on the other.

Because of the size and complexity of the schematics, it is impossible to include in-depth information on the automatic control systems in this text. Therefore, we will work with a typical description condensed from Chrysler's Auto Temp II information. The student can follow the vacuum system schematic shown

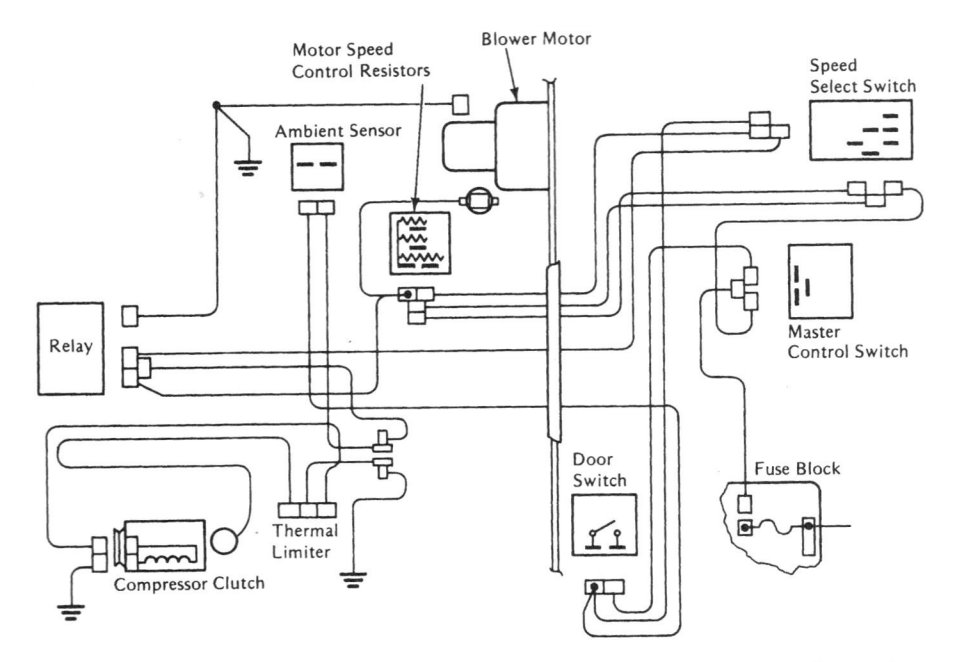

FIGURE 19–17 Typical General Motors air conditioning electrical schematic

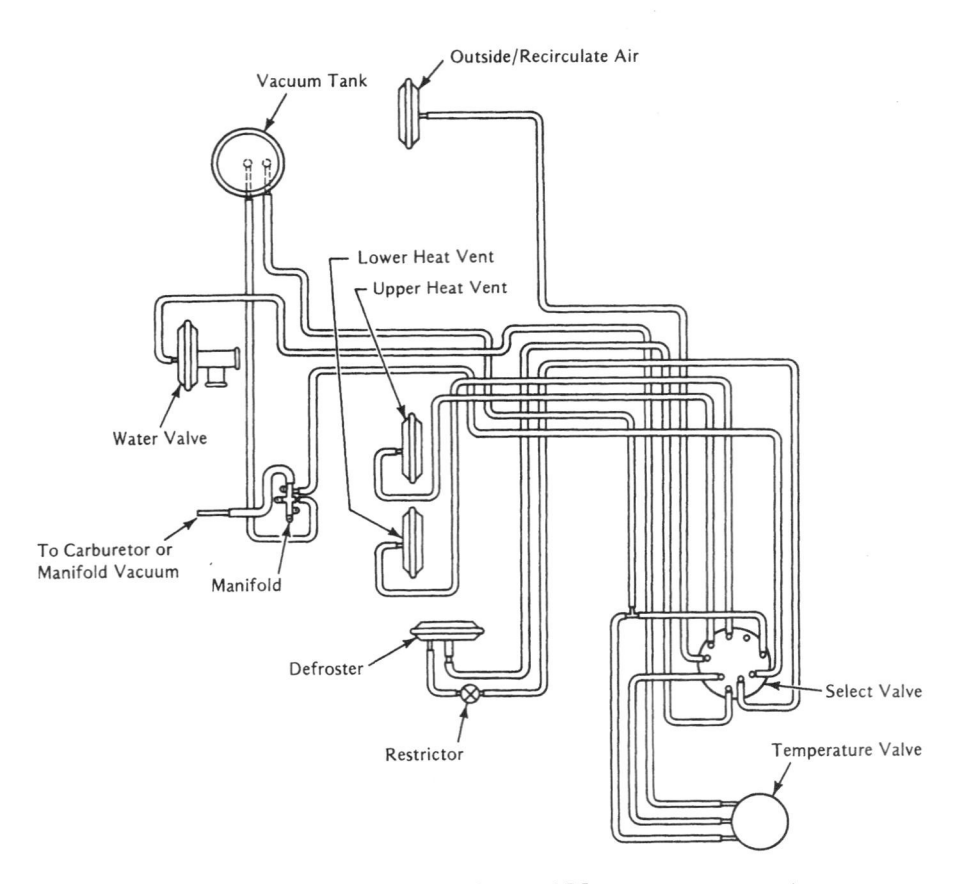

FIGURE 19–18 Schematic of a typical General Motors vacuum system

in Figure 19–19 and the wiring system schematic shown in Figure 19–20 with the text discussion.

Auto Temp II

The Auto Temp II package uses the same basic components as the standard heating and air conditioning system package. The controls, however, are different, and several sensors and components have been added. A brief description of the function of the various subsystems and components follows.

Lever Control. The lever control provides temperature selection from 65° to 85°F (18° to 29°C) in increments of 5°F (2.8°C). The lever controls a potentiometer (pot) that varies the electrical signal sent to the amplifier by the sensor string.

Push-Button Control. Six push buttons provide for OFF, VENT, LO-AUTO, HI-AUTO, LO-DEF, and HI-DEF. Briefly, the function of each is as follows:

OFF—The system is not operational.

VENT—The blower works on medium speed and the compressor is off. Otherwise, VENT is the same as LO-AUTO.

LO-AUTO—Maintains any selected temperature and automatically programs heating or cooling as required with the blower on any of four or five low speeds.

HI-AUTO—Same as LO-AUTO, except that the blower automatically operates on any two or three high blower speeds.

LO-DEF—Air blows from the defrost outlets at one of the two available low blower speeds.

HI-DEF—Same as LO-DEF, except the blower runs on high speed.

Sensor String. The sensor string consists of two sensors, in-car and ambient, connected in series with the control pot to send an electrical signal to the amplifier. This signal strength is constantly varied by the total resistance of the two thermistor sensors.

Compressor Switch. The vacuum-operated compressor switch closes to energize the clutch coil when a minimum of 2 inches Hg (6.75 kPa) is applied.

Amplifier. The amplifier converts signals from the sensor string into a single signal and sends it to the servo.

Servo. An electrical signal from the amplifier actuates an electric motor and gear train to position internal components as required to open and/or close the vacuum ports, change the blower speed, and position the "blend" door.

Cold-Engine Lockout Switch. The lockout switch prevents blower operation when the engine coolant temperature is below 125°F (52°C) if heat

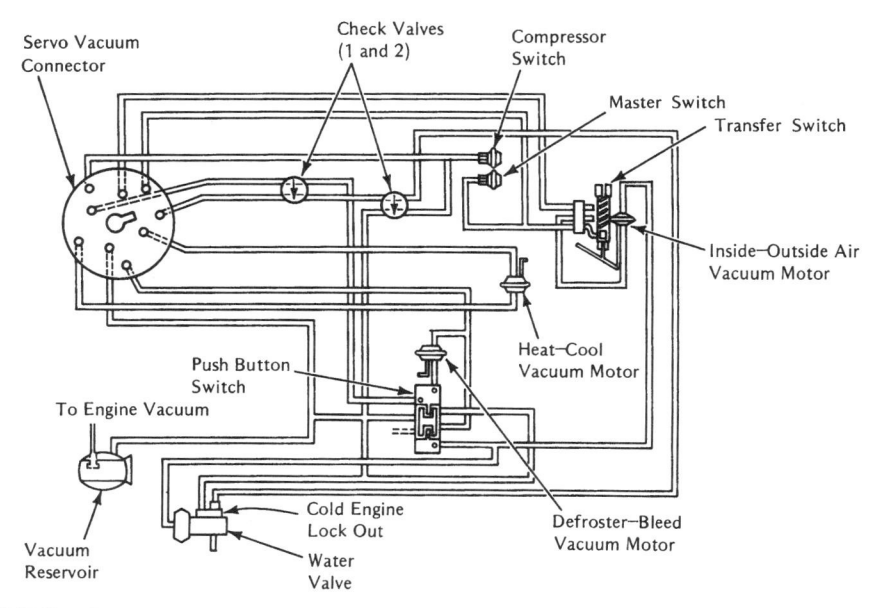

FIGURE 19–19 Auto Temp II vacuum circuit schematic

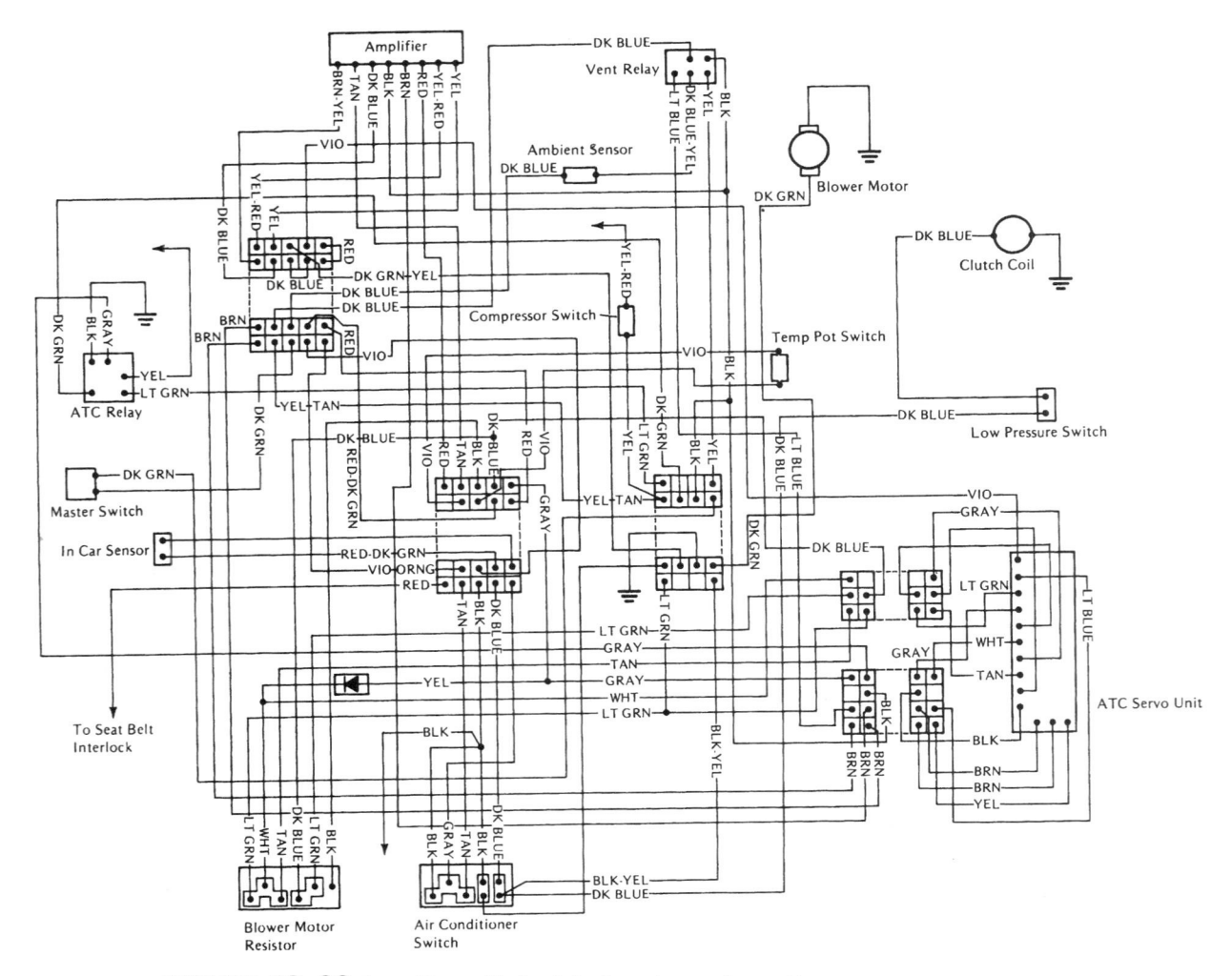

FIGURE 19-20 Auto Temp II electrical system schematic

is required when either AUTO push button is depressed.

Ambient Switch. The ambient switch prevents compressor operation if the outside ambient temperature is below 32°F (0°C).

Vacuum Transfer Switch. The vacuum transfer switch actuates the fresh-recirculate "mode" door in any of three positions: 0 percent, 20 percent, or 100 percent outside air.

SUMMARY

Unlike the manual air conditioning system, the automatic air conditioning system, the automatic air conditioning system does not utilize a cycling clutch for temperature control. Therefore, it has no thermostat. If less cooling is required, the blend-air door opens to admit hot air from the heater, or the fresh-air door opens to admit warm outside air (Figure 19-21). This method of blending cold, hot, and warm air has been used for years as a means of temperature and humidity control on automatic systems.

To dispel a myth, there has never been a reverse-cycle air conditioning system installed in a car. With readily available hot water, reverse-cycle air conditioning is not practical.

However, since the heater and air conditioner operate off the same control, many drivers think that theirs is a reverse-cycle system. Automatic systems are giving way to the cycling clutch system as automobile manufacturers attempt to squeeze every mile from a gallon of gasoline; if the compressor runs continuously, fuel consumption is high. But for those who are not

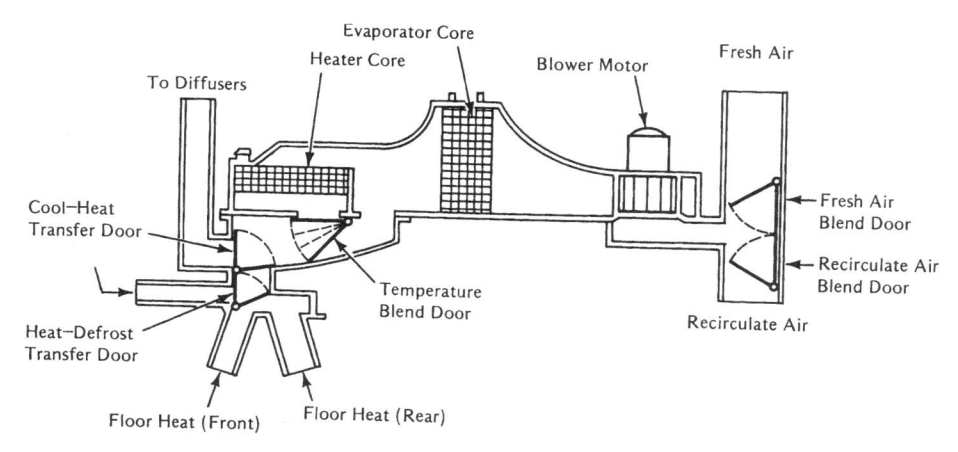

FIGURE 19–21 A typical heat-cool combination (factory-installed) duct system

concerned with the cost of fuel, the automatic system is still available.

PRACTICAL EXERCISE

This exercise is intended for classroom and shop discussion and activity. You will need a late-model vehicle with electrical schematics and component location manual.

1. Raise the hood of the vehicle. DO NOT start the engine.

2. In the schematic and on the vehicle, locate the major electrical components of the air conditioning system.

    a. Control panel
    b. Blower motor
    c. Compressor clutch
    d. Condenser fan, if equipped

3. Which of these components are shared by the heater system?

4. Can you find at least three other electrical control components in the air conditioning system and on the schematic? Briefly describe their function.

REVIEW

1. What is the comfort range for the human body? Temperature _____ °F _____ °C Humidity _____ % rh

2. Briefly describe the operation of the automatic temperature control system. _____

3. Can the automatic temperature control system be considered a reverse-cycle system? Explain your answer. _____

4. What is the purpose of the sensor string? _____

5. What is the purpose of the cold-engine lockout switch

    a. When cooling is required? _____
    b. When heating is required? _____

6. What is the most common type of heater in use today? _____

7. What is the purpose of the engine thermostatic switch for heater operation? _____

8. What is the purpose of the water control valve? _____

9. What are the two types of blower motor controls? _____

10. How does the blower motor speed control affect blower speed? _____

11. The purpose of the compressor is to:
    a. cause a change of state in the system.
    b. pump vapor refrigerant through the system.
    c. pump liquid refrigerant through the system.
    d. pump liquid and vapor refrigerant through the system.

12. "Mode" doors are used to:
    a. operate cables or vacuum controls.
    b. direct the flow of air.
    c. direct the flow of refrigerant.
    d. All of the above are correct.

13. What is the purpose of the thermostat?
    a. to cycle the clutch on and off
    b. to cycle the blower on and off
    c. to cycle the clutch and blower on and off
    d. None of the above is correct.

14. The compressor discharge switch prevents:
    a. compressor operation at low-ambient temperature.
    b. compressor operation at high-ambient temperature.
    c. compressor operation at low system pressure.
    d. compressor operation at high system pressure.

15. Most air conditioning systems depend on _____ -operated components for proper function.
    a. vacuum
    b. pressure
    c. temperature
    d. All of the above.

16. The engine thermostatic switch prevents heater operation until engine coolant temperature reaches _____ °F (_____ °C).

17. Approximately _____ percent of all cars manufactured today are equipped with an air conditioning system.

18. Technician A says that if the center of the compressor clutch is not turning, the thermostat is defective. Technician B says that if the center of the clutch is not turning, the clutch is defective. Who is correct?
    a. Technician A may be correct.
    b. Technician B may be correct.
    c. Either technician could be correct.
    d. Neither technician is correct.

19. All things in nature are composed of matter in one of three states: a. _____ b. _____ c. _____

20. Technician A says that heating may be accomplished by reverse-cycle air conditioning (heat pump). Technician B says that heating may be accomplished by use of a hot water heat exchanger (heater core). Who is correct?
    a. Technician A is correct.
    b. Technician B is correct.
    c. Either technician may be correct.
    d. Neither technician is correct.

◆ UNIT 20 ◆

SENSING AND CONVERSION DEVICES

OBJECTIVES

On completion of this unit you will be able to:

- Describe automotive sensing devices.
- Explain the system for converting analog and digital signals.

In the modern automobile, information relating to system operation must be available in order to properly control engine conditions, comfort systems, instrumentation systems, and safety systems. This information is obtained from sensors strategically placed to obtain the required information.

There are a number of different conditions to be sensed in modern vehicles. Included are

- Position
- Temperature
- Pressure
- Light ambient
- Speed (engine and wheel)
- Oxygen
- Air flow

Position sensors include

- Switches (ON-OFF)
- Resistive types (potentiometer)
- Optical sensors (light)

Temperature sensors include

- Resistive pickup types (thermistor)

Pressure sensors include

- Piezoelectronic sensors
- Capacitance types

Light sensors include

- Photocells

Speed sensors include

- Magnetic types (generator)
- Optical sensors (encoder)
- Hall effect sensors (pulse generating)

Oxygen sensors include

- S or R

Air flow sensors include

- Hot wire

POSITION-SENSING SWITCHES

The simplest type of position sensor is a switch. When the device to be maintained reaches a certain position, the switch is actuated. The change in switch positioning (open to close, or close to open) causes a change in voltage, which is sensed by the control module.

The switch may be connected to provide voltage to the circuit, or to provide ground to a circuit permanently connected to a voltage source (Figure 20–1).

Potentiometers

It is often necessary to continuously monitor the position of a system element, for example, the outside air-recirculate door of the air conditioning system. The ratio of fresh and recirculated air depends on the position of this door. If a potentiometer movable arm were connected to the door, the rotational position of the door could be determined by the voltage at the potentiometer output (Figure 20–2). The resister R_1 is used to protect the +5-volt regulator supply in case of a short circuit.

The output voltage to the control module is dependent on the position of the arm of the potentiometer, and therefore on the position of the outside air-recirculate door.

Light Sensors

The combination of a light-emitting diode (LED) and phototransistors can be used to determine the position of a rotating shaft, such as the steering shaft. The combination of mechanical movement of the encoder shutter with light from the LEDs and pickup by the light-sensitive transistors provides a no-physical-contact monitoring device.

Consider the optical steering position encoder shown in Figure 20–3. Two sets of photodiode and phototransistor combinations are included in the encoder. The position of the photodiode transistor sets and the slots in the shutter wheel are such that only one set may be active at a time. When LED A shines through a slot, turning on phototransistor A, LED B's light is blocked by the shutter wheel.

Positional information is obtained by processing the square waves of voltage taken from the photosensitive transistors. Three pieces of information may be obtained from this encoder:

1. Position of the steering wheel.
2. Direction of steering wheel motion.
3. Speed of steering wheel motion.

TEMPERATURE SENSORS

Temperature sensing is usually accomplished using a heat-sensitive resistor called a *thermistor*. The thermistor is held in contact with the medium to be sensed. The resistance of the thermistor varies with tempera-

FIGURE 20–1 Switch sensing

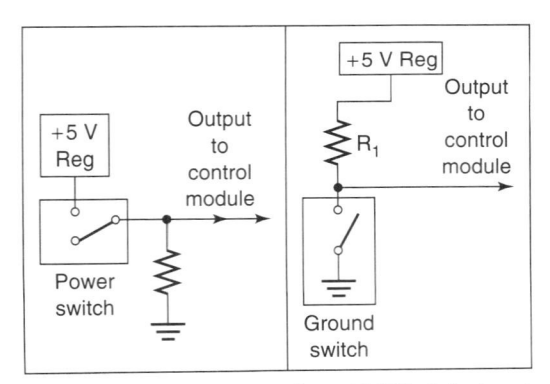

Output –0 V Switch open   Output 0 V Switch closed
   +5 V Switch closed   +5 V Switch open

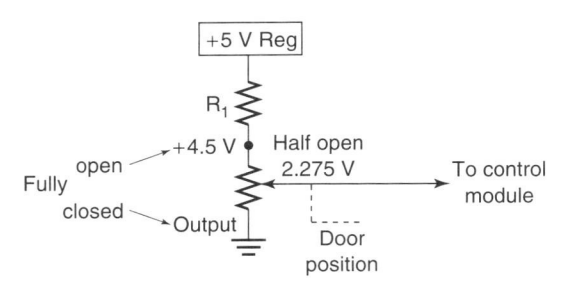

FIGURE 20–2 Position sensor (potentiometer)

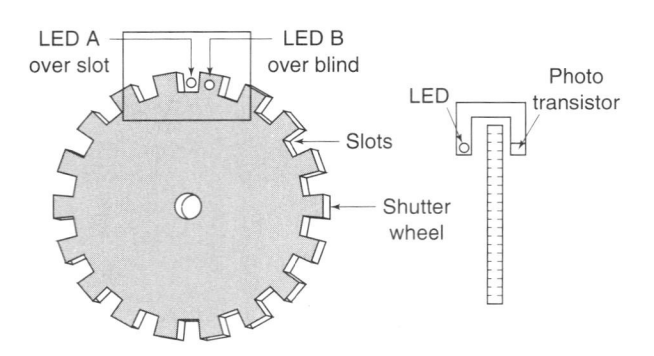

FIGURE 20–3 Optical steering wheel position encoder

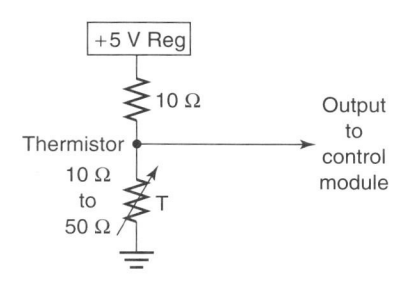

FIGURE 20–4 Temperature-sensing thermistor circuit

ture. Thermistors may have a positive temperature coefficient; a rise in temperature causes a rise in resistance. Thermistors are also available with a negative temperature coefficient; a decrease in temperature causes a rise in resistance. Temperature-sensing elements are usually connected in the system according to Figure 20–4. They usually have a negative temperature coefficient.

The series resistor at the +5-volt regulated supply output is a fixed resistor providing for voltage changes across the resistance of the thermistor, which varies with temperature.

PRESSURE SENSORS

There are basically two types of electronic pressure sensors found on the modern vehicle: piezoelectric and capacitance. The following is a brief description of the pressure sensors and their application.

Piezoelectric Sensors

A piezoelectric sensor has crystal material at its center and plates at either side of the crystal. It is similar in construction to a disc ceramic capacitor; a large-value disc ceramic capacitor produces a measurable voltage when subjected to a bending pressure.

One use of a piezoelectric sensor in the modern automobile is as a knock sensor. When an engine knock occurs, the shock wave causes a spike of voltage to be produced by the piezoelectric sensor. The voltage produced is proportional to the intensity of the shock wave. The output of the sensor is fed to the control module, where the signal is processed and a determination is made as to corrective action. A knock-sensor circuit is shown in Figure 20–5.

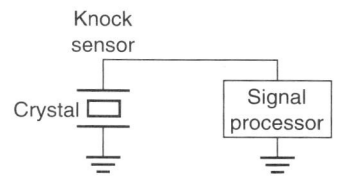

Oil Pressure Sensors. Another characteristic of a piezoelectric crystal that is useful in sensors is that the resistance through the crystal changes as pressure is applied to it. The piezoelectric crystal as a pressure sensor is similar to the standard change-in-resistance detectors. The oil pressure sensor circuit is shown in Figure 20–6.

Oil pressure exerts a force on the crystal, causing its resistance to vary. The voltage across the resistance of the crystal will change accordingly. This voltage reflects the oil pressure and will affect the control module action.

FIGURE 20–5 Piezoelectric sensor

Capacitance Sensors

A special type of sensor is used to indicate manifold air pressure and atmospheric air pressure. It is a variable capacitor (Figure 20–7).

As the plates are moved together the capacitance increases. When the plates are moved apart the capacitance decreases. The capacitance output at any time may be used to control the frequency of an inductor-capacitor oscillator. The frequency of the oscillator is given as:

$$f_0 = \frac{1}{2\pi \sqrt{LC}}$$

where f_0 = frequency of oscillation

L = coil inductance in henrys

C = Capacitance in farads

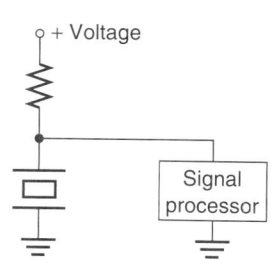

FIGURE 20–6 Piezoelectric variable resistance sensor

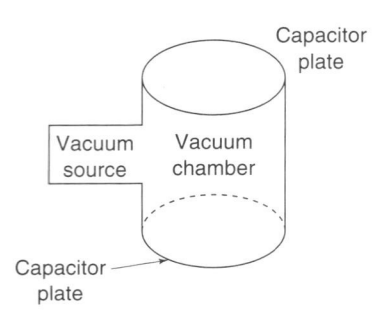

FIGURE 20–7 Pressure-sensitive capacitor

If the inductance of the coil is constant, the frequency of the oscillator is determined by capacitance. If the capacitance varies, the frequency varies:

- Larger capacitance, lower frequency
- Smaller capacitance, higher frequency

The output of the oscillator is fed through an overdriven amplifier to produce a square wave output. This square wave varies in frequency with pressure. Figure 20–8 represents a block diagram of a system generating a square wave output that varies in frequency with pressure changes.

LIGHT SENSORS

Light sensors are used to control headlights and rear-view mirror. The level of light is determined through the use of a photocell, also referred to as a photoelectric cell. A photocell is a special resistor that varies in resistance as light strikes its surface. In application, a photocell, when properly placed, can indicate an approaching vehicle's headlights and call for automatic dimming of your lights. The photocell may also be used to automatically call for change in the rear-view mirror to eliminate glare (Figure 20–9).

System operation of the light sensor is similar to that of the thermistor. A fixed resistor is connected in series with the photocell. When the resistance of the photocell changes because of a change in light on its surface, the voltage fed to the control module changes. A simple level detector (comparator) in the control module may be used to provide switching action for headlights or rear-view mirror.

FIGURE 20–8 Pressure-controlled signal generator

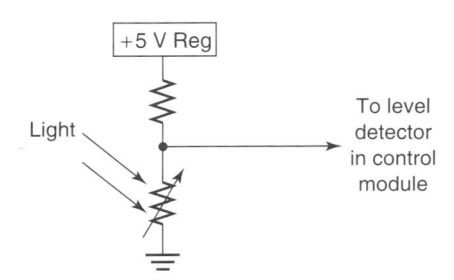

FIGURE 20–9 Light sensor

SPEED SENSORS

Magnetic Pickup Sensors (Reluctor)

Reluctor sensors (rotational) use a magnetic system to indicate position and speed. The sensor has a magnetic core in a coil. A rotating element turning and indicating the required measurement completes the magnetic path as teeth on the rotor pass the sensor. The rotating element is called a *reluctor*. Figure 20–10 shows a signal-generator system for a rotational speed sensor.

The sensor is a generator producing alternating current (ac) voltage. As the reluctor rotates, the magnetic field about the sensor coil increases and decreases. When a tooth is at the sensor coil center, the magnetic field is at maximum intensity. When a gap is at the sensor coil center, the magnetic field is at minimum intensity. The changing magnetic field induces a voltage in the magnetic sensor pickup coil.

The frequency of the voltage generated in the sensor coil is dependent on the rotational speed of the reluctor. The higher the speed, the higher the frequency.

Hall Effect Sensors

Electronic engine control finds use for a Hall effect device in the distributor. A rotating vane on the distributor shaft is used to expose and then shield a Hall effect device from a magnetic field. This shielding and exposure causes the Hall effect device to turn off and on accordingly. The device acts as a switch, producing the

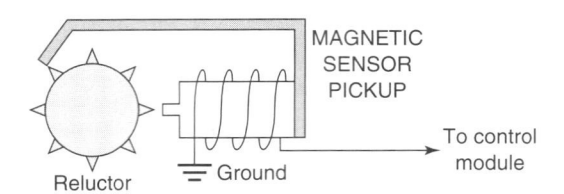

FIGURE 20–10 Magnetic pickup sensor (reluctor)

OFF-ON sequence. Electronic components associated with the switch provide for the sharp rise and fall of the leading and trailing edge on the output waveform. Engine speed and crankshaft positions are generated through the use of Hall effect devices. The system is shown in Figure 20–11.

OXYGEN SENSORS

Oxygen sensors are used in modern vehicles to determine the amount of oxygen in the exhaust gas. The sensor produces a variable voltage output from zero to approximately one volt. The sensor is made of platinum and zirconium dioxide, as shown in Figure 20–12.

The sensor is constructed and placed through the exhaust manifold wall in a manner that permits exhaust gases to pass over one platinum surface while the other platinum surface senses outside ambient air. The difference in oxygen content between the exhaust gas and outside air determines the voltage produced between the two platinum elements. The electrical circuit feeding the oxygen control signal to the control module is shown in Figure 20–13.

AIR-FLOW SENSORS (HOT WIRE)

The hot-wire air-flow sensor uses the principle of the Wheatstone bridge, also referred to as a resistance bridge, to determine air flow in a system. The bridge circuit is modified by the use of two elements sensitive to temperature. One is the thermistor in the cold-wire leg and the second is the hot wire in the hot-wire leg (Figure 20–14).

The initial condition of the bridge is that it should be in balance. The ratio of R_1 to R_T (thermistor) should equal the ratio R_{hw} (hot wire) to R_2. Under these conditions the bridge would be in balance and the voltages at point A and point B would be equal.

To measure air flow both the thermistor (R_T) and the hot wire (R_{hw}) must be placed in the air flow passage. The purpose of the thermistor is to compensate for resistance changes in the hot wire (R_{hw}) that are caused by temperature changes.

Consider temperature change only such as would be the case with no air flow. The resistance of the thermistor (R_T) will increase as the temperature is lowered. The resistance of the hot wire (R_{hw}) will decrease with a decrease in temperature. The bridge will remain in balance. Consider Figure 20–15, showing the effects of temperature change alone.

In Figure 20–15(a) the bridge is shown in its original, balanced condition. The voltage at point A is 2.5 volts, as is the voltage at point B. When the thermistor (R_T) and hot wire (R_{hw}) are placed in a cooler environment, the resistance of the thermistor (R_T) increases. At the same time the resistance of the hot wire (R_{hw}) decreases. Conditions are as illustrated in Figure 20–15(b). The resistance of the thermistor (R_T) increased to 200 Ω. The voltage at point A increased to 3.33 volts. The resistance of the hot wire (R_{hw})

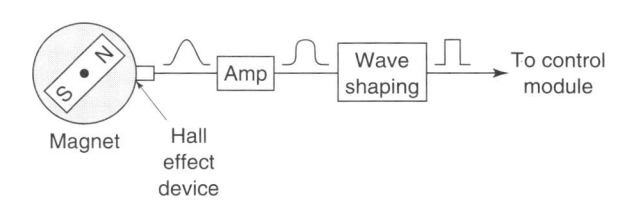

FIGURE 20–11 Hall effect signal sensor

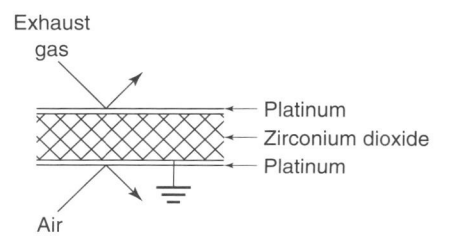

FIGURE 20–12 Oxygen sensor

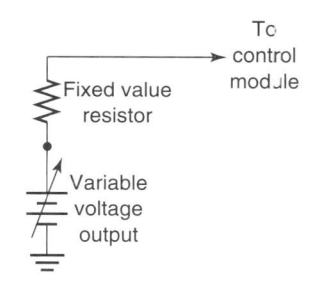

FIGURE 20–13 Oxygen sensor circuit

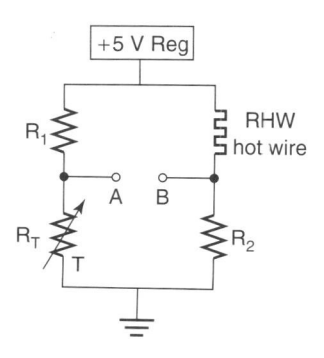

FIGURE 20–14 Hot-wire air-flow speed sensor

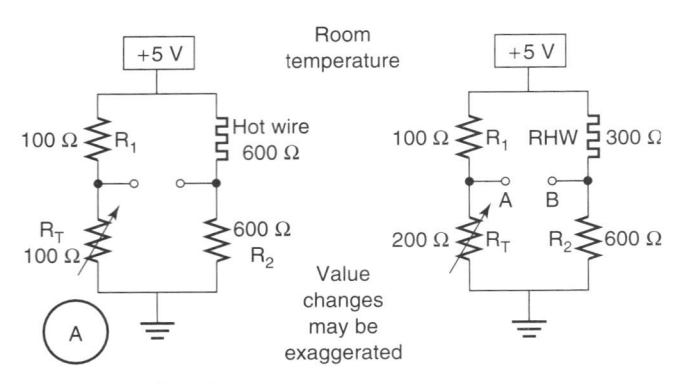

FIGURE 20–15 Balanced bridges

decreased to 300 Ω, and the voltage at point B increased to 3.33 volts. Under these conditions the bridge is still in balance.

If the fan were turned on to provide air flow across the hot wire (R_{hw}), it would be cooled. The greater the air flow, the greater the cooling. The resistance of the hot wire (R_{hw}) would decrease to below 300 Ω, unbalancing the bridge. The difference output between point A and B terminals is amplified and used as an indication of air-flow speed.

CRASH SENSOR

The crash sensor is an example of a special switch sensor. A special deceleration device allows for switch closure and inflation of an air bag. Figure 20–16 illustrates the deceleration control device. When at rest, the magnet holds the gold-plated steel ball at the one end of a tube far away from the gold-plated contact points.

Under extreme deceleration, such as would be the case in a collision, the steel ball is pulled away from the magnet and moved to the contact points, completing a circuit (Figure 20–17). The air bag will deploy. For safety reasons, a number of circuits must be completed before the air bag is released. Extreme caution must be exercised by the automotive electronics technician to guard against unintentionally inflating an air bag. Manufacturer's precautions for any particular car line must be followed to deactivate the air bag circuit for certain electrical and electronic troubleshooting and repair procedures.

DIGITAL CONVERSION CIRCUITS

The on-board computer (microprocessor) operates using digital information at its input. After combining and manipulating the information received, it produces digital information and presents this information at its output.

The sensors that produce the information used by the on-board computer produce the information in the form of analog voltages (measured voltages). These measured voltages must be converted to a digital signal in order to be processed in the computer.

The output of the computer is digital information. This must be converted to analog voltages in order to produce moving action, changes of position, speed of rotation, and so on, that are required in the output devices.

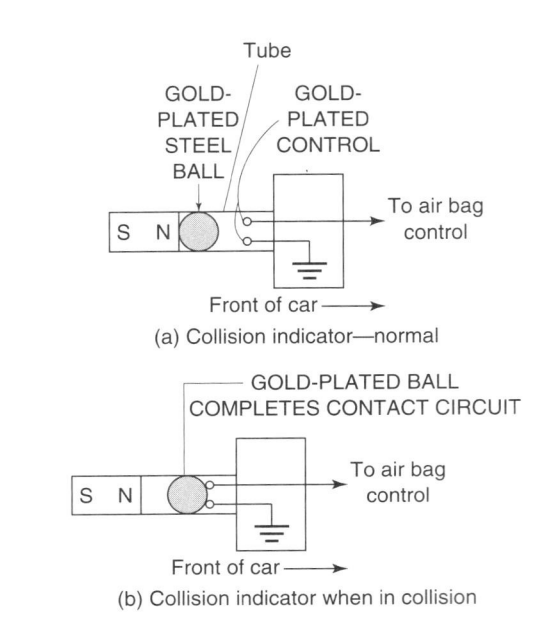

FIGURE 20–16 Collision indicator: (a) normal and (b) extreme deceleration

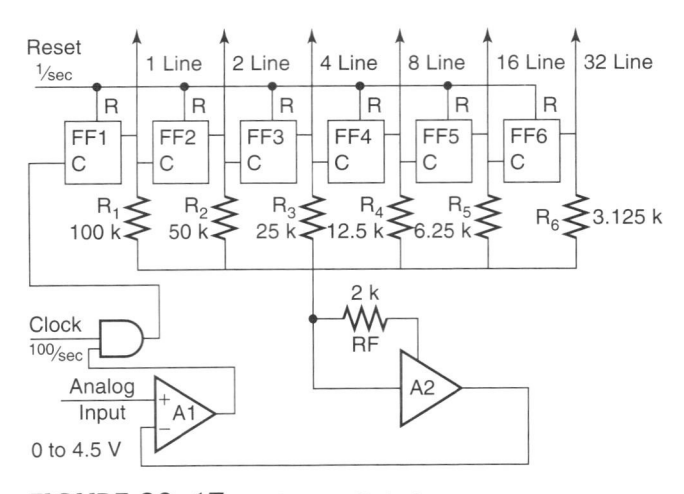

FIGURE 20–17 Analog-to-digital converter

Analog-to-Digital Converters

The input devices to the computer are analog-to-digital converters (ADC). Two main types of ADCs are in use. They are:

- dc voltage to digital
- Frequency to digital

dc Voltage to Digital

In actual practice an eight-bit digital counting system is used. Scaling is also used in the actual system. A count of 150 (01101001) could represent 2.5 volts. A count of 198 (01000011) could represent 3.3 volts, and a count of 75 (11010010) could represent 1.25 volts.

For explanation, let us use a six-bit counter. The maximum count is 64 (111111). For simplicity's sake a count of 64 will represent 6.4 volts. A count of 50 (010011) will represent 5.0 volts, whereas a count of 12 (0011001) represents 1.2 volts.

Consider the system diagram shown in Figure 20–17, an analog-to-digital converter. A slow repetition rate of 1 complete count per second is provided. The clock pulses are at 100 per second. The six flip-flop circuits across the top of Figure 20–17 form a standard counting circuit. Once every second a reset pulse returns the count to zero.

Clock impulses, when available at the C input to FF_1, will provide for counting in the circuit. When clock pulses are not available at the C input to FF_1, counting will stop.

When a flip-flop is in its active state (1) it is assumed to have +5 volts at its Q output. Note the scaling resistor R_6 at the output of FF_6, representing a count of 32 as 1/32 the size of scaling resistor R_1, representing a count of 1. The application of +5V to R_6 has 32 times the effect of the output of A_2 as the application of +5V to R_1.

Amplifier A_2 is used as a summing amplifier. The output of the amplifier is equal to the sum of the voltage inputs from the flip-flops times the feedback resistor, divided by the input scaling resistors (Figure 20–18).

The output of the amplifier A_2 is fed to the comparator A_1. If the analog input is greater than the output of A_2, the output of the comparator A_1 will be positive (+). If the output of the amplifier A_2 exceeds the analog input, the comparator output immediately drops to zero.

The AND gate has two inputs, the output of the comparator and the clock pulses. So long as the comparator output is positive (+), the clock pulses will get through the AND gate. Should the comparator output drop to zero, no further clock pulses will get through the AND gate and counting will stop.

Consider an analog input of 3.5 volts. With our selected scaling an output count of 35 is required. In digital binary code this is 110001.

1. A reset pulse returns all FF to zero. The input to A_2 is 0 volts and the output is 0 volts.

2. Counting starts as clock pulses get through the AND gate. Since A_1 output is positive (+), analog input voltage is greater than A_2 output.

3. Counting continues until FF_6 is in the (1) state. FF_1, FF_2, FF_3, FF_4, and FF_5 are at zero.

The output of A_1 is as shown in Figure 20–19; with 3.2 volts at the comparator inverter input and 3.7 volts at the analog input, the comparator output is positive

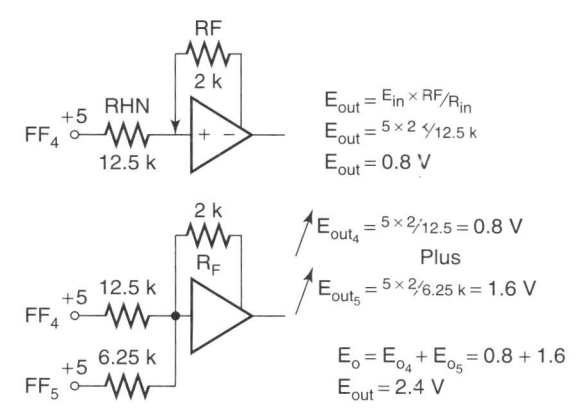

FIGURE 20–18 Amplifier output

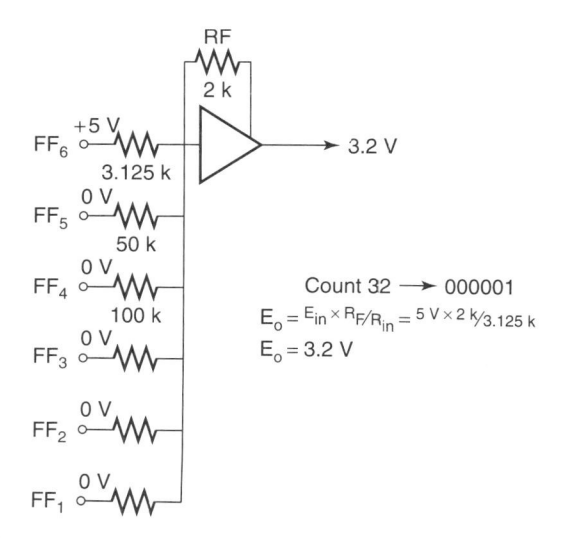

FIGURE 20–19 Amplifier output at count 32

(+). The AND gate continues to pass clock pulses and counting continues.

When the count reaches 34 (010001) the output of the amplifier A_2 is as shown in Figure 20–20. On the next count, 35 (110001), the amplifier input becomes 3.5 volts, as shown in Figure 20–21.

The comparator output switches to zero and digital counting stops at 110001 or 35, representing 3.5 volts. In the on-board computer this digital number may be sent to a memory location for storage and then on to further calculations.

Frequency-to-Digital Conversions

Frequency-to-digital conversions (Figure 20–22) are usually based on counting for a specific period of time. The clock or timing pulses come from the on-board computer. In this explanation timing pulses of 1 second on and 1 second off will first be considered. As frequencies at the input to a counting circuit increase, the counting period may decrease. Keep in mind that a binary eight-bit counter can only count to 256.

The clock time input to the AND gate is positive (+) for one second and zero for one second. When the clock input is positive (+) the squared signal from the overdriven amplifier may pass through the AND gate and drive the counter up. After one second the clock input decreases to zero and counting stops. The count is equal to the frequency.

A read pulse follows the clock or count timer. The read pulse activates the output gates of the counter feeding the frequency count to a memory location on the eight output leads. These lines may be part of a multiplexed system where this count is on the line for a small part of one second. The read pulse is followed by a reset pulse, which returns all flip-flops to zero.

For higher frequencies at the input the clock or count time may be reduced. For a one-second count the eight-bit binary counter can count up to 256. With a clock or count time of 0.1 seconds the same counter, counting at 256, indicates a frequency of 2560. With a clock or count time of 0.01 seconds the counter, at 256, indicates a frequency of 25,600, and so on.

Digital-to-Analog Converters

One of the devices found at the output signal process area of the control module is a digital-to-analog (D/A) converter. Here the output commands from the computer are converted from digital binary 0s and 1s (0 volts or +5 volts) to analog voltages. The analog voltages can provide for the moving of solenoids, the control of fan speed, control of brakes, and so on.

The digital-to-analog converter operation is exactly opposite to the analog-to-digital converter, though internally, some of the processes may seem the same.

Remember in the analog-to-digital converter that the counter section provided for scaled inputs to an operational amplifier. A similar process takes place in a digital-to-analog converter. Consider the converter shown in Figure 20–23. Again, for simplicity, only six digits are shown.

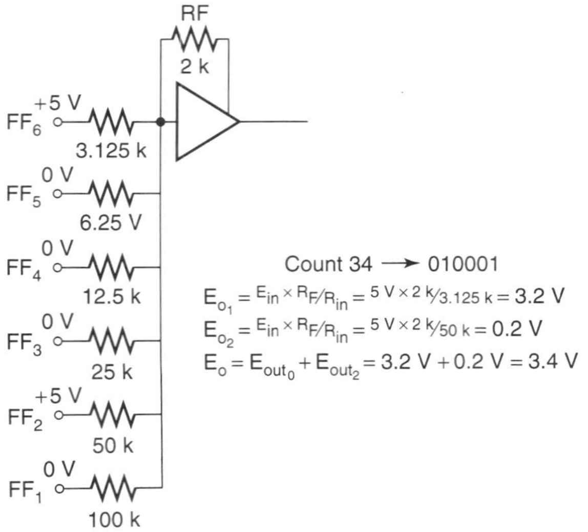

FIGURE 20–20 Amplifier output at count 34

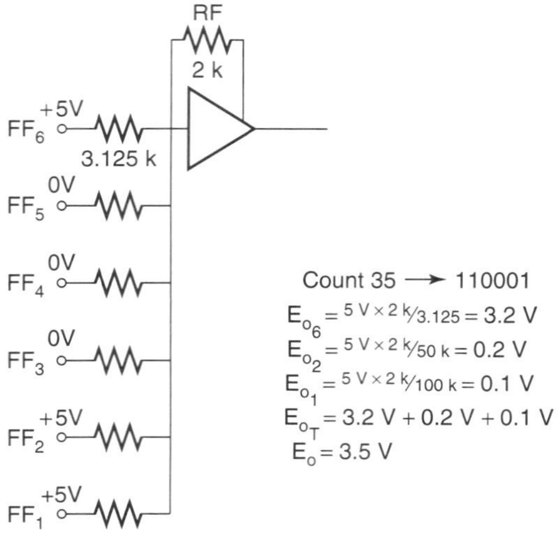

FIGURE 20–21 Amplifier output at count 35

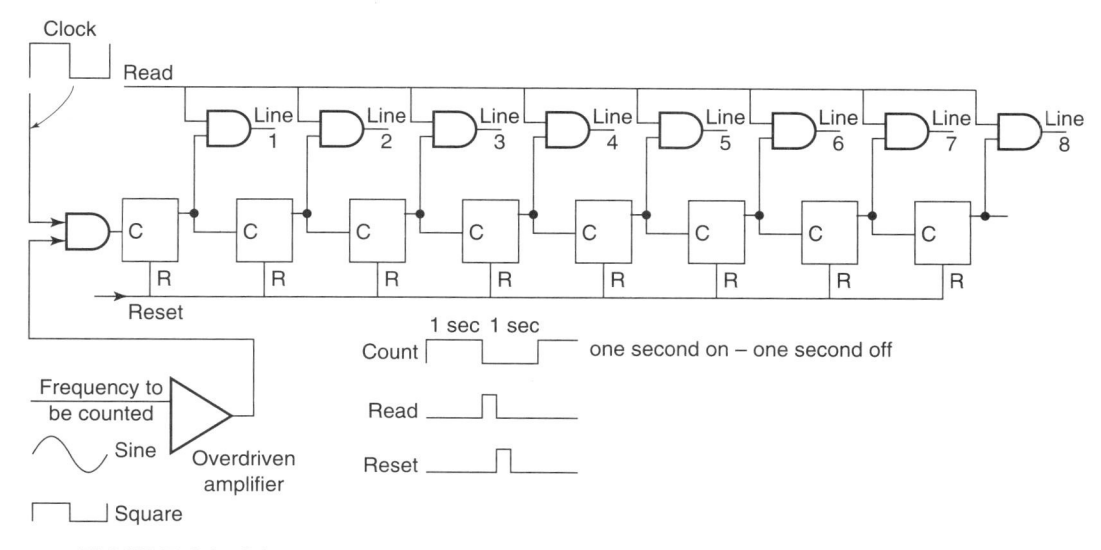

FIGURE 20–22 Frequency-to-digital converter

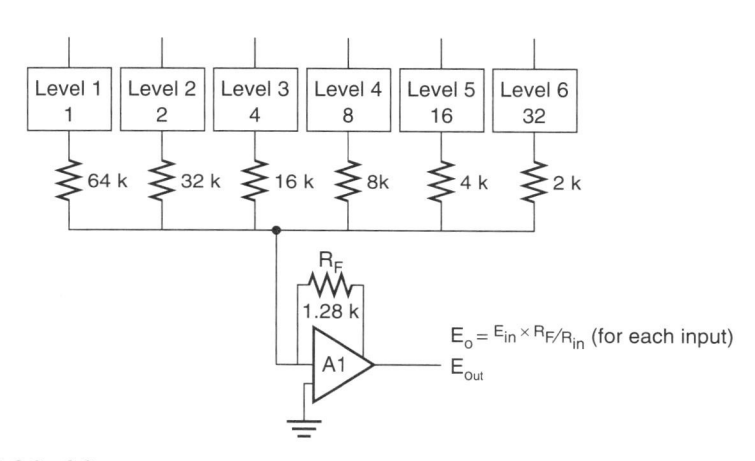

FIGURE 20–23 Digital-to-analog converter

$E_o = E_{in} \times R_F/R_{in}$ (for each input)

The output lines shown (1, 2, 4, 8, 16, and 32) represent the digital output levels. A digital count of 26 would have logic output (+5 volts) on lines 2, 8, and 16. This could be shown as 010110.

The output of amplifier A_1 would be as shown in Figure 20–24. The output of the digital-to-analog converter is 2.6 volts, representing the digital number 26.

$E_{o_T} = E_{in} \times RF/R_{in} + (\ ) + (\ )$
$E_{o_5} = 5\,V \times 1.28\,k/4\,k = 1.6\,V$
$E_{o_4} = 5\,V \times 1.28\,k/8\,k = 0.8$
$E_{o_2} = 5\,V \times 1.28\,k/32\,k = 0.2$
$E_{out} = 1.6 + 0.8 + 0.2$
$E_{out} = 2.6\,V$

FIGURE 20–24 DC amplifier output

SUMMARY

Many sensing devices are included in the modern automobile. The information sensed is converted to a proper electrical signal and transferred to a location where appropriate action is indicated. The modern technician must be familiar with the location of these devices, how they operate, and where the information they produce is used. The technician must also understand and perform proper maintenance.

PRACTICAL EXERCISE

This exercise is intended to stimulate classroom and shop discussion and participation. A late-model vehicle with an appropriate service manual for electrical service and a component locator diagram will be needed.

EXERCISE

1. On the electrical schematic, locate the outside/recirculate air door. (See note 1.)
    a. Is there a door position sensor?
2. On the component locator diagram, locate the:
    a. air door.
    b. sensor.
3. On the electrical diagram, locate the knock sensor. (See note 2.)
    a. Is there a knock sensor in the vehicle? If so, locate it.
    b. Locate the knock sensor on the component locator diagram.
4. Repeat the exercise to locate the oil pressure sensor. (See note 3.)
5. Repeat the exercise to locate other sensors. (See note 4.)

NOTES

1. The outside/recirculate air door is to be found in the heater and/or air conditioner duct system.
2. Not all vehicles have a knock sensor.
3. Unless mechanical, most vehicles have oil pressure sensors.
4. Others sensors may include low fuel, door ajar, engine coolant, and others.

QUIZ

1. If the answer to Step 1(a) was "yes," what type sensor is it? _____
2. Were the air door and sensor (Step 2) where you expected to find them? Explain. _____
3. If the answer to Step 3(a) was "yes," what type sensor is it? _____
4. What other sensors did you find in Step 5? _____

REVIEW

(T) (F) 1. The output of the switch sensor shown in Figure 20–1 is either +5 volts or 0 volts.

(T) (F) 2. In Figure 20–2, resistor R_1 is used for current limiting.

(T) (F) 3. In Figure 20–3, the phototransistor sends a signal to the LED.

(T) (F) 4. The thermistor produces a signal because its resistance changes with temperature.

(T) (F) 5. The knock sensor produces a signal because its resistance changes under pressure.

(T) (F) 6. The piezoelectric oil pressure sensor produces a signal because its resistance varies with pressure applied.

(T) (F) 7. The oxygen sensor produces a voltage at its output terminals.

(T) (F) 8. Under extreme deceleration the sensor in Figure 20–16 provides +12 V to ground to the air bag control circuit.

(T) (F) 9. In Figure 20–20, what is the output wave form from the overdriven amplifier? ____

(T) (F) 10. The output of the dc amplifier in Figure 20–24 would (increase) (decrease) if feedback resistor R_F were increased in value.

UNIT 21

IGNITION SYSTEMS

OBJECTIVES

On completion of this unit you will be able to:

- Explain how and where ignition timing signals are generated.
- Explain the function of
  - Distributors.
  - Transistorized ignitions.
  - Hall effect sensor systems.
  - Distributorless ignitions.
  - Reluctor systems.

The ignition system of a motor vehicle provides the electric spark that ignites the fuel/air mixture in a cylinder. The spark must be provided at precisely the proper time and at a sufficient amplitude.

Many changes have taken place in ignition systems over the years. Ignition timing control was rather simple in early vehicles, where timing was changed (advanced) with only one consideration: engine speed. In present-day electronics a number of factors are considered in computing the timing of ignition spark.

It is interesting to note that some modern ignition systems use multiple coils to provide spark to individual cylinders. The old four-cylinder Model T Ford used a four-coil pack.

Unit 8 presented a basic approach to the effect of inductance (a coil of wire) in a dc circuit. In Figure 21–1, a switch is shown controlling the connection of a battery to a coil. When the switch is closed, the current builds up in the coil. The length of time it takes for the current to build up to maximum is determined by the time constant of the circuit. As the current builds up through the coil, so does the magnetic field about the coil. The time for current buildup is an important consideration as engine speed is increased.

While the switch is closed, current flows through the coil, and a magnetic field builds up to a maximum value shown at 5 TC in Figure 21–1. The current is constant after 5 TC, and so is the strength of the magnetic field.

When the switch is first closed, the magnetic field about the coil tries to increase immediately, but a back emf is produced in the coil by the increasing magnetic field cutting through the coil. The maximum back emf is slightly under 12 volts when the switch is closed, which allows for current flow. The current then increases to maximum in five time constants.

At this time the back emf is zero, since the current is no longer changing (zero change in magnetic field). When the switch is opened, the magnetic field must decrease to zero rapidly. This rapid change in magnetic field induces a high voltage in the coil winding.

In practice, it is necessary to place a capacitor across the switch as shown in Figure 21–1 in order to keep arcing from occurring as the switch is opened. Even with the capacitor across the points, the voltage induced (back emf) in the primary of a standard oil field ignition coil is in the order of 300 volts at the opening of the points.

MOTOR VEHICLE IGNITION

The points and capacitor ignition system includes the primary and secondary coil as shown in Figure 21–2.

Unit 21 IGNITION SYSTEMS

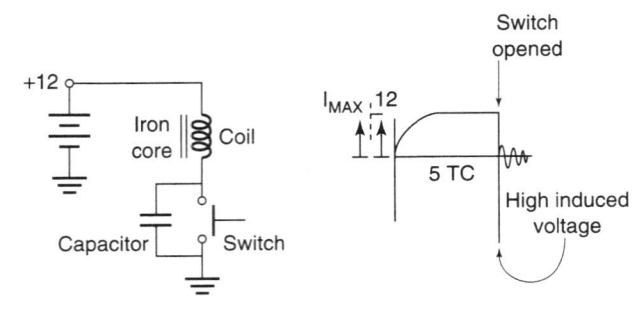

FIGURE 21-1 Current and voltage of a coil

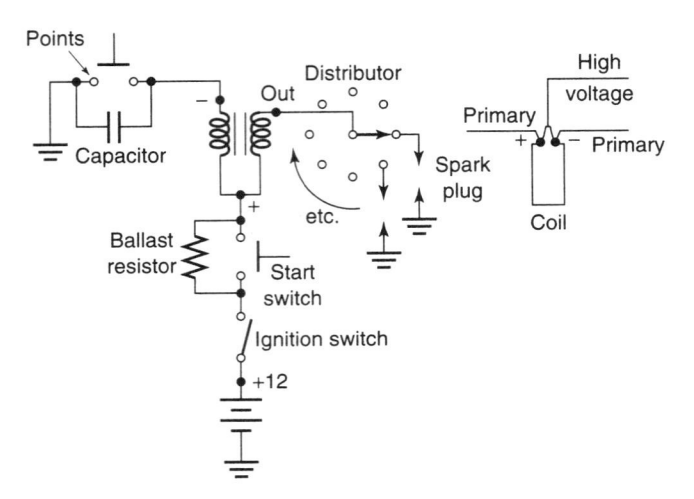

FIGURE 21-2 Conventional ignition

The secondary coil has many times more turns of wire than the primary coil. For the purpose of explanation, the primary coil is given with 300 turns; the secondary has 100 times as many turns—30,000.

$$300 \times 100 = 30{,}000$$

When the points open, the magnetic field collapses, inducing 300 volts of bemf in the primary:

$$300 \text{ volts}/300 \text{ turns} = 1 \text{ volt per turn}$$

Conventional Ignition

The same magnetic field that cuts through the primary coil cuts through the secondary coil. Recall that the secondary has 100 times as many turns as the primary (30,000 turns). At one volt per turn, 30,000 volts will be induced in the secondary. (The number of turns on the primary coil and turns ratio varies with manufacturer.)

This voltage is sufficiently high to jump the gap in the spark plugs, igniting the fuel/air mixture. The actual voltage rises to about 10,000 where the spark jumps the spark plug gap. With current now flowing, the voltage stops rising and drops down to the firing level.

The ignition system shown in Figure 21-2 contains components that provide for efficient operation during start and run periods of motor vehicle operation. The battery voltage is approximately 7 volts during start because of battery internal voltage drop caused by high starting current. The total 7 volts is applied to the ignition coil primary through the start switch, which is closed during starting.

When the engine is running and the start motor is disengaged, the battery voltage returns to 12 volts (13.5 volts, alternator voltage). The start switch is open. Ignition coil primary current must flow through the ballast resistor and coil primary in series. About 6 volts appears across the ballast resistor, leaving 7 volts across the coil primary. The coil is designed to operate at 7 volts. With this design "hot" spark is produced during start and run conditions.

Ignition Waveform

The waveform shown in Figure 21-3 is that of a standard ignition system. The waveform is numbered to show times where specific actions occur.

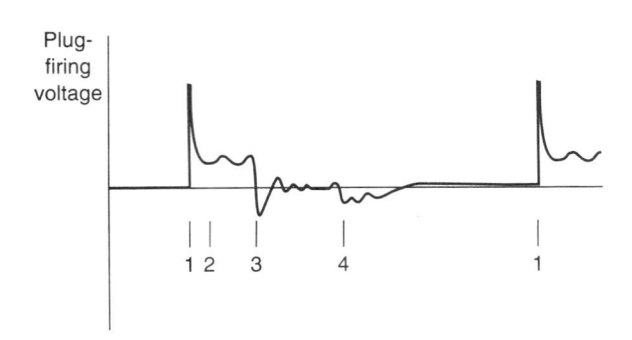

FIGURE 21-3 Ignition waveform

They are:

1. Points open; high voltage is induced in the secondary coil. At the peak, a spark jumps the gap in the spark plug.
2. With the spark in the plug, current flows in the secondary circuit. The rate of change in the magnetic field is reduced, as is the secondary voltage, which drops down to the level maintaining spark.
3. Spark is maintained until time 3, when the energy stored in the coil magnetic field is dissipated. The spark extinguishes. There is a slight rise in voltage before dropoff, due to decreased current and circuit oscillations controlled by the coil inductance and the combination of connected and distributed capacitances.
4. The points close, applying power to the coil primary. Current builds up in the coil. Note the voltage is in the opposite direction when voltage is applied to the coil at 4 compared to when the points open, removing supply voltage from the primary as at 1.
5. The points again open and the process repeats.

DISTRIBUTORS

As its name implies, the *distributor* distributes the ignition spark to the spark plugs at the proper times and in the proper sequence to ensure maximum torques from the engine. The job of the distributor is also to trigger the exact instant that the spark is to occur. Its job is very exacting—it must trigger and distribute voltages that can be up to 30,000 volts and must do so at exact time intervals. Slight irregularities in the operation of the distributor caused by either mechanical or electrical misadjustments can result in poor engine performance.

Conventional Distributor

The breaker points in the distributor, shown in Figure 21–4, open and close the circuit to the primary winding of the ignition coil. At the instant the points close, current begins to flow in the primary circuit and a magnetic field builds up around the primary coil. When the points open, the current flow stops and the magnetic field collapses rapidly. This sudden collapse builds up a very high voltage in the secondary winding of the coil. This high voltage is connected by a wire to the center terminal on the distributor cap. This terminal, in turn, is connected through a rotating contact to the rotor inside the distributor. The rotor turns with the distributor shaft and distributes, in the proper order, the high voltage to the spark plug. Figure 21–5 shows a typical distributor for a 6-cylinder engine, and Figure 21–6 shows the top view of the breaker plate used in another popular 6-cylinder distributor.

An eight-cylinder engine has an eight-sided cam; a four-cylinder engine, a four-sided cam. A change in dwell angle will change the opening time of the points, and this, in turn, will be reflected in the ignition timing.

The parts that normally wear and require service are:

1. The breaker points.
2. The rubbing block.
3. The capacitor.
4. The cam.
5. The shaft bearings.

The standard point–capacitor ignition system served the automotive vehicle faithfully for a good number of years. A major development came with the introduction of the power transistor (late 1950s) to the system. Two of the drawbacks to the point–capacitor system are:

1. Current draw is limited.
2. Points become pitted (high resistances).

TRANSISTORIZED IGNITION SYSTEM

A diagram of a transistorized ignition system is shown in Figure 21–7. The ignition coil primary current is controlled by the turning on and off of the power transistor. When the points are closed, the base of the transistor is connected to ground. The collector is connected to ground through the coil. Making the base voltage like the collector voltage turns the transistor on. When the points are open, the only connection to the base is to the emitter through a resistor. The base of the transistor is at the same voltage as the emit-

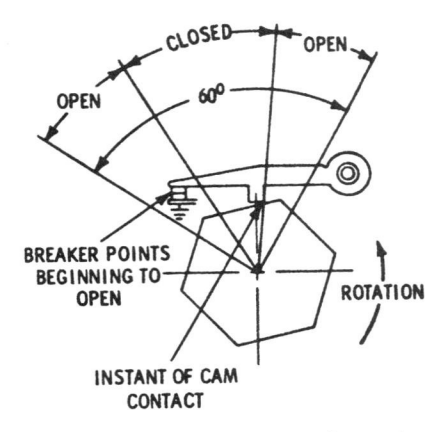

A. Points at the instant of opening.

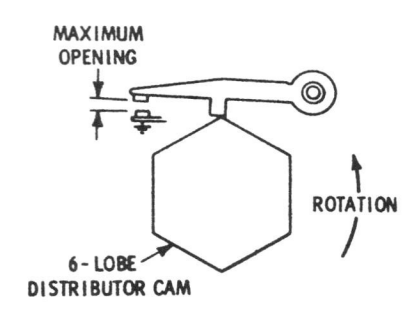

B. Points at maximum separation.

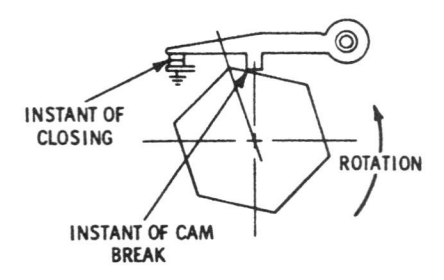

C. Points at the instant of closing.

FIGURE 21-4 The cycle of breaker point action in a six-cylinder engine

ter. The transistor turns off. The transistor is a PNP. Positive battery voltage is connected to the emitter through the switch. Note the use of the ballast resistor during run to maintain proper voltage (approximately 7V).

Advantages

1. Only the low value of base current flows through the points. This means low point arcing and wear.
2. The transistor is capable of carrying higher current than the points (hotter sparks).

The addition of the transistor improved ignition performance and decreased maintenance problems, but still included a source of mechanical failure—the points.

Breakerless Ignition

In the early breakerless ignition systems, the points condenser plate and cam were replaced with a star wheel and magnetic pickup system located in the distributor.

Reluctor Pickup

The reluctor and the rotating star wheel change the magnetic path, providing sharp changes in the total magnetic field. The change in magnetic field induces a voltage in the coil. The induced voltage pulse may be shaped, and then used to turn a transistor on and off to provide the ignition pulse, as shown in Figure 21-8. Mechanical vacuum advance was still used with the early reluctor pickup system. In later vehicles, advance is accomplished in the associated computer system. The system diagram is shown in Figure 21-9.

Hall Effect Ignition

The *Hall effect* is the change that occurs when a current flow is subjected to a magnetic field. A Hall effect switch is a combination of a Hall effect pickup, amplifier, Schmitt trigger, and output transistor in a small three-terminal package. The Hall effect device may be used to provide trigger output to a transistorized ignition system.

The Hall effect sensor is mounted in the point–capacitor area of the distributor (the points and capac-

TRANSISTORIZED IGNITION SYSTEM 257

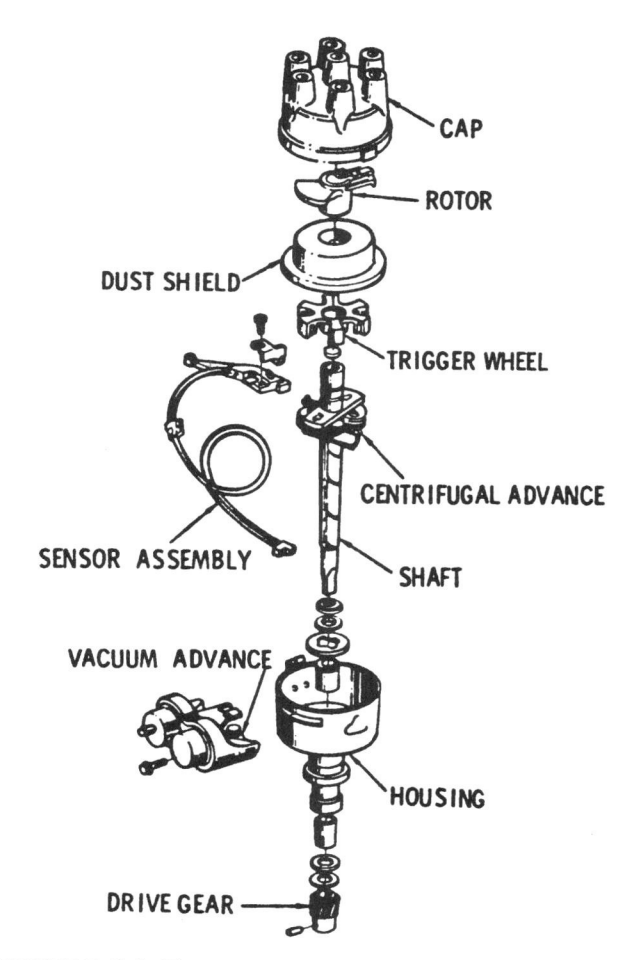

FIGURE 21-5 A typical distributor for a six-cylinder engine

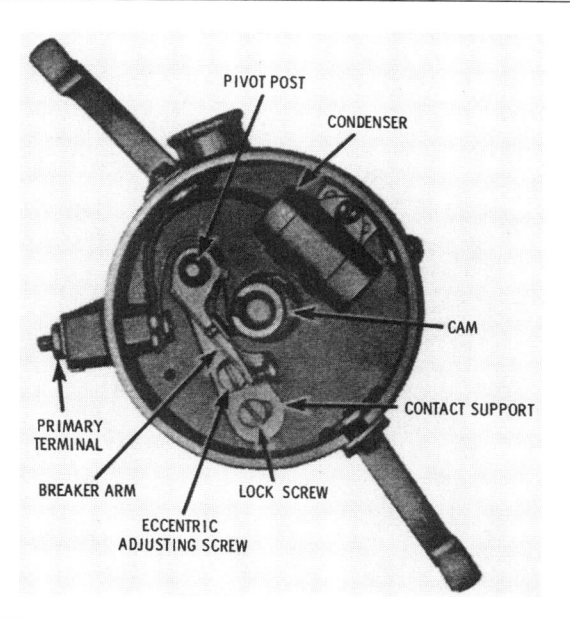

FIGURE 21-6 Top view of the breaker plate in a six-cylinder distributor.

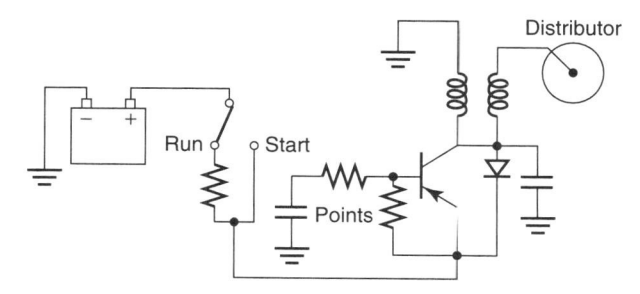

FIGURE 21-7 Transistorized ignition system

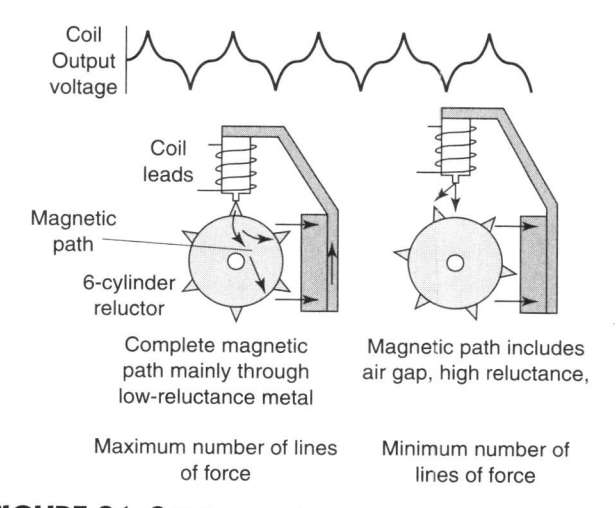

FIGURE 21-8 Reluctor pickup

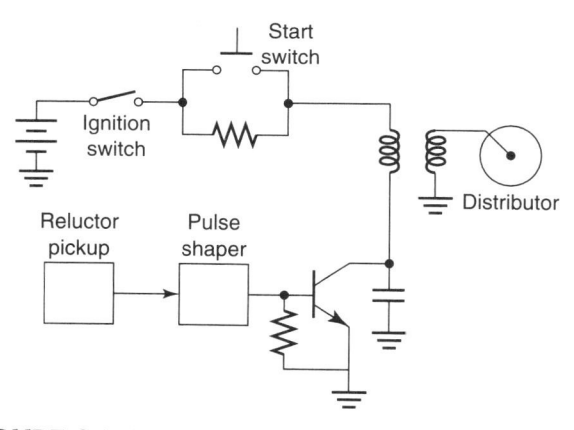

FIGURE 21-9 Magnetic pickup breakerless ignition

itor are no longer needed). A magnet is fixed in a position across from the Hall effect device. The shutter wheel rotates, and a shutter blade blocks and opens the space between the magnet and the Hall effect device. The shutter wheel and blades provide a low-reluctance path for the magnet's field, keeping the field from the Hall effect device. The Hall effect device produces square wave output pulses. These pulses are used in the electronic ignition system. Mechanical vacuum advance was used to control spark timing in early Hall effect systems. Hall effect pickup for a four-cylinder engine is shown in Figure 21-10.

On later-model vehicles, spark advance is controlled by the electronic control assembly. The indication of cylinder position is taken from the reluctor pickup or Hall effect pickup in the distributor. Proper timing of the spark is then computed in the electronic control module and ignition module. Actual spark is produced and distributed as in earlier systems (Figure 21-11).

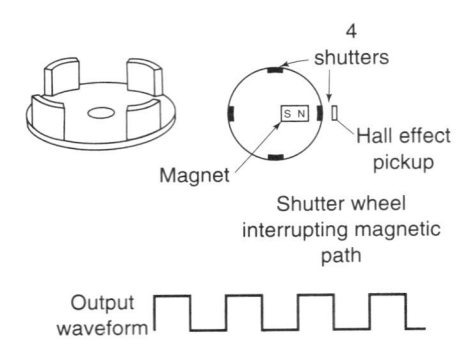

FIGURE 21-10 Hall effect pickup (distributorless ignition)

Distributorless Ignition System (DIS)

The distributor provides two main functions in the modern motor vehicle.

1. It determines the position of the cylinder for proper timing of the spark.
2. It distributes the spark to the proper cylinder.

The two main functions of the distributor can be provided by other means.

1. Cylinder position can be determined by crankshaft or cam position.
2. Increasing the number of coils used eliminates the need for mechanical distribution of the spark.

The position of the pistons may be taken from a vaned rotor mounted behind the pulley on the crankshaft. Hall effect pickup provides signal output indicating crankshaft rotational position and speed.

A second sensor, the cylinder identification sensor, usually installed in the bore originally provided for the distributor, is used to indicate the position of the engine cam and thereby may be used to indicate the position of piston #1.

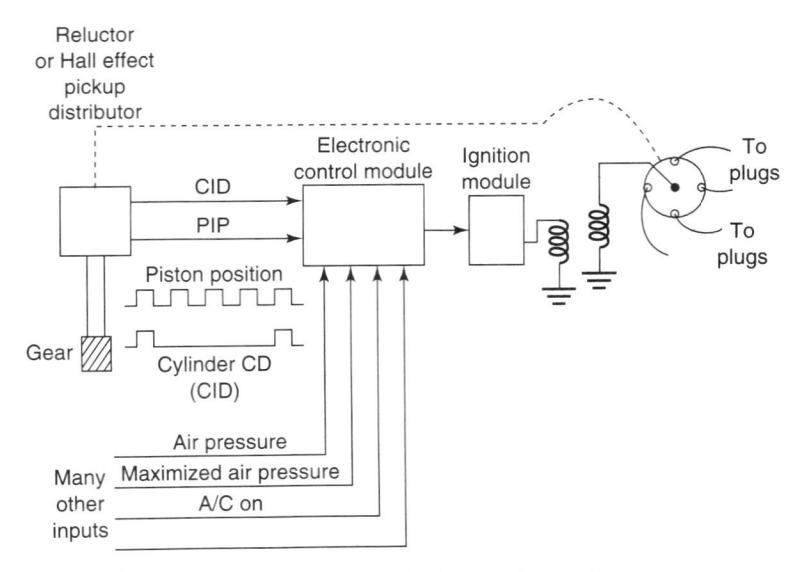

FIGURE 21-11 Distributor ignition with electronic control

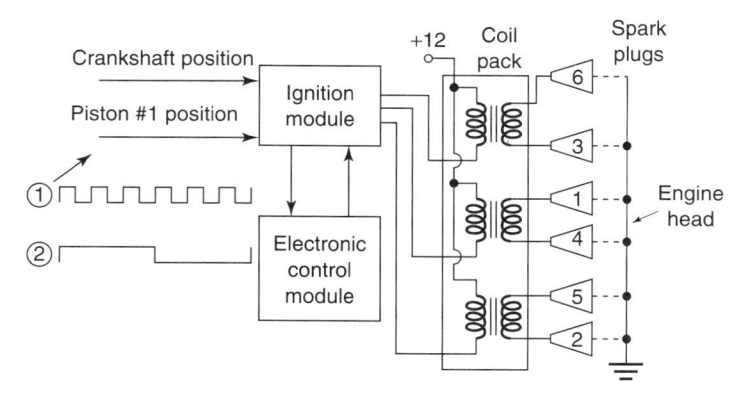

FIGURE 21-12 Ignition system components (six cylinder)

The combination of piston #1 position and the position of all other pistons is used to determine firing order: timing and sequence. The system for one six-cylinder engine is shown in Figure 21-12.

Observe the coil pack at the top right of Figure 21-12, the distributorless ignition system (DIS). This is a major change from distributor-type ignitions. There are three coils in this pack, one coil for every two plugs; two plugs are fired at the same time. One plug fires on the compression stroke while the second is on the exhaust stroke.

The plug that fires on the compression stroke ignites the fuel/air mixture. The plug that fires on the exhaust stroke simply completes the electrical path for the other plug that is firing.

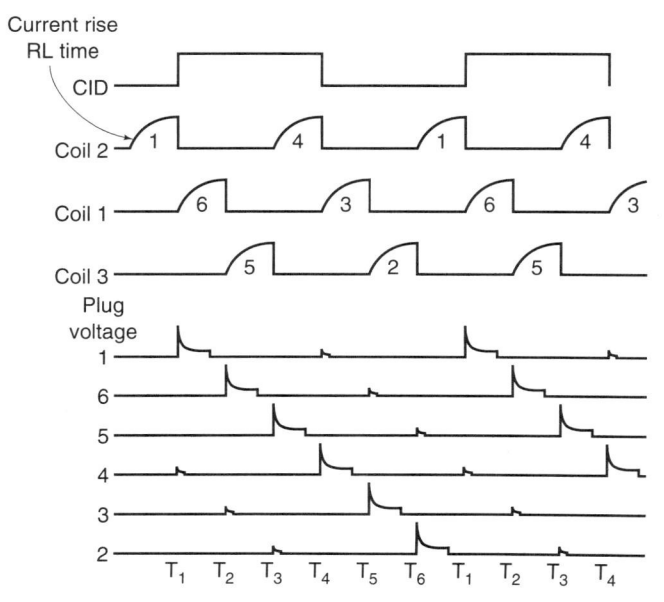

FIGURE 21-13 Coil and plug voltages

The amplitude of the voltage at the plug under compression is much higher than that of the plug on exhaust. Observe the amplitude of the voltage at plugs #1 and #4 at T_1 in Figure 21-13. Plug #1 is firing, as is plug #4. The two plugs are in series; the same current is flowing through both plugs. The higher voltage is at plug #1, since the resistance of the compressed air/fuel mixture (valves closed) is much higher than that of the exhaust gas (valves open). Since the resistance is higher at the plug under compression, the power (I^2R) is higher at that plug. The higher-power plug is the "hot" spark plug. Hot firing of plugs continues with plugs #6, #5, #4, #3, and #2.

When plug #4 fires, the "hot" spark will be at plug 4 and the "cool" spark at plug 1 (see Figure 21-13, plugs 4 and 1 at T_4). Keep in mind that the time of the firing is controlled by the electronic engine control module. Crankshaft and cam position provide information as to which cylinder is to be fired.

A distributorless ignition system (DIS) for a four-cylinder vehicle requires two coils, whereas four coils are required for an eight-cylinder vehicle. Future design may provide for a single coil for each cylinder.

Single-Sensor Electronic Ignition

With the single-sensor electronic ignition, a trigger wheel is mounted directly behind the pulley at the crankshaft end. The trigger wheel is a 35-tooth wheel with a tooth at every 10 degrees (10°), as shown in Figure 21-14. The space at the "missing" tooth provides for indication of top dead center (TDC) of #1 piston.

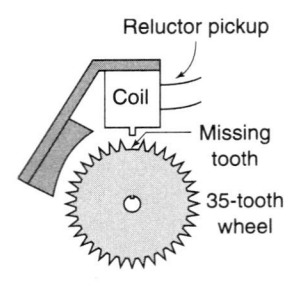

FIGURE 21–14 Trigger wheel

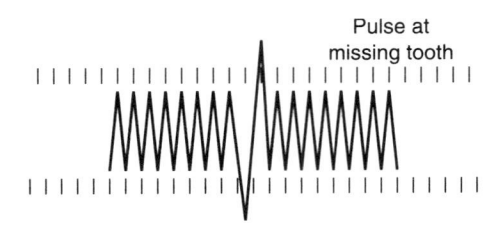

FIGURE 21–15 Reluctor output trigger wheel

A variable-reluctance sensor, as described earlier, is used as the pickup device. The output of the sensor is a sine wave, as shown in Figure 21–15.

The location of the missing tooth pulse indicates piston #1's position. The frequency of sine wave output of the star wheel/reluctor sensor is used as the engine speed indicator. Once the position of piston #1 is determined, the remaining piston positions can be determined by their relation to the 35 other pulses from the reluctor's pickup. Ignition timing is still controlled from the computer system, which considers many other factors along with piston number and position. Firing of plugs may be with one ignition coil for each pair of plugs or, in the future, with one ignition coil for each plug.

SUMMARY

The modern ignition system is an important device in the motor vehicle. One of the factors providing for maximum engine performance is ignition. Optimum maintenance of this system is the technician's responsibility.

PRACTICAL EXERCISE

This exercise is intended to stimulate group activity and discussion in the classroom and shop. A late-model vehicle with appropriate manuals for electrical schematics and component locations are required.

1. Locate the ignition system diagram on the electrical schematic.
2. Locate the device(s) on the component locator diagram.
3. Discuss and answer the following questions.

QUIZ

1. What type of signal pick-up device is used to initiate ignition timing?
2. What type of ignition system do you find—distributor or distributorless?
3. Describe the system (How many cylinders? Straight or "V" design?).
4. What is the firing order?

REVIEW

(T) (F) 1. The back electromotive force (bemf) of a coil is usually equal to the applied electromotive force (emf).

(T) (F) 2. It takes five time constants for current to reach maximum in a coil.

(T) (F) 3. There is higher voltage in an ignition coil secondary because it is wound with larger wire.

(T) (F) 4. Although the ignition coil could provide higher output, the spark plugs usually fire at around 10,000 volts.

(T) (F) 5. When the spark plug fires, the voltage output of the ignition coil stops rising.

(T) (F) 6. It is possible to get more energy out of the secondary of a coil than is provided to the primary.

(T) (F) 7. Use of a transistorized ignition system increases the pitting of the points.

(T) (F) 8. The star wheel/reluctor combination of a breakerless ignition system produces an output by capacitive pickup.

(T) (F) 9. Cylinder position can be determined by the position of the crankshaft.

(T) (F) 10. More than one coil is required in a distributorless ignition system.

◆ UNIT 22 ◆

COMPUTER CONTROLS

OBJECTIVES

On completion of this unit you will be able to:

- Explain on-board computer operation.
- Describe multiplexing, data transfer, and converters.
- Explain parallel to series and series to parallel transfer.
- Describe memory cells and sample and hold devices.

The modern motor vehicle is controlled by several computer systems. Information gathered from sensors throughout the vehicle is processed in one of the computer systems to control engine operations. The computer provides for efficient use of fuel as well as limiting exhaust emissions. Many other on-board devices are controlled by other computers.

Within the computer a system of data transfer called *multiplexing* is used. Data transfer takes place on eight lines, called a *bus*. Analog information taken into the computer is converted into an eight-bit digital code. Along with other inputs the information is multiplexed into memory, multiplexed out of memory into the microprocessor, and multiplexed out of the microprocessor to the output section. There it is converted back to analog form for use by an operating system (air/fuel mixture control, for example). If multiplexing were not used in computers, computers would have to be much larger than they are.

There is also a movement toward the use of multiplexing in modern vehicles for simple controls such as turning devices on and off from a remote location. This form of multiplexing is straightforward and is covered in this unit, though it may not be utilized in motor vehicles at present. However, a technician needs to understand multiplexing in order to have a basic understanding of computers and how they function.

MULTIPLEXING

The term *multiplexing* is not new to electronics. The term, as it implies, refers to the process of combining several measurements for transmission over a signal wire or link by time division or frequency sharing. It provides a means of transmitting two or more messages in either one or both directions over the same transmission path.

Multiplexing for automotive electrical control, however, is relatively new. Presently used on only a few car lines, it is expected that multiplexing will be expanded in the future for use on most car lines.

The multiplexing systems covered first in this unit should be considered typical and not as an example of any particular operating system. We intend this discussion to be only an illustration of how multiplexers operate. A pulse width of 125 microseconds (μs) has been chosen for this discussion of a simple multiplexing system. This pulse time, also, is not related to any particular operating system.

Multiplexing is a system of control where a limited number of wires is used to remotely control the turning on and off of a multitude of devices. Multiplexing is also used to transfer digital data from sensors to the on-board computer and from the computer to the points to be controlled.

A SIMPLE MULTIPLEXING SYSTEM

In the modern automobile, multiplexing finds application for the control of electrical devices in the rear of the vehicle from the front. Consider a number of individual items at the rear that may be controlled from the front area—things such as tail lights on/off, left or right rear window up or down, backup lights on/off, rear window wiper on/off, rear window washer on/off, rear window defogger on/off, left or right brake/turn signal light on/off. Other devices may also need control, but they are not necessarily included in this limited explanation.

In present vehicles most load devices require at least one wire for proper operation. The wires must be sized to carry the current drawn by the individual loads. The cost of this wire system is important when considering how practical multiplexing might be. Consider:

1. The cost of copper wire, one heavy wire for each item to be controlled. Overall vehicle weight is an important factor in fuel economy.
2. The cost of manufacturing multiwire cable from the vehicle front to the rear.
3. The cost of installation of the cable from the vehicle front to the rear.

With multiplexing, one wire is required for current. A second wire is used for the control signal, and a third for timing synchronization. All control is initiated and maintained with the three wires from the front to the rear of the vehicle. Depending on system design, a fourth wire may be used for a reset circuit.

Operation

One system of multiplexing uses a series of voltage pulses at specified times as the basis of control. The time matching of two pulses, one from the vehicle front section and one generated in the rear section, will cause a specific system to be turned on or off.

Consider the pulse system shown in Figure 22–1. It shows pulses of voltage that are intended to turn on the backup lights and the rear window wiper. No other voltage pulses are present, so no other electrical activity is being called for from the vehicle front.

In Figure 22–1 a signal sequence is shown that repeats itself after eight time periods. The signal from the vehicle front indicates that:

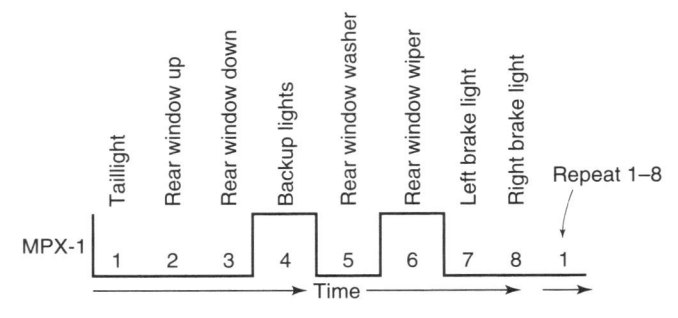

FIGURE 22–1 Simple multiplex signal

1. Tail lights are not required.
2. Rear window up not required.
3. Rear window down not required.
4. Backup lights are required.
5. Rear window washer not required.
6. Rear window wiper required.
7. Left brake light not required.
8. Right brake light not required.

This signal from the vehicle front is connected as one input to eight two-input AND gates located at the vehicle rear. Also located in the vehicle rear is a pulse-generating device, which is timed (synchronized) to the vehicle front's pulse generator.

Observe Figure 22–2, showing the output pulses from the pulse generator. The same pulses are produced at pins 1 through 8 at the same time in the front and rear. They are synchronized.

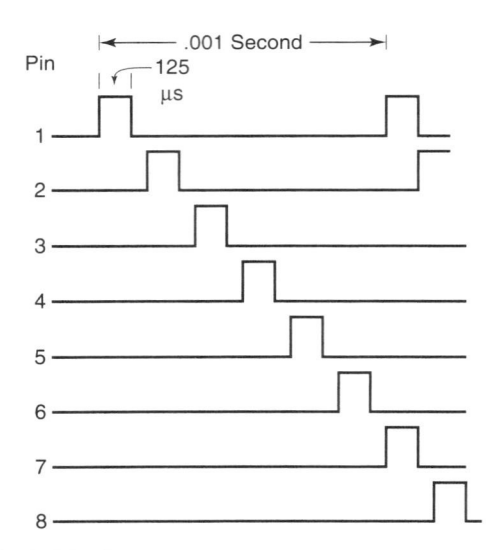

FIGURE 22–2 Pulse generator output

A SIMPLE MULTIPLEXING SYSTEM

With the control switches at the vehicle front, only the rear wiper and backup light switches are closed, as shown in Figure 22–3. Only pulses 4 and 6 will be fed to the AND gates at the vehicle rear. The output from the rear pulse generator produces pulses at the AND gate input in time sequence with its front pulse generator; only gates 4 and 6 will have pulses at their two inputs. AND gates 4 and 6 will produce output pulses. AND gates 1, 2, 3, 5, 7, and 8 will not have outputs.

The diode, resistor, and capacitor combination at the AND gate outputs are required to maintain the dc voltage level between pulses. If the output frequency were 1000 pulses per second, the diode–resistor–capacitor trio would be selected, so the capacitor would discharge in approximately 1/100th second. As long as 1000 pulses come along every second the capacitor remains charged. If 1/100th second goes by without pulses, the capacitor will have time to discharge.

The output from AND gate 4 will turn on field effect transistor (FET) 4, allowing current to flow through the backup lights. The output from AND gate 6 will turn on FET 6, allowing current to flow through the rear wiper motor so it will operate.

Brake and Turn Signals

When the turn signal switches are not activated, and the brake pedal is depressed, pulses are fed through the turn signal switch to the control line, and then to the AND gates in the rear. The brake lights will light.

When a turn is selected, left, for example, the turn switch is activated. The switch connection is changed

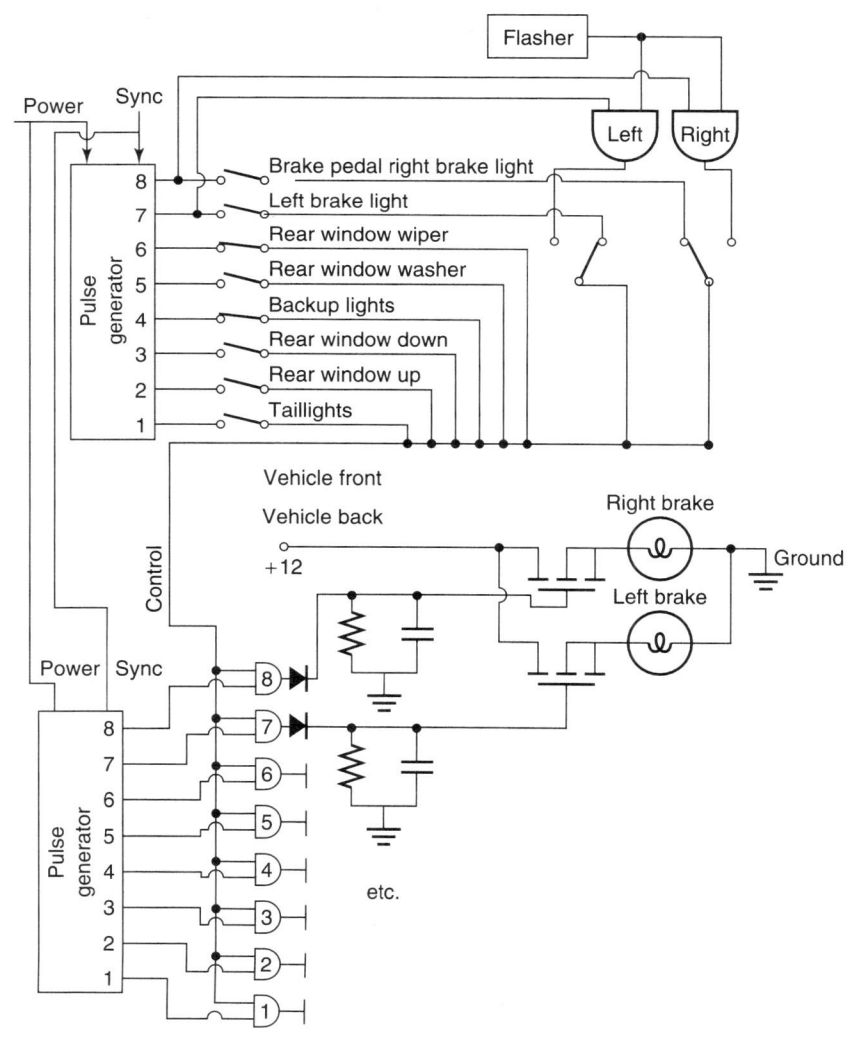

FIGURE 22–3 Front to rear multiplex

Unit 22 COMPUTER CONTROLS

to pick up the output of the left turn AND gate. The AND gate inputs on this left brake pulse from pin 7 and the output of the turn flasher. These are shown in Figure 22–4. When both inputs are positive (+) there will be an output from the gate. It will be a series of approximately 500 pulses for 1/2 second and zero output for 1/2 second. This signal will be fed through the turn signal switch of the control wire to the AND gate at the rear. Each of the 500 pulses will match the pulses from pin 7 of the pulse generator at the rear. The left brake lamp will be on for approximately 1/2 second and then off for approximately 1/2 second, and so on.

Although it is not necessary to have a complete understanding of how individual integrated circuits function internally, it is of interest in some cases. In the case of the pulse generator a simple system is used. It consists of three flip-flop circuits and eight three-input AND gates. The flip-flop circuits are connected as shown in Figure 22–5.

This is a simple countdown circuit producing output waveform A through F as shown in Figure 22–6. These outputs are ANDed in combinations of three, as shown in Figure 22–7. The final outputs are shown in Figure 22–2.

In Figure 22–6, if waveforms A, C, and F are combined in a three-input AND gate the only time an output will be produced is during time period 1. With B, C, and F the output will be during time period 2. The ANDed combination, as shown in Figure 22–7, will provide individual pulses 1 through 8 in sequence. If a greater number of pulses in sequence are needed, another flip-flop may be added. Sixteen time-sequenced pulses would be obtained. Sixteen four-input AND gates would also be required.

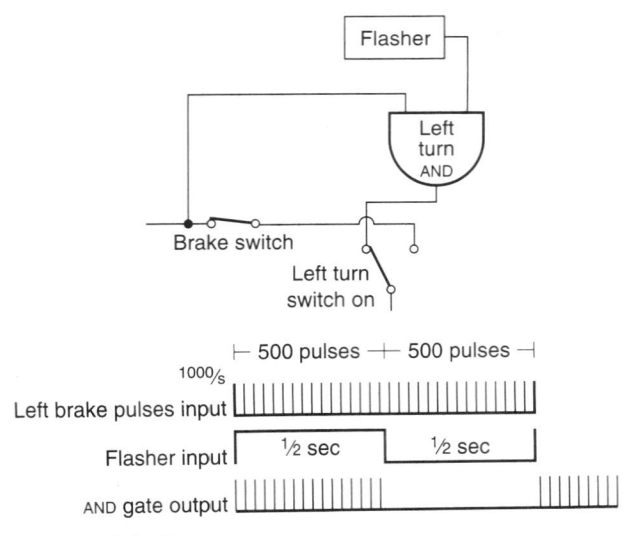

FIGURE 22–4 Left turn signal

MULTIPLEXING OF DIGITAL ENGINE DATA

There are two other procedures of multiplexing used to transfer coded data, particularly as related to a computer. The two methods are parallel and series transfer. Again, the system shown here for data transfer and computer or electronic control module operation is typical, but not necessarily an example, of any particular operating system.

Data is obtained from sensors found at specific locations around the engine. Sensors were covered in Unit 20 of this text. The data signal that is generated by

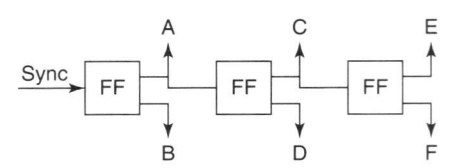

FIGURE 22–5 Countdown circuit

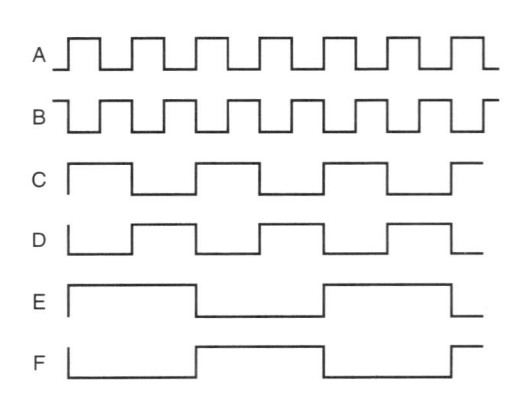

FIGURE 22–6 Countdown circuit output

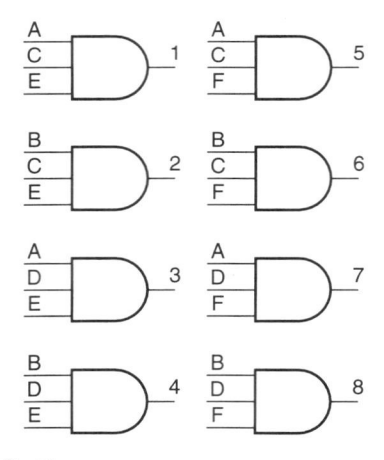

FIGURE 22–7 AND gating for timing pulses

a sensor is an analog signal. Figure 22–8 shows an engine coolant temperature (ECT) sensor (A).

The analog signal is fed to an analog-to-digital converter (B) (Figure 22–8). The converter produces an eight-bit digital output equivalent to the analog input (see Unit 13). The eight-bit digital signal is then fed to the parallel-to-series converter (Figure 22–9). This device provides a means of storing the digital information coming in on the eight lines and allows the individual bits of data to be fed out on a single data line in sequence.

The serial data, engine coolant temperature, is fed on the line during the time period allotted for it. It is converted from serial data back to parallel data, and is then fed to the proper memory area in the control module.

The output of unit one is then isolated from the data line, and the output of unit two is fed to the line. This output data will be fed to the control module and stored in its proper location. This process continues for the eight bits of information, and the data line will be connected to the engine rpm parallel-to-series converter, and so on. There may be 20 or more pieces of information fed to the control module on the same wire.

Memory Cell

There are a number of methods available that will provide acceptable memory in a computing system. The automotive electronics technician will seldom, if ever, have to get down to the memory cell level when troubleshooting. In fact, neither does the mainframe computer technician. It is well, though, to have some basic understanding of what a memory cell is and how it functions.

Figure 22–10 shows a memory cell. Components shown as R_1 and R_2 are simply FETs fixed biased so that they function as resistors. The input data is either 0 volts for a data 0 or +5 volts for a data 1.

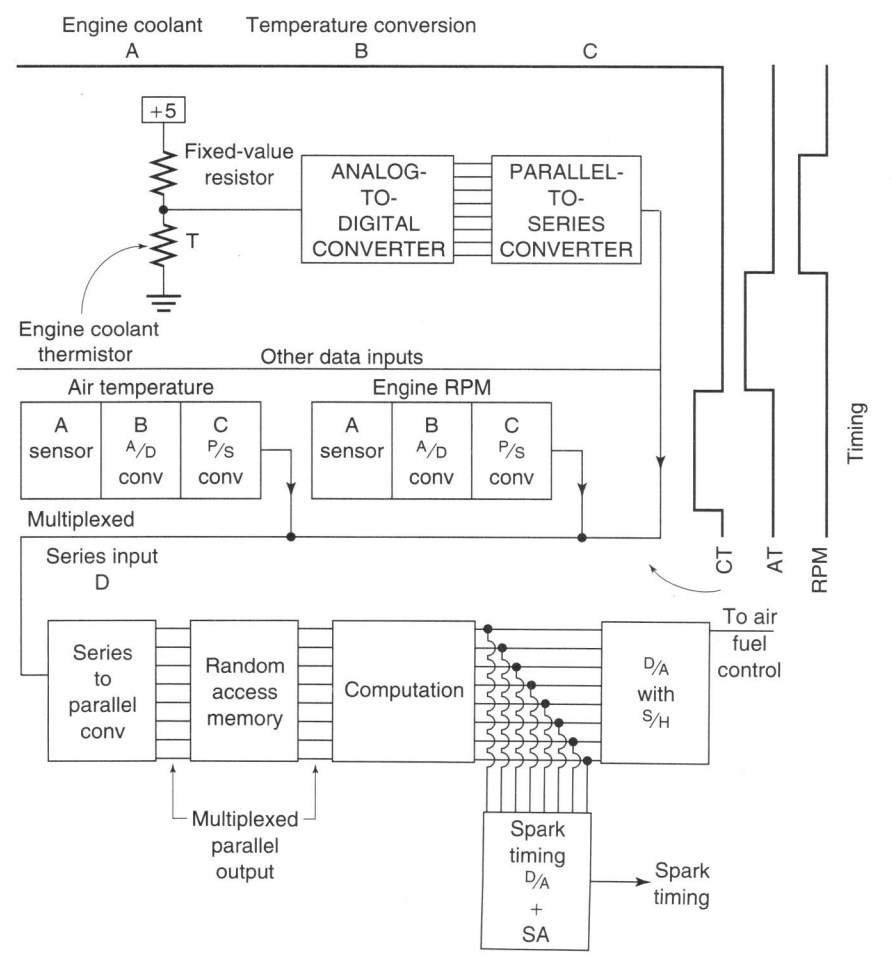

FIGURE 22–8 Data sensing, transfer, computation, and transfer

268 Unit 22 COMPUTER CONTROLS

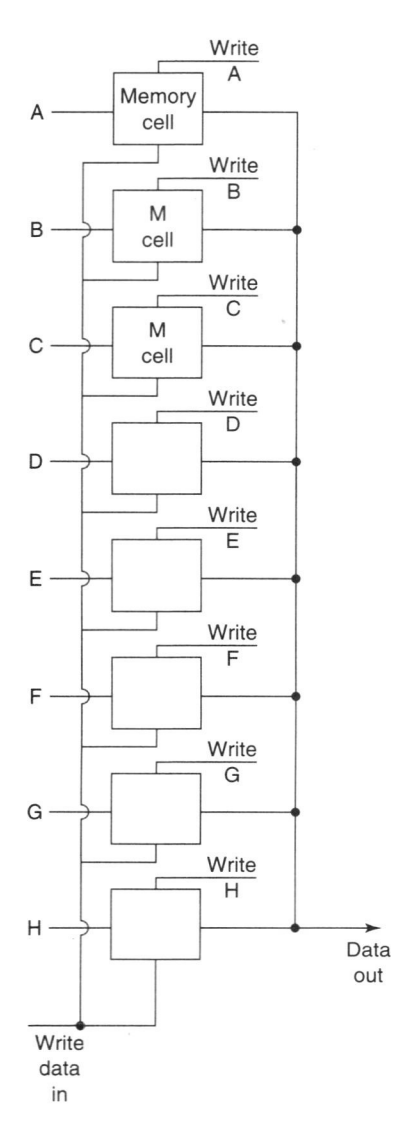

FIGURE 22-9 Parallel-to-series converter

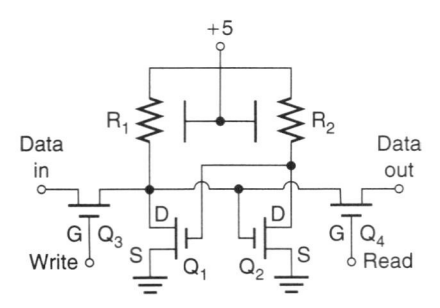

FIGURE 22-10 Memory cell

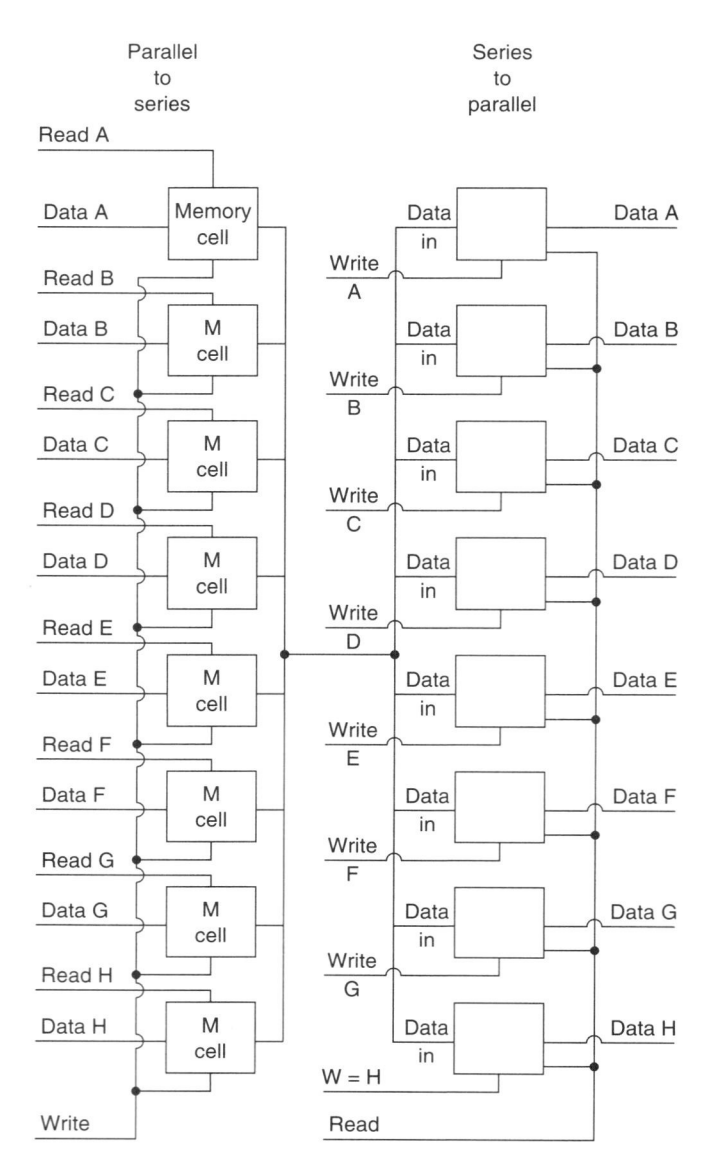

FIGURE 22-11 Parallel-to-series and series-to-parallel converters

Data are written into memory when +5 volts appears at the gate of Q_3. Positive (+) voltage (+4.5 to 5.0 volts) at the transistor gate turns the transistor on. No voltage (0 to +0.5 volts) at the transistor gate turns them off.

Data are read out of memory when +5 volts appears at the gate of Q_4. The information contained in memory will remain in memory after being read. Memory data can change only when a new data bit is written in (+5 volts at Q_3 gate).

Operation

As an example, a one (1) will be stored in the memory cell. This means that +5 volts (one) is available at the data-in terminal. When the write pulse comes along

(Figure 22–10), the +5 volts is fed to the gate of Q_2; Q_2 turns on. Current flow through Q_2 and R_2 causes the voltage across R_2 to rise above +4.5 volts. A voltage of less than +0.5 volts is fed to the gate of Q_1; Q_1 turns off. Since there is no current through Q_1 and R_1, the voltage drain at Q_1 is +5 volts. The drain of Q_4 is connected to the drain of Q_1. When the read pulse, +5 volts, is applied to the gate of Q_4, the voltage at the drain of Q_4 will be available at its source terminal, which is the memory cell output.

When data bit zero (0) is to be stored in the memory cell, 0 volts is available at the data-in terminal. Where the write pulse appears, Q_3 conducts. Zero volts is fed to the gate of Q_2; Q_2 stops conducting. There is no current flow through R_2. The drain of Q_2 moves to +5 volts, which is fed to the gate of Q_1; Q_1 turns on. The drain voltage of Q_1 drops to zero. The data output will be zero (0) when the next write pulse is applied to the gate of Q_4.

PARALLEL-TO-SERIES AND SERIES-TO-PARALLEL CONVERTERS

The parallel-to-series converter and the series-to-parallel converter are similar in construction. In fact, it can be shown that the only difference is in the timing of read and write pulses and the input and output data lines.

Parallel to Series

Observe the left side of Figure 22–11, the parallel-to-series converter. The individual data lines A through H contain the eight (8) bits of information to be multiplexed to the control module (computer).

The write inputs are connected together for the eight memory cells. With the application of a write pulse the eight bits of data, 0s or 1s, will be temporarily stored in memory. Note that the read pulses follow each other in sequence. The output of each memory cell will be fed on to the data line in sequence.

Series to Parallel

The serial data line is connected to the data input terminal of each memory cell. The write pulses follow each other in sequence in the series-to-parallel converter. As the individual data pulses arrive they will be written into memory cells A through H in sequence. The application of a single read pulse transfers the eight bits of data out of the memory cells to the control module.

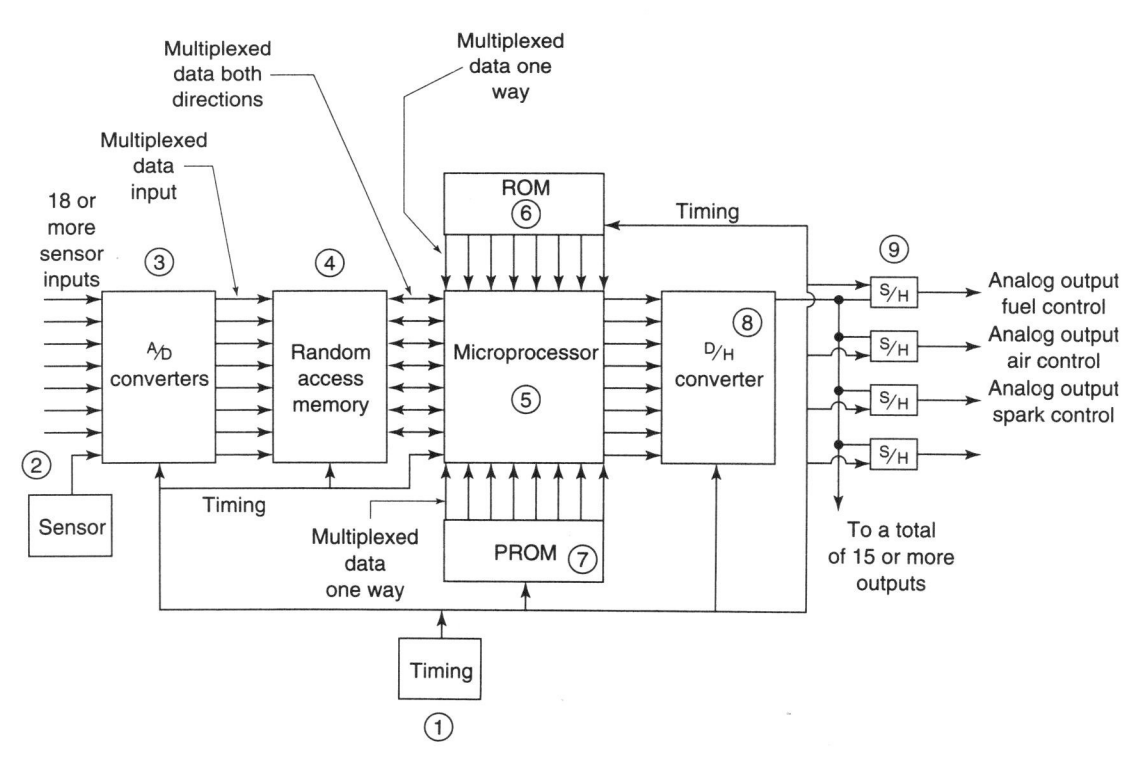

FIGURE 22–12 Motor vehicle computer control

In normal operation it takes a longer time period to transfer data in serial than in parallel (the time for eight pulses rather than one). Parallel, however, does require the use of eight lines rather than one for serial transfer.

Present automatic computers use parallel transfer with the multiline bus. The system for data sensing, transfer, computation, and transfer, (Figure 22–8) becomes as shown in Figure 22–12. The system shown in Figure 22–12 is closely related to that used in the modern motor vehicle. A clock generator (1) produces timing pulses that control the time of operations within the computer.

Information is taken from individual sensors (2) in the form of analog signals. The analog voltages are fed into the electronic control module (computer). At the input, A-to-D converters (3) change the analog voltages to digital equivalents several thousand times each second. The digital data are multiplexed into the random access memory (RAM) (4).

The microprocessor (5) manipulates and computes using the RAM data and information taken from read-only memory (ROM) (6). The microprocessor temporarily stores information in RAM (4) and also provides output information that is used in engine control.

The output of the microprocessor is fed to the digital-to-analog (D/A) converters (8), where the digital signals are converted to voltage usable in operational control. Simple and hold (S/H) devices (9) are used to maintain the outputs while multiplexing takes place.

SUMMARY

The modern motor vehicle includes many sensor devices feeding signals to one or more computers that provide for high-performance engine operation. The technician's responsibility is to master proper diagnosis and troubleshooting procedures in order to perform occasional maintenance to the sensing devices necessary to maintain efficient operation of the vehicle.

PRACTICAL EXERCISE

This exercise is expected to stimulate classroom and shop activity and discussion. A late-model vehicle with appropriate electrical schematics and component locator diagrams are required.

1. On the electrical schematic, locate all the components—lamps, locks, switches, and so on, that are located in the rear of the vehicle.
    a. Taillamps
    b. Stop (brake) lamps
    c. Backup lamps
    d. License plate lamp(s)
    e. Trunk lock (manual/electric)
    f. Trunk lamp
    g. High-mount stop lamp
    h. Rear side marker lamps
    i. Trunk-lid pull-down motor (if equipped)
    j. Antenna motor (for rear-mount antennas)
    k. Trunk lid antitheft switch
2. Locate the electrical cable(s) serving the rear of the vehicle.
3. Consider the manufacturing cost:
    a. Cost of material
    b. Cost of construction
    c. Cost of installation

QUIZ

1. Do you believe that a system that requires only one wire for battery power, one wire for control, and one wire (or fiber optic) for a signal unit would be less expensive to produce than the present system?

REVIEW

(T) (F) 1. Multiplexing is a term used to indicate the transfer of more than one piece of information on a single line.

2. In Figure 22–3 there will be an output from the right turn AND gate when brake input is (positive) (zero).

3. In Figure 22–3 the right brake FET is maintained in the ON condition during pulses by the:
   a. #1 and #2 AND gate.
   b. lamp supply voltage.
   c. RC network.
   d. Pickup from the brake radiation.
4. The countdown circuit of Figure 22–5 is made up of:
   a. AND gates.
   b. exclusive OR gates.
   c. flip-flops.
   d. one-shot multivibrators.
5. The gate circuits shown in Figure 22–7 are:
   a. 3-input OR gates.
   b. 3-input NAND gates.
   c. 3-input AND gates.
   d. 3-input NOR gates.
6. (Parallel) (Series) data transfer can take place on a single line.
7. The date stored in a memory cell (may) (may not) be changed.
8. When feeding data out on a serial transfer line, the read pulses occur (at the same time) (in sequence).
9. When feeding data into a serial-to-parallel converter, the write pulses occur (at the same time) (in sequence) (Figure 22–11).
10. When reading data out of a serial-to-parallel converter, the read pulses occur (at the same time) (in sequence) (Figure 22–11).

APPENDIX A

POWERS OF TEN

The use of the power of ten provides for mathematical operations using very large or very small numbers. In electronics most calculations are held to three significant digits, that is, in the number 3,262,137,812, only the 326 and the position of the decimal would be important in most situations. Electronic components and measurements are usually only accurate to three significant digits.

Considering three-significant-digit use in electronics makes the powers of ten very important. The number 3,262,137,812 becomes 3.26×10^9. The decimal is simply moved to the point providing a single digit whole number. The number of places the decimal has been moved provides the power of ten. For numbers less than one, say, 0.0000000192, this becomes 1.92×10^{-8}.

Numbers with a power of ten may be added if the numbers have the same power of ten.

$$\begin{aligned} 2.13 \times 10^4 \\ +1.33 \times 10^4 \\ \hline 3.46 \times 10^4 \end{aligned}$$

If the powers of ten are different, they must be made the same before adding.

EXAMPLE 1

Add (2.14×10^6) plus (1.21×10^5).

Solution: Change 1.21×10^5 to 0.121×10^6.

$$\begin{aligned} 0.121 \times 10^6 \\ +2.14 \times 10^6 \\ \hline 2.26 \times 10^6 \end{aligned}$$

The same conditions exist in subtraction.

273

EXAMPLE 2

Subtract 8.37×10^5 from 1.28×10^6.

Solution: Change 1.28×10^6 to 12.80×10^5.

$$12.80 \times 10^5$$
$$-8.37 \times 10^5$$
$$4.43 \times 10^5$$

When powers of ten are used in multiplication, the numbers are multiplied and the powers of ten are added.

EXAMPLE 3

Multiply (3.11×10^4) times (2.06×10^2)

Solution:

$$3.11 \quad \times 10^4$$
$$\times 2.06 \quad \times 10^2$$
$$6.4066 \times 10^6 \quad \text{or} \quad 6.41 \times 10^6$$

EXAMPLE 4

Multiply (2.17×10^{-4}) times (4.23×10^3).

Solution:

$$2.17 \quad \times 10^{-4}$$
$$\times 4.23 \quad \times 10^3$$
$$9.1791 \times 10^{-1} \quad \text{or} \quad 9.18 \times 10^{-1}$$

In division, the number is divided by the divisor and the powers of ten are subtracted.

EXAMPLE 5

Divide (6.11×10^6) by (2.07×10^3).

Solution:
$$(6.11 \times 10^6) \div (2.07 \times 10^3) = 2.95 \times 10^3 \quad \text{or} \quad 2950$$

EXAMPLE 6

Divide (3.78×10^5) by (2.16×10^{-3}).

Solution:

$$(3.78 \times 10^5) \div (2.16 \times 10^{-3}) = 1.75 \times 10^8$$

Powers Of Ten

10^{10}	10,000,000,000
10^9 Giga	1,000,000,000
10^8	100,000,000
10^7	10,000,000
10^6 Mega	1,000,000
10^5	100,000
10^4	10,000
10^3 Kilo	1,000
10^2	100
10^1	10
10^0	1
0	0
10^{-1}	0.1
10^{-2}	0.01
10^{-3} Milli	0.001
10^{-4}	0.0001
10^{-5}	0.00001
10^{-6} Micro	0.000001
10^{-7}	0.0000001
10^{-8}	0.00000001
10^{-9} Nano	0.000000001
10^{-10}	0.0000000001

◆ APPENDIX B ◆

PERIODIC TABLE

Appendix B PERIODIC TABLE

Group	I Mono-Valent	II Bi-Valent	III Tri-Valent	IV Tetra-Valent	V Penta-Valent	VI Hexa-Valent	VII Hepta-Valent	VIII			0 Inert
1	1 H Hydrogen										2 He Helium
2	3 Li Lithium	4 Be Beryllium	5 B Boron	6 C Carbon	7 N Nitrogen	8 O Oxygen	9 F Fluorine				10 Ne Neon
3	11 Na Sodium	12 Mg Magnesium	13 Al Aluminum	14 Si Silicon	15 P Phosphorus	16 S Sulfur	17 Cl Chlorine				18 Ar Argon
4 A	19 K Potassium	20 Ca Calcium	21 Sc Scandium	22 Ti Titanium	23 V Vanadium	24 Cr Chromium	25 Mn Manganese	26 Fe Iron	27 Co Cobalt	28 Ni Nickel	
4 B	29 Cu Copper	30 Zn Zinc	31 Ga Gallium	32 Ge Germanium	33 As Arsenic	34 Se Selenium	35 Br Bromine				36 Kr Krypton
5 A	37 Rb Rubidium	38 Sr Strontium	39 Y Yttrium	40 Zr Zirconium	41 Nb Niobium	42 Mo Molybdenum	43 Tc Technetium	44 Ru Ruthenium	45 Rh Rhodium	46 Pd Palladium	
5 B	47 Ag Silver	48 Cd Cadmium	49 In Indium	50 Sn Tin	51 Sb Antimony	52 Te Tellurium	53 I Iodine				54 Xe Xenon
6 A	55 Cs Cesium	56 Ba Barium	57-71* La Lanthanum	72 Hf Hafnium	73 Ta Tantalum	74 W Tungsten	75 Re Rhenium	76 Os Osmium	77 Ir Iridium	78 Pt Platinum	
6 B	79 Au Gold	80 Hg Mercury	81 Tl Thallium	82 Pb Lead	83 Bi Bismuth	84 Po Polonium	85 At Astatine				86 Rn Radon
7	87 Fr Francium	88 Ra Radium	89 **Ac Actinium								

* LANTHANIDES

58 Ce Cerium	59 Pr Praseodymium	60 Nd Neodymium	61 Pm Promethium	62 Sm Samarium	63 Eu Europium	64 Gd Gadolinium	65 Tb Terbium	66 Dy Dysprosium
67 Ho Holmium	68 Er Erbium	69 Tm Thulium	70 Yb Ytterbium	71 Lu Lutetium				

**ACTINIDES

90 Th Thorium	91 Pa Protoactinium	92 U Uranium	93 Np Neptunium	94 Pu Plutonium	95 Am Americium	96 Cm Curium	97 Bk Berkelium	98 Cf Californium
99 Es Einsteinium	100 Fm Fermium	101 Md Mendelevium	102 No Nobelium					

INDEX

air conditioning
  electrical aspects, 232
  mechanical aspects, 231–232
  protective switches, 232–233
  system, 231
    automatic, 234
    schematics, 233
  physics, 231
air flow sensor, 245
air induction system, 219
alloy, 11
alternating current, 147
alternator, 36, 147, 151–152
  diodes, 153
  output, 153
  rotor, 152–153
  stator, 152–153
  tester, 65–66
aluminum, 13
ambient switch, 233–234
ammeter, 58
  connecting to circuit, 25, 58, 82
  use of, 63–64
Ampere, 1
amplifier circuits, 141
analog meters, 57, 62
analogy of electricity, 15
AND gate, 266
antennas, 224
  AM/FM, 224
  AM/FM/CB, 225

antennas, *continued*
  cellular phone, 225
  CB, 224
antenna systems, 224
antilock brake system, 2
antitheft alarm system, 199
atomic structure, 12
Automatic
  air conditioning control, 234–238
  door locks, 214
  headlamp control, 162
  highway control, 6
  steering control, 7
automotive
  battery, 76
  electrical systems, 5
  history, 4

backup lamps, 166
balance
  bridge, 245–246
  in, 16
bar magnet, 50
battery, 73
  acid, 77–78
    neutralizer, 77–78
  automotive, 76
  cable, 27–28
  care, 80
  cells
    in parallel, 75

279

battery, *continued*
    in series, 75
  charging rate, 83
  chemical action, 78
  components, 77
  electrolyte, 79
  internal resistance, 74
  jumper cables, 65, 67
  power
    ratings, 80
    transfer, 76
  safety, 76
  self discharge, 82
  side post connected, 76–77
  tester, 65–66
  testing, 80, 81
    capacity, 82
    maintenance free, 82
  top post connected, 76–77
bemf polarity, 98
Benz, 3
bipolar transistor, 130
Bohr's Law, 12
brake and turn signals, 265
brake light, 172
breakerless ignition, 256–257
  magnetic pickup, 257
bridge rectifier, 129
bulb, 166
  trade jargon, 167
  types, 165–166
bunch stranding, 30

cables, wires and, 31
capacitance, 87
  and inductance, 87
  in ac circuits, 150
  sensor, 243
calculator solutions, 44–45
capacity test, 82
capacitor
  construction, 91
  electrolytic, 91
  rating, 91
car, electronic, 1
  history, 2
  of today, 4
  of tomorrow, 5
cell (battery)
  in parallel, 75
  in series, 75
  primary, 73–74
  symbol, 74
cell (memory), 267–268
cellular telephone, 223
charging
  rate table, 83
  system, 194

circuit
  breaker, 160, 162
  flip-flop, 143
  gate, 140
  parallel, 41
  resistance, 27
  resistive ac, 149
  series parallel, 45
  tank, 150
  voltage, 27
circular mil, 28–29
clearance lamps, 166
clocks, 221
closed circuit, 18
collision indicator, 246
color coding
  resistors, 121
  wire and cable, 31, 121
comfort systems, 229
common, 18
common ground, 19
complete electrical circuits, 35
compound, 11
compressor, 231
  discharge switch, 232–233
computer controls, 263
concentric stranding, 30
conical stranding, 30
connectors and connections, 31
convenience lamps, 173
converters
  parallel to series, 268–269
  series to parallel, 268, 269–270
coolant temperature system, 192–193
copper, 12
cornering lamps, 166
Coulomb, 2, 18
countdown circuit, 266
courtesy and convenience lamps, 173
crash sensor, 246
current, 88
  and resistance, 21
  and voltage relationship, 92
  flow, 88

dash
  instruments, 189
  light, 174
data sensing, 267
delay off, headlamp, 162
depressed park system, 180
diagnostic computer, 6
diameter mil, 28–29
digital
  conversion circuits, 246
    analog to digital, 247
    counting, 143
    dc voltage to digital, 247

digital, *continued*
    frequency to digital, 248
    digital to analog, 248
  counting, 143
  engine data, 266
  meters, 61
dimmer switch, 162
diode, 153
  silicon, 127
  testing, 128
  uses, 129
  zener, 129
DIS, 258
distributor, 255
  conventional, 255
  ignition, 258
    with electronic control, 258
distributorless ignition system, 258
dome lamps, 173
door
  ajar warning system, 202
  duct, 229–230
  light, 174
  locks, 213–214
duct
  door, 229–230
  system, 238–239
Duryea, 3
dynamic electricity, 14

Earth as a magnet, 50
electrical
  circuits, 35
  testing, 62
electricity
  analogy, 15
  dynamic, 14
  static, 14
electric
  clocks, 221
  power and energy, 111
  door locks, 213
  energy, 113
  fuel pump, 219–220
  measuring, 114
  overdrive, 219
  test equipment, 57
    meters, 57
electrolyte, 79
electrolytic capacitor, 91
electromagnets, 52
  coil strength, 54
electron, 11, 12
  behavior of, 14
electronic
  car, 1
  control unit, 218
  cruise control, 222

electronic, *continued*
  entry system, 214, 216
  fuel injection system, 217
  ignition, 259
    single sensor, 259
electropneumatic speed control, 220–221
engine thermostatic switch, 229–230
eye protection, 84–85

fiber optics, 174
front side marker lamps, 168
field effect transistor, 130, 135
flasher
  hazard, 172
  turn signal, 170
flip-flop
  circuit, 143
  JK with truth table, 143
  wave form, 143
fluid level warning systems, 183
four way power seat, 205–206
front seat, 205
  back lock, 207
  power, 205
fuel
  delivery system, 218
  level system, 191
  pump, 219

gate circuit, 140
  AND gates, 141, 266
  OR gates, 141
gauges, 194, 195, 196
  and lamps, 196, 197
  electromagnetic, 190
  thermoelectric, 190
generated electromotive force, 147
generating ac voltage, 147
generator, 16
  rotating coils of, 148
ground, 18
grounding, personal, 14

half wave rectifier, 129
hall effect
  ignition, 256
  pickup, 258
  sensor, 244
harness wiring, 27–28
hazard
  flasher, 172–173
  lamps, 171
  warning symbol, 145
headlamp, 159
  aiming, 160, 163
  automatic control, 162, 164
  circuit protection, 160

headlamp, *continued*
  construction
    rectangular, 160
    round, 160
  delay off, 162
  dimmer switch, 162, 164
  on warning system, 202
    quick check, 202
  switchers, 162
  systems, 161
  twilight sensing, 162, 164
heaters, 229
high resistance, 21
hood open warning system, 203
horn, 197
  button, 198
  relay, 199
  without relay, 197
  with relay, 198
horsepower, 113
hot wire sensor, 245
hydrogen, 12
hydrometer, 79

ignition
  breakerless, 256
  conventional, 254
  Hall effect, 256
  Kettering's, 3
  motor vehicle, 253
  system, 253
    components, 259
    transistorized, 255
  waveform, 254
induced polarity, 152
inductance, 92
  capacitance and, 87
  rating, 92
insulation, 30
insulator, 14
integrated circuit, 144
interlocks, relays and, 212

jumper
  cables, 65, 67
  wire, 65, 67
jump starting, 83

Kettering, 3
key in warning system, 201
  quick check, 201
keyless entry system, 214

lamps, 166, 194, 197
  backup, 166
  brake, 172
  clearance, 166
  convenience, 173
  cornering, 166

lamps, *continued*
  courtesy, 17
  dash, 174
  dome, 173
  door, 174
  hazard, 171
  license plate, 165
  map, 174
  parking, 165
  side marker, 165
  tail, 165
  test, 68
  turn signal, 167
large scale integration, 144
Lazer optical system, 7
LED, 133
left hand rule, 53
Leland, 3
license plate lamp, 165
light
  emitting diode, 133
  sensors, 242, 244
lighting, 159
load, 81
  parasitic, 81
  sneak, 82
  phantom, 81
  test, 81
lodestone, 50–51
low fluid level warning system, 183
low resistance, 21
LR time constant, 92
LSI, 144

magnetic
  fields, 50
    effects of, 51
  flux, 51
  lines of
    flux, 51
    force, 51
  pickup, 257
    Hall effect, 258
    sensor, 244
  poles, 50–52
  polarity, 50
magnets
  and magnetism, 49
  artificial, 49
  atomic arrangement, 51
  bar, 50
  Earth, 50
  electro, 52
  natural, 49
maintenance free battery, 82
map light, 174
Matter, 11
measuring electric power, 114
memory cell, 267–268

metal oxide semiconductor field effect transistor, 135
meters
    analog, 57
    ammeter, 58
    digital, 61
    electric, 57
    ohmmeter, 59
    voltmeter, 59
mil/foot, 28–29
moon roof, 212
MOSFET, 135
motor
    control, 105
        pulse width modulation, 106
        series resistive, 106
    operated door locks, 214
    principles, 100
    speed control, 229–230
multiplexing, 263
    brake and turn signals, 265
    digital engine data, 266
    front to rear, 265
    simple system, 264
        signal, 264
        operation, 264
        pulse generator output, 264

N-channel enhancement, 135
neutron, 11
nondepressed park system, 177
nuclear power plant, 14
nucleus, 11

Ohm, 2, 24
ohmmeter, 59
    use of, 26, 64
Ohm's Law, 24
    using, 26
oil pressure system, 195–196
Omega, 22
one shot multivibrator, 142
open circuit, 18
operational amplifier circuits, 141
oxygen sensor, 245

parallel
    circuits, 41–43
    to series converters, 268–269
parasitic load, 81
parking lamps, 165
P-channel enhancement, 135–136
personal injury warning symbol, 145
phantom load, 82
photoconductive cells, 133
phototransistors, 134
photovoltaic cells, 134
piezoelectric sensor, 243
Pope, 3
position sensing switches, 242

pot, 24
potentiometer, 24, 242
power
    reclining seat back, 208
    tailgate window, 210
    three phase, 149
    transfer chart (battery), 76
    window, 208
        adjustment, 210
        electrical testing, 212
        motor testing, 211
        relays and interlocks, 212
pressure
    sensitive capacitor, 243–244
    sensor, 243
proton, 11

Radar braking system, 7
radios
    AM, 222
    CB, 223
    FM, 222
    FM stereo, 222
    and sound systems, 221
        cassette player, 222
        compact disc, 222
RC time constant, 89
rear window
    defogger, 185
    deicer, 185
    grid system, 185
    wipers, 185
rear view mirror, 217
reclining seat back, 208
rectangular headlamp, 160
relays, 97, 212
    and interlocks, 212
    and solenoids, 97
reluctor, 244
    output, 260
    pickup, 256, 257
    trigger wheel, 260
Resistance, 22, 25
    determining, 43, 44
    length of wire, 29
    mil/foot, 28
    of wires and cables, 27
    stranded wire, 30
    values, 22
resistive
    ac circuits, 149
    parallel components, 40
resistor, 22
    color code, 23
    variable, 24
    wire, 35
reverse
    bias diode, 127–128
    warning system, 202

rheostat, 174
   schematic symbol, 174
   speed control, 107
rotor, 152
round headlamp, 160
Route guidance system, 8
Rice, 3

safety
   glasses, 84–85
   systems, 177
schematics, 117
schematic
   interpretation, 121
   sequence and arrangement, 123
   symbols, 117–120
seat back
   lock, 207
   reclining, 208
seat belt warning system, 186
self discharge, 82
semiconductor, 14, 127
   devices, 127, 139
   integrated circuits, 139
sending unit, 191–192
sensing and conversion devices, 241
sensor
   air flow, 245
   capacitance, 243
   crash, 246
   light, 242, 244
   hall effect, 244
   hot wire, 245
   magnetic pickup, 244
   oxygen, 245
   piezoelectric, 243
   position, 242
   pressure, 243
   speed, 244
   systems, 219
   temperature, 242
series-parallel circuits, 45–46
series to parallel converters, 269–270
shunt
   circuit, 58
   operation, 58–59
side marker lamps, 165, 168
silicon diode, 127
   testing, 128
sine wave, 149
single sensor electronic ignition, 259
six way power seat, 207–209
sneak
   circuit, 39–40
   load, 82
solenoid, 98
   operated door locks, 214
specific gravity test, 81

speed controls, 219
   electrical, 221
   electropneumatic, 220
speedminder, 199
speedometer, 198–202
   calibration check, 199
   conversion tables, 201
starter
   control circuits, 105
   current meter, 63–64
   drives, 103, 105
   motor, 102
      circuits, 103
      control, 105
      update, 105
stator, 153
static electricity, 14
   warning symbol, 145
step speed control, 106–107
stereophonic radio, 222–223
stranded wire, 20
sulfuric acid, 76
   neutralizing, 77
sunroof, 212
switches, 39, 162

tachometer circuit, 143
tailgate
   lock, 214
   window, 210
tail lamps, 165
tank circuit, 150–151
temperature sensor, 242
testing a diode, 128
test instruments
   ammeter, 58
   lamp, 68
   ohmmeter, 59
   other, 65
   voltmeter, 17, 59
thermistor, 243
thermometer, 79–80
thermostatic switch, 229–230
three phase power, 149
   delta, 149–150, 153
   wye, 149–150, 153
torque, 101
transistor, 130
   bipolar, 130
   data sheet (typical), 134
   field-effect, 130, 135
   ignition, 255
transistorized ignition system, 255, 257
trigger wheel, 260
trunk
   lid closing system, 215
   lock-release system, 214
   open warning system, 203

turn signal, 167
    flasher, 170
twilight sensing, 162
two way power seats, 205–206

unbalanced, 16

vacuum system, 234, 236–238
variable resistor, 24
Vehicle proximity system, 8
Volta, 2
voltage, 18, 25, 88
    limiter, 189–190
    regulator, 155
voltage and current, 15
voltmeter, 17, 59
    connecting to circuit, 26, 62
    internal resistance, 59–60
    loading, 59–60
    use of, 62, 63

Watt, 2
Waveform, ignition, 254
window
    adjustment, 210
    motor testing, 211
    power, 208
    safety, 208
    tailgate, 210
windshield
    washers, 183–185
    wipers, 177, 182, 185
wire resistors, 35
wires and cables, 31
wire wound resistor, 36, 38
wiring
    harness, 124
        routing, 125
    circuits, 117

zener diode, 129
    voltage regulator, 129